Lacquer Chemistry and Applications

Lacquer Chemistry and Applications

Rong Lu

Tetsuo Miyakoshi

Department of Applied Chemistry,
School of Science and Technology,
Meiji University, Japan

ELSEVIER

AMSTERDAM • BOSTON • HEIDELBERG • LONDON • NEW YORK • OXFORD
PARIS • SAN DIEGO • SAN FRANCISCO • SINGAPORE • SYDNEY • TOKYO

Elsevier
Radarweg 29, PO Box 211, 1000 AE Amsterdam, Netherlands
The Boulevard, Langford Lane, Kidlington, Oxford OX5 1GB, UK
225 Wyman Street, Waltham, MA 02451, USA

Notices

Knowledge and best practice in this field are constantly changing. As new research and experience broaden our understanding, changes in research methods, professional practices, or medical treatment may become necessary.

Practitioners and researchers must always rely on their own experience and knowledge in evaluating and using any information, methods, compounds, or experiments described herein. In using such information or methods they should be mindful of their own safety and the safety of others, including parties for whom they have a professional responsibility.

To the fullest extent of the law, neither the Publisher nor the authors, contributors, or editors, assume any liability for any injury and/or damage to persons or property as a matter of products liability, negligence or otherwise, or from any use or operation of any methods, products, instructions, or ideas contained in the material herein.

ISBN: 978-0-12-803589-4

British Library Cataloguing in Publication Data
A catalogue record for this book is available from the British Library

Library of Congress Cataloging-in-Publication Data
A catalog record for this book is available from the Library of Congress

For information on all Elsevier publications
visit our website at http://store.elsevier.com/

Contents

Preface

Lacquer is the only natural product that is polymerized by an enzyme. It has been used as a coating material with a beautiful and durable surface in Asian countries for thousands of years. Although there are many species of lacquer trees growing in the world, in the long history and experience of lacquer science, there have been only three kinds of lacquer trees that are able to produce lacquer sap. The first one is *Toxicodendron vernicifluum* lacquer trees, which grow in China, Japan, and Korea, and whose main liquid component is urushiol; the second is *Toxicodendron succedanea*, which grow in Vietnam and Chinese Taiwan, and whose main liquid component is laccol; and the third is *Gluta usitata*, which grow in Myanmar, Laos, Cambodia, and Thailand, and whose main liquid component is thitsiol. Although the human lacquer culture including the cultivation of lacquer trees, harvesting, and the use of lacquer sap has a history of more than a thousand years, the application of modern scientific and technological methods to study lacquer chemistry is very short.

After the retirement of Professor Kumanotani, one of the pioneers of lacquer research, from the University of Tokyo, we started a study of lacquer chemistry in the late 1970s at Meiji University as a continuous research project. During more than 30 years of research in lacquer chemistry, we have published many research papers and a book entitled "Progress of Lacquer Chemistry" in Japanese in 1999. Of course, we also note that Oda et al. published a book entitled "Lacquer—the Science and Practice" in Japanese in 1999, Gan Jinghao from Fujian Normal University of China published a book entitled "Lacquer Chemistry" in Chinese in 1984, and Zhang Feilong, a member of the Xi'an Raw Lacquer and Coating Research Institute published a book entitled "Chinese Lacquer: Technology and Conservation" in Chinese in 2010. However, a book about lacquer in English, the language of international communication, has not been published yet. Yet, with advances in modern separation and analysis technology, the research in lacquer chemistry has made great progress. Therefore, in order to facilitate the worldwide lacquer exchange among lacquer researchers, we would like to publish this book entitled "Lacquer Chemistry and Application" in English, describing our comprehensive results in lacquer research so far. The purpose of this book is to reveal the history of lacquer and recent advances in the research on lacquer including component analysis, relationship between structure and biological activities of lacquer polysaccharides, mechanisms of enzymatic catalysts, modification and hybrids of lacquer sap, synthesis of lacquer, mechanism of lacquer allergy, analysis of lacquer films, lacquering technology, and application of lacquer coating material.

Lacquer chemistry is the subject of research into the characteristics of lacquer substances. During the development of lacquer studies, different researchers have obtained results with theoretical contradictions. When encountering such situation, the authors only summarize the lacquer characteristics according to our own research results.

Rong Lu, Tetsuo Miyakoshi
Japan

CHAPTER

Introduction

1

1.1 LACQUER DEFINITION

Lacquer is a naturally occurring product that is polymerized by laccase enzyme. It is an Asian specialty and often called oriental lacquer. Lacquer has been used in Asian countries for thousands of years as a durable and beautiful coating material [1,2]. Cultural treasures and lacquerwares have maintained their lovely surfaces without losing their original beauty for more than 2000 years [3,4]. Lacquer is a sap tapped from a lacquer tree. Like the rubber tree, when the bark on the trunk of lacquer tree is scratched, a milky white sap is exuded. This exuded sap slowly trickles into a container, which may be made of shells, leaves, or bamboo. After the containers are collected, the sap is made into lacquer. Lacquer is called "Shengqi" in Chinese; "Urushi" in Japanese; "Sơn mài" in Vietnamese; "Thitsi" in Burmese, "Marea" in Cambodian; and "Rak" in Thai. Lacquer is considered to be a mystic paint.

Lacquer is a special natural product in Southeast Asia that was considered to be a mysterious coating material used in human life for thousands of years. In modern society, with the progress of studies on composition analysis, people have marveled at the ingenious composition of natural products and felt the charm of lacquer and lacquerwares.

Lacquer trees are a special tree species that grows in nature based on the principle of natural selection. Their leaves receive the sun's rays and use photosynthesis to produce lacquer sap. Lacquer has become a useful material for human life, due to the ingenuity of humanity, which continuously attempts to improve its lifestyle and culture. Lacquer is not necessary for human survival, but is collected by humans for its special functions. However, this behavior does not contradict the principle of biological coexistence in nature; therefore, lacquer culture has continued for thousands of years, and will continue to be refined in the future.

Lacquer sap is generated by leaf photosynthesis in the presence of sunlight, air, water, and soil. The yield is low and tapping is labor intensive, so the efficiency of collection is very low, making lacquer a rare material with a very high price. Furthermore, due to the constant pursuit of aesthetic qualities by humans, lacquer application has become an art and craft, and that also increased its value. Many scientists have focused on its super durability that allows it to survive thousands of years. In 1878, Ishimatsu Sadama, then a student at the University of Tokyo, was the first to try to elucidate the composition of lacquer. From that time to now, nearly 135 years,

Lacquer Chemistry and Applications. http://dx.doi.org/10.1016/B978-0-12-803589-4.00001-8

international lacquer research has expanded, a large number of research papers have been published, and lacquer chemical research has made great progress.

Lacquer chemistry is the study of the properties of lacquer sap. In other words, lacquer chemistry separates, purifies, and analyzes the components of lacquer sap. Of course, based on the analyzed results of lacquer sap, understanding of the plant physiology of lacquer trees is also important. As the discovery of a new organic carbon bonds continues, study of the enzymatic polymerization of urushiol, its homologs, and its derivatives is also important. During the process of conversion of urushiol to a copolymer by enzyme oxidation, lacquer polysaccharides in the lacquer sap spread to form a network of polymerization. With the glycoprotein as an emulsifier and reactive stabilizers, enzymatic polymerization progresses smoothly and creates an insoluble lacquer film. The drying mechanism has a great effect on polymer synthesis, especially the organic synthesis of urushiol.

Modern humans seek a safe, healthy, and comfortable life. Lacquer is the only self-catalyzing natural polymer. After polymerization of the urushiol catechol core by the laccase contained in the lacquer sap, the unsaturated side chains of urushiol are autoxidized. During the drying process, no organic solvent evaporates, only water. Because of the self-drying system, natural lacquer can be called a solvent-free and high-solid content component-type coating material. Therefore, the studies on lacquer sap and film are ongoing. Due to different allocations of lacquer resources, the research on lacquer chemistry is carried out in several countries and areas. Although the results obtained from different research groups have some differences, they can gradually correct and confirm each other to a unified understanding close to the truth. Many research results have been reported, and several books including those written in Chinese [5,6] and Japanese [7–9] have been published. However, with the development of chemical science and advanced technology of instrumentation and equipment, research in lacquer chemistry has made considerable progress. In order for readers to have a comprehensive understanding of the changes in lacquer chemical research, combined with the research of our laboratory, this book introduces the up-to-date achievements in lacquer chemistry research involving component analysis, the relationship between structure and biological activities of lacquer polysaccharides, mechanisms of enzymatic catalysis, immobilization of laccase, analysis of lacquer films, and the standard pyrolysis GC/MS data of lacquer sap, mechanism of lacquer allergies, synthesis of lacquer, and lacquerware technology.

1.2 LACQUER TREE

1.2.1 SPECIES AND DISTRIBUTION

Lacquer trees belong to the genus *Toxicodendron* (formerly *Rhus*) of the family Anacardiaceae, and have more than 73 genera and 600 species all over the world. Most of them grow in the subtropical region of Southeast Asia [5–7]. It is a Cenozoic tertiary relic species, hermaphrodite or dioecious, and has a wide variety of types. Although there are many species in the world, however, only a few kinds of lacquer

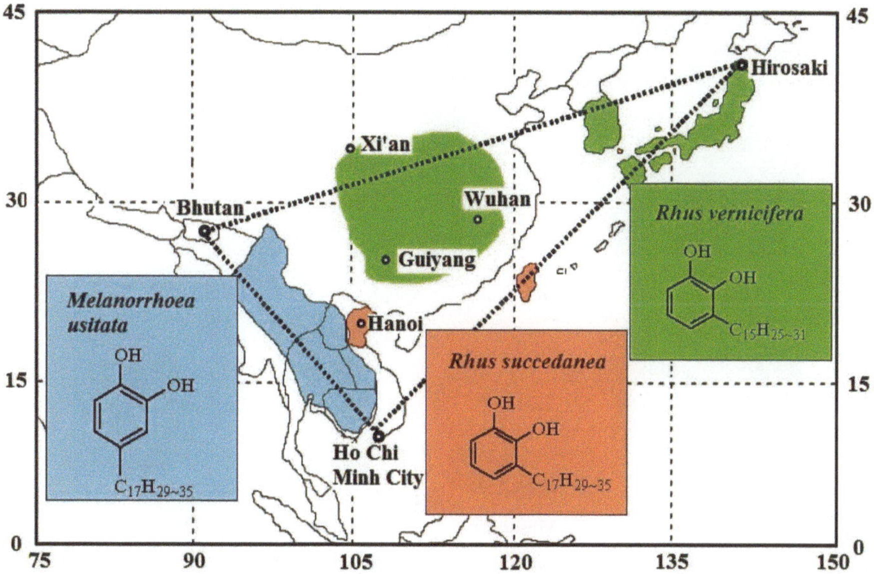

FIGURE 1.1

World distribution of lacquer trees. Urushiol is obtained from *T. vernicifluum*, laccol from *T. succedanea*, and thitsiol from *G. usitata*.

trees, which grow in the evergreen forest of East Asia, are able to produce lacquer sap. In Asian countries, lacquer sap is obtained from the lacquer trees of *Toxicodendron* (*Rhus*) *vernicifluum* in China, Japan, and Korea, *Toxicodendron* (*Rhus*) *succedanea* in Vietnam and Taiwan, and *Gluta* (*Melanorrhoea*) *usitata* in Myanmar, Cambodia, Lao, and Thailand (Figure 1.1).

The family was named *Ana* (looks like) and *Cardiac* (heart) because the seed of the lacquer tree looks like a heart [8]. The seed has an ectocarp, mesocarp, and a wax layer, as shown in Figure 1.2.

It was formerly called *Rhus* from the Greek "Rhudd," meaning "red," because the leaves of lacquer trees become red in late autumn (Figure 1.3). The constituents and properties of the lacquers differ not only with the species, but also with the age of the tree, location, and season of collection. The saps of lacquer trees are composed of catechol derivatives (60-70%), water (20-30%), plant gum (4-10%), glycoproteins (3-5%), and the enzyme laccase (1.5-2%) [9]. The catechol derivative of *T. vernicifluum* lacquer is urushiol, of *T. succedanea* lacquer is laccol, and of *G. usitata* lacquer is thitsiol [8,10].

1.2.2 BIOLOGICAL SPECIFICITIES

The early family name of the lacquer tree was *Rhus* from Greek; however, plant taxonomy found a kind poison in the lacquer tree that can cause human skin allergies, so it was separated from *Rhus* and named *Toxicodendron*. The scientific name of

FIGURE 1.2

Lacquer seeds. (a) Seeds, (b) Dry seeds, (c) Ectocarp, mesocarp, wax-pericarp, and dewaxed seed.

FIGURE 1.3

Lacquer trees in (a) July and (b) December.

the lacquer tree is *Toxicodendron vernicifluum* (Stokes) Barkley. Later, it was found that the allergen is the lipid component of lacquer sap. *Rhus vernicifluum* (or *Rhus vernicifera*) is usually used in the literature.

The lacquer tree is a tall, broad-leaved, deciduous tree. *T. vernicifluum* species can grow up to 5-20 m, *T. succedanea* are shorter at 2-5 m, but *G. usitata* can grow up to 30 m. It usually grows at an altitude of 400-2800 m. Lacquer trees always produce new leaves in spring, flourish in summer, begin to shed their leaves in late autumn,

FIGURE 1.4

Cross section of lacquer tree stem.

and are completely defoliated in winter. During growth and development, a lacquer tree requires more moisture and a warm climate, so July to October is the best season to tap lacquer sap. Each part of a lacquer tree contains a certain amount of sap. The fresh sap is white, and after contact with the air, the sap is oxidized to a deep color.

Lacquer trees have a woody stem, and the stem structure is similar to that of dicotyledons. From the outside to inside, the stem consists of bark, phloem, and xylem, as shown in Figure 1.4. The bark is a phellogen composed of dead cells which are suberized that make it air- and water tight. The second phloem is formed from the vascular cambium outward, and it consists of phloem rays, sieve tubes, companion cells, phloem parenchyma, sclereid (stone) cells, and lacquer canals. It plays a role in transporting nutrients during the growth of lacquer trees. The vascular cambium is composed of several layers of neatly arranged meristematic cells, and outward and inward layers produce secondary phloem and secondary xylem, respectively.

A detailed discussion of the botany of the lacquer tree can be found in textbooks. Here we simply introduced the structure of the stems to help the readers understand the key points of lacquer chemistry.

The organs of the lacquer tree closely related to lacquer chemistry are the laticiferous canals. A laticiferous canal is an intercellular canal, an ellipse, or round hole in the phloem of stem, root, and leaves, the place of secretion and storage of lacquer sap. The laticiferous canals can be divided into four types according to the shape: single tube, single branch, parallel tubes, and nets (Figure 1.5). The diameter of a laticiferous canal is about 85-350 μm.

The bark structure of a lacquer tree is shown in Figure 1.6. It shows wood rays (R), a radial arrangement of ray parenchyma cells; stone cells (SC), a kind of sclerenchyma cell; and laticiferous canals (L) in the bark, respectively. A drop of lacquer sap (Sap) also can be observed.

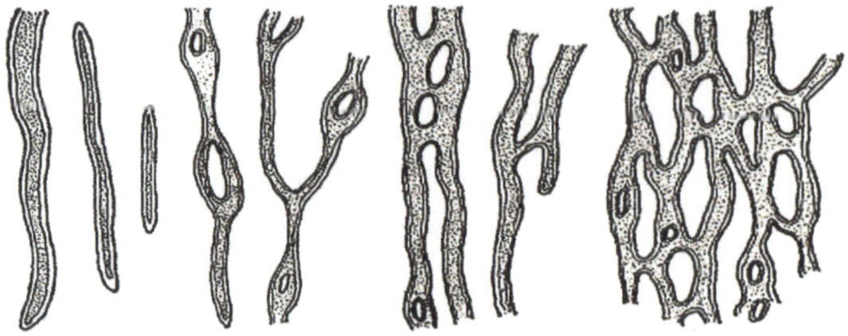

FIGURE 1.5

Schematic diagram of laticiferous canals.

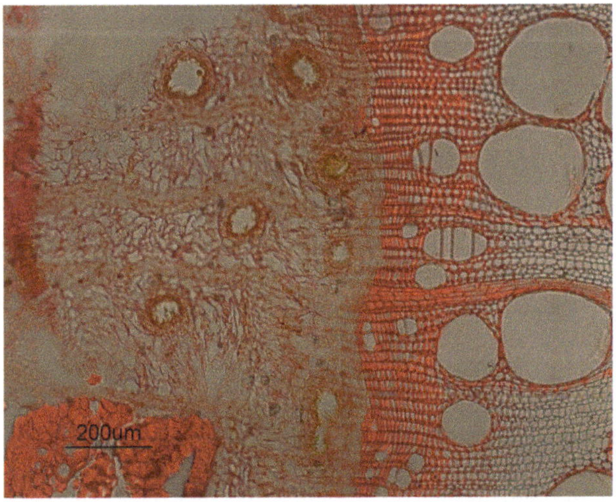

FIGURE 1.6

Bark structure of lacquer tree.

The raw lacquer is produced by laticiferous canals. Although laticiferous canals are found across the various organs of the lacquer tree, most of them exist in stem phloem. Lacquer is synthesized mainly in the plastid and endoplasmic reticulum in a cell, after intracellular translocation, and the sap is stored in laticiferous canals. If the tree is hurt, it will secrete sap to protect the wound, similar to an animal's wound protection through blood coagulation.

The quantity and diameter of laticiferous canals differs in different species of lacquer trees, and affects the yield of lacquer sap. In general, a tree with more laticiferous canals per unit area, bigger diameter, and fewer stone cells yields a higher amount of lacquer sap.

There are many variations in lacquer trees due to their wide geographical distribution, and different "species" names are also found in the literature. However,

Table 1.1 Classification of lacquer trees

Original area	Old name	New name	Main component
China, Japan, Korea	Rhus vernicifera	Toxicodendron vernicifluum	Urushiol
Vietnam, Taiwan	Rhus succedanea	Toxicodendron succedanea	Laccol
Myanmar, Lao, Thailand, Cambodia	Melanorrhoea usitata	Gluta usitata	Thitsiol

during the in-depth study of lacquer, understanding of lacquer trees has increased, and a correct classification has been obtained, as summarized in Table 1.1.

Different species of lacquer trees have different film-forming components. In this book, the main component is named according to its chemical structure. Urushiol is a mixture of 3-substituted catechol derivatives with $n=15$ carbon chains with 0-3 olefins, and found in sap tapped from a *T. vernicifluum* lacquer tree. Laccol is a mixture of 3-substituted catechol derivatives with $n=17$ carbon chains with 0-3 olefins, and found in the sap tapped from a *T. succedanea* lacquer tree. Thitsiol is a mixture of 4-substituted (with a little 3-substituted) catechol derivatives with $n=17$ carbon chains, and found in the sap tapped from a *G. usitata* lacquer tree (Table 1.1).

1.2.3 COLLECTION OF LACQUER

When a lacquer tree is subjected to mechanical injury, that is, when the laticiferous canal is cut, it will exudate lacquer sap. Whether a lacquer tree can be tapped to collect lacquer sap or not, depends not only on the size of tree but also on its developmental status. Generally, lacquer sap can be tapped only from an adult tree. The seedling and juvenile tree have immature laticiferous canals, and if a tree of this age were to be tapped to collect sap, not only will it produce less sap and of poor quality, but it will also affect the normal growth of the trees. Of course, if a lacquer tree is too old, the yield also will be reduced.

The tapping method differs with the type of lacquer production. Figure 1.7 shows the horizontal type, V type, egg type, and vertical or oblique types of lacquer sap tapping.

Each tapping type also has several variants, as shown in Figure 1.8. In general, the horizontal type is popular in Japan, and is also sometimes used in Vietnam; V type and egg types are popular in China, Vietnam, and Myanmar; vertical and oblique types are popular in Thailand and Myanmar.

The lacquer tree is a deciduous tree, and the leaves will completely falloff in winter. This phenomenon is called "falling in winter." Different kinds of lacquer trees shed their leaves at different times. The leaf growth period also is different in different tree species, with the longest one being 7 months, and the shortest only 5 months. The largest harvest is collected during the leaf heyday, and this is also associated with promoting assimilation of laccase. Collection is carried out between

FIGURE 1.7

Lacquer tree tapping methods.

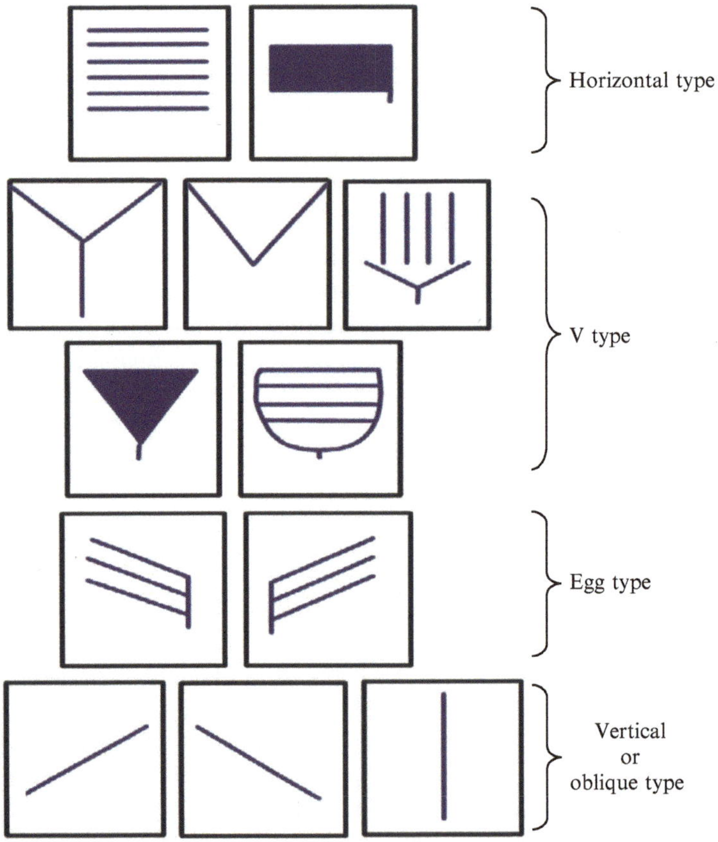

FIGURE 1.8

Schematic diagram of tapping types.

June and October. On sunny days, the best tapping time is from dawn to sunrise. At this time, due to the weak transpiration of lacquer trees and high relative humidity of the environment, the secretion of lacquer sap is fast and continues for a long time. A cloudy and hot day is best for tapping. Lacquer sap is secreted evenly from morning till night, and can be tapped throughout the day. Because the lacquer sap will contain too much water, a rainy day is not suitable for collection of lacquer sap.

1.2.4 YIELD AND QUALITY

Lacquer is an excellent coating material widely used in the chemical and petrochemical industries, shipbuilding, and crafts. However, the yield of lacquer is not very high for various reasons. Because the collection of lacquer mainly involves a manual rather than a mechanical operation, and lacquer allergy, lacquer trees grow in inconvenient places, lacquer sap is expensive.

Many factors affect the yield and quality of lacquer sap. Meteorological conditions, growth environment, tree species, tree age, collection season, and tapping techniques will affect the yield and quality. Long periods of sunshine and high temperature can increase the lacquer yield, while increased rainfall, low temperatures, and short periods of sunshine will reduce the yield. Because the production of lacquer sap is closely related to the tree's assimilative capacity and bark thickness, the species in which bark thickness increases relatively quickly, such as HuoYanZi (Chinese lacquer tree), sap can be collected after 5 years growth. On the other hand, species in which the bark thickness increases relatively slowly, such as GaoBaChi (Chinese lacquer tree), need up to 10 years growth before sap can be collected.

The collection of lacquer is usually divided into three harvests. The first collection period is from about the middle June to early July, when the lacquer sap harvested has a lot of water and a high viscosity. The second collection is about from the middle of July to late August. During this period, because the life activity of a lacquer tree is strong, the sap secretion also is very strong, and because the lacquer tree transpires strongly, the physiological water makes up less of the sap content, so the yield is very high and the quality is very good. The third collection period is from early September to early October. During this period, the decrease in temperature and transpiration weakens the laticiferous canal cell activity, decreasing the secretion, and the harvested lacquer sap has a light color and low lipid concentration.

In the growth environment, the soil is a very important factor, and generally yellow or black soil is best. Because the altitude affects the distribution of light, heat, and water, the growth of trees of the same species in different altitudes has different qualities. Lack of water in soil and too low humidity in the growth environment affect the yield and quality of lacquer sap. It has reported that ionization of saturated urushiol in alkaline solution is favorable in the presence of water molecules. However, too much moisture increases the yield of lacquer sap, but decreases the quality. The relationship between sap quality and growing altitudes of lacquer trees are summarized in Table 1.2.

Table 1.2 Relationship between sap quality and altitude of lacquer trees

Height (m)	Lipid%	Monoene%*	Diene%*	Triene%*	Saturated%*	Water%	Oil%
550	67.5	25.63	5.56	67.61	1.19	16.8	1.4
620	65.7	19.93	9.84	69.02	1.23	16.8	1.2
630	70.6	19.99	9.76	69.12	1.12	10.2	0.3
640	68.1	18.94	9.27	70.79	1.01	13.8	1.3
650	62.5	19.29	10.44	69.37	0.91	18.4	1.1
820~860	63.0	21.74	11.59	65.44	1.24	21.0	1.2
950~1000	62.9	18.82	11.90	68.18	1.10	16.7	1.0

Content of monoene, diene, triene and saturated urushiol refers to the percentage of the total proportion of lipid component.

Table 1.3 Relationship between sap quality and growth mountains slope of lacquer trees

Slope orientation	Lipid (%)	Water (%)	Oil (%)
East	68.9	17.2	1.99
Southeast	71.4	16.6	2.10
South	67.2	22.9	1.47
West	64.2	25.5	1.27
Northwest	67.7	22.8	1.33
North	68.8	16.6	1.12
Northeast	70.8	18.2	3.48

It can be observed from Table 1.2 that the optimum altitude is 620-650 m. Of course, different species of lacquer tree prefer different altitudes; for example, the Chinese GaoBaChi grows best at an altitude of 930 m, HuoYanZi at 960 m, and DaHongBao at 820-852 m.

The slope of a mountain also obviously affects the quality of lacquer sap because the slope affects the duration and intensity of sunshine on a lacquer tree. In general, lacquer sap obtained from lacquer tree grown on a mountain facing the sun has a bright color, high lipid content, less water, and strong smell, and is good-quality lacquer. On the contrary, lacquer sap obtained from a lacquer tree grown on the back of a mountain has a dull color and is rather muddy due to the higher water and lower lipid content. The smell is not strong and the quality is not as good. Table 1.3 summarizes the different qualities of lacquer sap obtained from trees grown on different mountain slopes.

It can be seen that the sunny slopes facing East, Southeast, and Northeast have a higher lipid component and lower water concentration, and the mountain slope with its back to the sun of west and south have a lower lipid component and higher water concentration. These characteristics are related to sunshine duration.

The same lacquer tree will produce lacquer sap of different qualities at different ages. In general, during the period of vigorous growth of a lacquer tree, that is, the age of 8-9 years, the average lipid component is more than 70% and the water is less than 20%. From 8 to 15 years, the total lipid component tends to be reduced and the water content increased, as summarized in Table 1.4.

Tapping techniques also affect lacquer quality. Different qualities of sap can be obtained from the same lacquer tree by different tapping techniques. In general, a skilled lacquer farmer can tap an incision of moderate width and suitable depth, and such an incision does not damage the xylem. The lacquer sap obtained has a clear layer, clean and fresh scent, high lipid component, low water concentration, and high yield. On the other hand, when done by the low-skilled lacquer farmer, the operation is rough and should the xylem be cut, water in the xylem infiltrates the lacquer sap, so the sap obtained has relatively high water and low lipid component concentrations. The sap layer is not clear and the quality is poor.

Table 1.4 Relation of lacquer tree age and sap quality

Age (years)	Lipid (%)	Water (%)	Oil (%)
8	72.3	15.0	2.0
9	71.9	14.9	0.8
10	68.3	17.4	0.7
11	64.1	22.3	0.8
12	69.3	17.7	0.9
13	69.0	20.0	0.4
14	61.2	26.0	0.6
15	63.8	27.1	0.6
25	67.8	21.4	2.3

Stimulants also affect the yield and quality of lacquer sap. It has reported that a stimulant used to stimulate a lacquer tree can increase the yield 25-40%, delay the leaves falling 7-15 days, and accelerate healing of the incision. 2-Chloro-ethyl phosphonic acid (commercial name: ethephon) and calcium carbide are often used to treat lacquer trees, and they produce plant stimulants of ethylene or acetylene, respectively, during the decomposition process. Ethephon is relatively stable below pH3, but gradually decomposes when the pH is more than 4. The reaction equation is as follows:

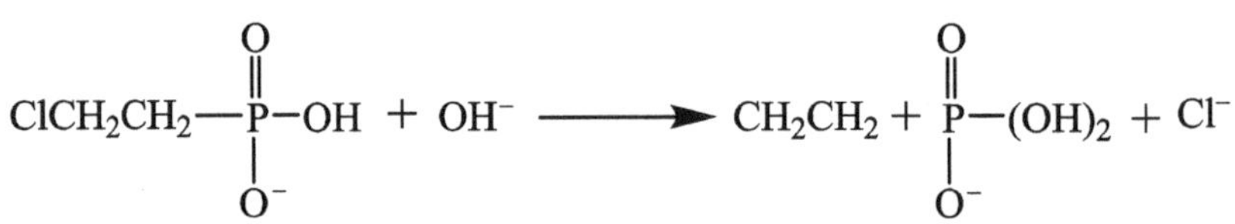

Calcium carbide decomposes to acetylene in the presence of water.

$$CaC_2 + H_2O \rightarrow C_2H_2 + CaO$$

The use of ethephon is reported in rubber production technology. It has two types: a water type and an oil type. Generally, the concentration of the water soluble type is 8% and the dose is 2-3 ml. There are three treatment methods for a lacquer tree. (1) Smearing: the rough skin is gently scraped to form a band around about 1/3 of the tree circumference and 3 cm wide with a spatula in the middle or middle-upper trunk, then the water-type or oil-type ethephon is smeared on the band. (2) Root injection method: a ditch is dug about 15 cm deep in the ground about 50 cm from the root on the side with more incisions in the lacquer tree, and then water-soluble ethephon is poured in evenly. (3) Hole injection method: a hole about 3 cm deep is made in the side of lacquer tree with more incisions (make two holes for a lacquer tree that has a trunk diameter over 20 cm), and then water-soluble ethephon is poured in.

Table 1.5 Components and characteristics of control lacquer sap compared with ethephon-treated sap

Items	Control sap	Treated sap	Remarks
Solid content (%)	87.50	86.00	Drying for 1 h at 15 °C
Lipid content (%)	82.00	80.25	Tested using Xylene
Gum content (%)	3.75	4.25	Tested using Xylene
Nitrogenous (%)	1.75	1.5	Tested using Xylene
Touch free dry (h)	1.5 ~ 2	3.5 ~ 4	Dried at 23 °C, 80%RH
Film impact strength	≦ 20	≦ 20	cm × kg
Adhesion	3 level	3 level	Zoned circle method
Hardness	≧ 0.56	≧ 0.56	Film/glass

The characteristics of lacquer sap collected from a lacquer tree treated with ethephon are not obviously different from the natural production, as summarized in Table 1.5.

It has been reported that the bark of the rubber tree is prematurely aged by ethephon, and maybe the main reason as it causes the death of the bark. Although premature bark aging was not found in lacquer trees, stone cells appear early compared with the control. However, it is still not clear whether the laticiferous canals cells are damaged or not during treatment with ethephon process.

REFERENCES

[1] J. Hu, Conservation of water-logged lacquerwares unearthed in China, in: Conservation and Reatoration of Cultural Property: Conservation of Far Eastern Objects, Tokyo National Research Institute of Cultural Property, Tokyo, Japan, 1980, pp. 89–112.

[2] O. Vogl, J. Bartus, M. Qin, J. Mitchell, Progress in Pacific Polymer Science 3, Springer-Verlag, Berlin, 1994, p. 423.

[3] O. Vogl, Oriental lacquer, poison ivy, and drying oils, J. Polym. Sci. A: Polym. Chem. 38 (2000) 4327–4335.

[4] E.J. Kidder, Ancient Peoples and Places, Thames & Hudson, Japan, 1959.

[5] J. Gan, Lacquer Chemistry, Chinese Academy of Science, Beijing, China, 1984.

[6] F. Zhang, Chinese Lacquer: Technology and Conservation, Science Publisher, Beijing, China, 2010.

[7] T. Terada, K. Oda, H. Oyabu, T. Asami, Lacquer—the Science and Practice, Rikou Publisher, Tokyo, Japan, 1999.

[8] T. Miyakoshi, K. Nagase, T. Yoshida, Progress of Lacquer Chemistry, IPC, Japan, 1999, p. 43.

[9] J. Kumanotani, Talk About Lacquer, Bungeisha Publisher, Tokyo, Japan, 2012.

[10] R. Oshima, J. Kumanotani, Structural studies of plant gum from sap of the lac tree, *Rhus vernicifera*. Carbohydr. Res. 127 (1984) 43–57.

CHAPTER

History of lacquer chemistry

2

Lacquer chemistry, which is independent of oil and other fossil resources, uses only the natural cycles of plant materials to make various craft and industrial products. The polymerization of lacquer sap is catalyzed by the oxidation of lipid components (urushiol, laccol, and thitsiol) by laccase contained in the sap, followed by a coupling reaction as well as an autoxidation reaction on the long aliphatic unsaturated side chain. During the drying process, no organic solvent evaporates, only water. In recent years, new concepts of green and sustainable chemistry have been instituted and are considered in lacquer chemistry. Lacquer is a natural resource as long as lacquer trees are cultivated (Figure 2.1).

2.1 BRIEF HISTORY OF LACQUER CHEMISTRY

Like other disciplines, the study of lacquer chemistry has a long development and has been repeatedly revised, mainly in the following points:

(1) The research has been carried out in several different countries and areas due to different lacquer resource allocations. Although the results from different laboratories exhibit certain differences, they could provide mutual correction and then gradually be unified. For example, in the latter half of the nineteenth century, *T. vernicifluum* lacquer sap was studied in Japan, and almost at the same time, France discovered this resource in its colony, Vietnam, and analyzed *T. succedanea* lacquer sap independently. Therefore, georelationships, production, and utilization are the power to promote the development of lacquer chemistry.

(2) Lacquer chemistry developed along with the development of chemistry and biology as well as with the improvement of instrumentation and equipment technology. In the primary stage of lacquer chemistry research, in terms of studies of composition separation, structure analysis, and mechanisms, especially the catalytic mechanism of laccase, all reflected the phenomenon and qualitative research due to the basic theories of chemistry and biology. For example, the mechanism of active copper center of laccase gradually became clear after the development of spectroscopy technology.

(3) The study of lacquer chemistry occurs more in the laboratory and less in the wilderness or plant due to limitations of lacquer resources. Because lacquer is a

Lacquer Chemistry and Applications. http://dx.doi.org/10.1016/B978-0-12-803589-4.00002-X

(a) (b)

FIGURE 2.1

Lacquer tree seedlings (a) and forest (b).

special Asian product not naturally available in Europe and America, European and/or American studies should only use extraction vacuum lacquer samples provided by Japan. However, because lacquer composition is easily changed (e.g., oxidized), some results of research on laccase are quite different from those of natural laccase, and thus the reliability of the results obtained is poor. For these reasons, fungal laccase instead of lacquer laccase has been used to study the catalytic mechanism. Although the results can be published and compared, they are distorted. Another example of the effect of the distribution of resources is the French loss of its Eastern colonies, making studies of lacquer chemistry in France increasingly rare.

(4) Applied research needs to continue and improve. Our lab is continuously engaged in the reforms and improvements of lacquer, such as blends, hybrids, and microdispersion. Of course, synthesis of lacquer sap also needs to continue. It is also necessary to do interdisciplinary and international research in lacquer chemistry.

The systematic study of lacquer has long been carried out in Japan. In 1878, the Japanese researcher Ishimatsu Sadama, a student at the University of Tokyo, tried to separate lacquer sap with ethanol [1]. He added anhydrous ethanol to the lacquer sap, then mixed and filtered it. The filtrate was evaporated, and an "alcohol-soluble" residue was obtained. The residue was treated with hot water, stirred, and filtered, the filtrate was evaporated to obtain a "gum" and the matter insoluble in hot water was considered the residue. The residue included bark, cellulose, and a mud-like matter (glycoprotein), and the "gum" was like an ordinary gum (gum Arabic) according to element analysis. The alcohol-soluble portion was the main component of lacquer sap that had a brown color and was rather viscous. Although Ishimatsu attempted to determine the molecular formula of the alcohol-soluble residue by elemental analysis, the results were not satisfactory. The results of Ishimatsu are summarized in Table 2.1.

Five year later, urushiol, the main component of lacquer, was investigated by Yoshida, who was one generation younger than Ishimatsu, in the same laboratory

Table 2.1 Compositions of lacquer sap (separated by Ishimatsu sadama)

Alcohol soluble	58.24
Gum (alcohol insoluble, water soluble)	6.34
Residue (both alcohol-, water insoluble)	2.24
Moisture and other volatile matters	33.18
Total	100

at the University of Tokyo [2]. He used acetone as the separation solvent to instead of anhydrous alcohol. Because the acetone-soluble portion has a –COOH group according to a Litmus test, Yoshida named the acetone-soluble portion "urushi acid" and gave it the molecular formula of $C_{14}H_{18}O_2$ from element analysis results. He also pointed out that the part insoluble in ethanol contained tree gum and an enzyme that made the lacquer self-drying, and found that the enzyme is not stable, but subject to thermal decomposition. Almost at the same time, in 1890, a French researcher, Gabriel Bertrand, studied Vietnamese lacquer and suggested that the acetone-soluble component was a kind of polyphenol and named it "laccol." The alcohol-insoluble water-soluble material was a "gum" containing a small amount of nitrogen. It was not clear whether this nitrogen belonged to an enzyme protein or not, but because it can catalyze laccol to dry and solidify, he named it "laccase" [3–6]. In fact, the acidic property of lacquer is due to the phenolic hydroxyl group, which was later confirmed by Miyama, and the main lipid component of lacquer was named "urushiol" [7]. In 1907, Majima found that the urushiol from *T. vernicifluum* tree grown in Japan is a hydroxyl compound with a divalent benzene nucleus due to the *O*-dihydroxy benzol reaction. He found that urushiol is not composed of a single substance, but a mixture having the basic chemical structure shown in Scheme 2.1. The side chain is a straight 15-carbon chain mixture with saturated and unsaturated monoenes, dienes, and trienes in the carbohydrate structures [8–12].

Lacquer saps from *T. succedanea* and *G. usitata* trees were also studied by Majima. He concluded that laccol has 17-carbon side chain combination on the 3-position of a catechol ring mixture with saturated and unsaturated monoenes, dienes, and trienes. On the other hand, thitsiol has a 17-carbon side chain combination on the 4-position of a catechol ring mixture with saturated and unsaturated monoenes, dienes, and trienes, as shown in Scheme 2.2.

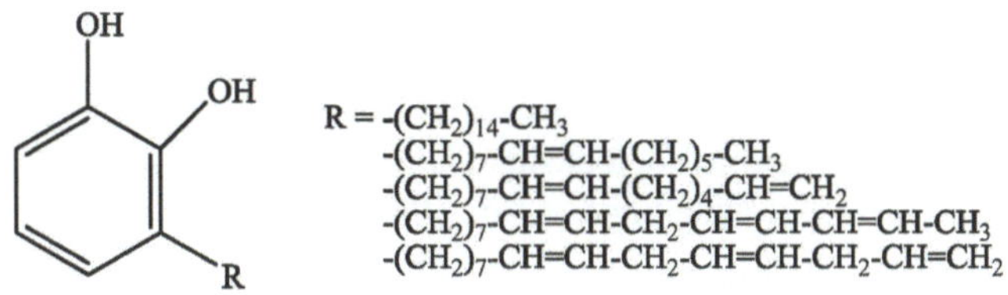

SCHEME 2.1

Urushiol structure (Majima).

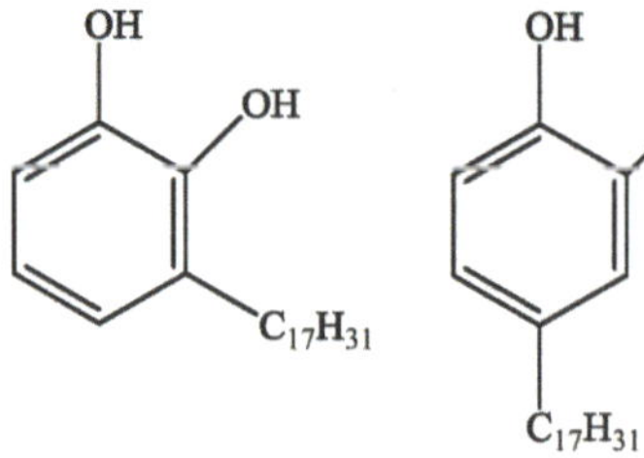

SCHEME 2.2

Laccol and thitsiol structures (Majima).

Majima and his coworkers published many research papers on lacquer chemistry. They revealed a part of the urushiol structure by carbonization, nitric acid oxidation, methylation, acetylation, potassium permanganate oxidation, and ozone oxidation methods. They also successfully synthesized the hydrogenation of urushiol.

In China, Xu Shanxiang was the first to study lacquer chemistry in 1926. He reported physical and chemical properties, use, and drying behaviors of lacquer [13]. After that, polychrome formulas of lacquer [14], morphological and physiological bases of the lacquer tree [15], and lacquer toxicity [16] were reported.

It is worth mentioning here that Bertrand and Brooks studied the main component of lacquer sap from *Gluta laccifera* trees grown in Cambodia and Thailand, and named it "moreacol." However, due to insufficient research, it is considered to be the same structure as the thitsiol described by Majima.

About 30 years later, in 1954, Dawson et al. examined the numbers and positions of the double bonds in the side chains of urushiol by an ozonolysis method and ultraviolet spectroscopy [17–19]. They compared the structure of poison ivy and lacquer urushiol, analysis of -ene numbers, and locations on the side chain. They found that a double bond exists at the poison ivy urushiol side chain terminal (14′) without a conjugated double-bond structure; in contrast, the lacquer urushiol side chain has a conjugated double-bond structure at C-13′ and about 50% trienes in the lacquer urushiol side chain. The conjugated structure has a significant impact on drying polymerization. The results are shown in Scheme 2.3.

Based on their results, Dawson suggested naming the urushiol components as following:

Urushiol—natural material, mixture
Hydrourushiol—saturated side chain
Urushenol—one double bond at side chain (monoene)
Urushedienol—two double bond at side chain (diene)
Urushetrienol—three double bond at side chain (triene)

While the number and position of double bonds in the side chain of urushiol were gradually being revealed, Kumanotani et al. recognized the importance of the geometric structure of the double bond [20,21]. In 1959, Kumanotani began to study

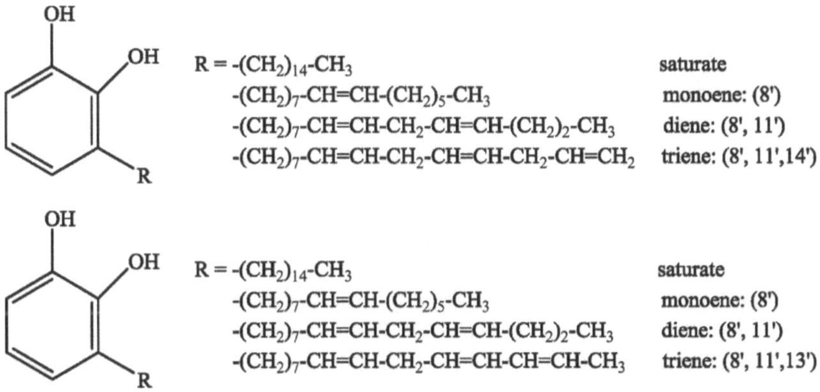

SCHEME 2.3

Urushiol structures of poison ivy (above) and *T. vernicifluum* (below) (Dawson).

natural lacquer chemistry at the University of Tokyo. He systematically separated *T. vernicifluum* lacquer sap with acetone, and obtained urushiol, gum, glycoprotein, enzymes (including laccase, peroxidase, and stellacyanin), and water. The concentrations of each composition were calculated and are summarized in Table 2.2.

Kumanotani revealed not only the double-bond positions, but also the geometric structure of the urushiol side chain using modern instrumental analytic techniques such as gel permeation chromatography (GPC), high-performance liquid chromatography (HPLC), and capillary gas chromatography (GC). Because an urushiol quinone radical was found in the drying process, Kumanotani deduced that the mechanism of lacquer drying is the polymerization of urushiol by the laccase catalyst [22]. After the composition and drying mechanism of lacquer were revealed, Nagase [23] reported the practical application of a synthetic paint that contained lacquer sap, and Terada [24] attempted to synthesize an artificial lacquer film by mixing 3-alkenylcatechol with the laccase from *Pycnoporus coccineus*. Selectively introducing a steric structure and considering the olefin position in the side chain, we have reported the synthesis of 3-[(8*Z*,11*E*,13*Z*)-8,11,13-pentadecatrienyl]catechol, 3-[(8*Z*,11*E*,13*E*)-8,11,

Table 2.2 Composition and properties of lacquer sap

	Urushiol	Gum	Glycoprotein	Laccase	Water
Molecular weigh	310	~84,000	(8000)n	~120,000	18
Polar groups	OH, –O–	OH, –O–, COOMetal (Ca, Na, Mg)			OH
Sugar content (%)		100	10	45	
Protein content (%)		0	90	55	

13-pentadecatrienyl]catechol, and 3-[(8*Z*,11*E*)-8,11,14-pentadecatrienyl]catechol [25], which are the main components of urushiol, and also attempted to prepare lacquer film using synthetic urushiol derivatives. Thus, the composition, structure, and properties of lacquer as a natural polymer have been studied for 135 years by many researchers.

2.2 SEPARATION OF LACQUER SAP

Lacquer sap just tapped from lacquer tree is a milky-white water-in-oil emulsion. It will turn brown (*T. vernicifluum*, *T. succedanea*) or black (*G. usitata*) when exposed to the air for about 5 minutes or more. It can be divided into many components by acetone and water, as shown in Figure 2.2.

Sap from different species of lacquer trees, the same species grown in different locations, or collected at different season has different compositions. Lacquer sap collected from *T. vernicifluum* lacquer trees grown in China, Japan, and Korea, from *T. succedanea* lacquer trees growth in Vietnam, and from *G. usitata* lacquer trees grown in Myanmar and Thailand have been analyzed in detail in our laboratory. The results showed that *T. succedanea* sap has relatively more water and gum, correspondingly less lipid component, and *G. usitata* sap has less acetone powder (gum and nitrogen-containing material) compared with *T. vernicifluum* sap, as summarized in Table 2.3.

We have tried to photograph fresh lacquer sap in our laboratory, but a photograph does not reveal the internal structure because the sap is transparent. After trying many dyes, we found that methylene blue stains lacquer sap, and a photomicrograph of fresh lacquer sap is shown in Figure 2.3A.

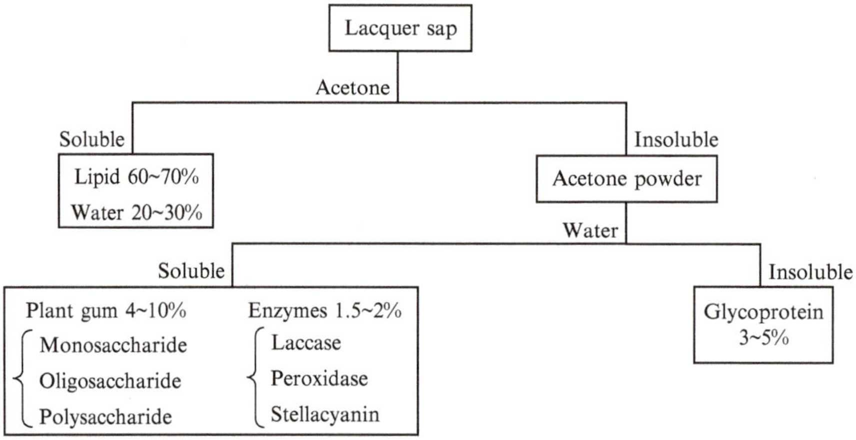

FIGURE 2.2

Separation of lacquer sap.

Table 2.3 Compositions of lacquer sap by tree species and growing region

Origin/species	Lipid (%)	Water (%)	Gum (%)	Nitrogen-containing material (%)
Japan/T. vernic.	71.5	18.3	7.5	2.6
China/T. vernic.	66.0	24.6	6.0	0.3
Korea/T. vernic.	70.7	20.4	6.2	2.7
Vietnam/T. suc.	47.2	34.9	15.0	2.8
Myanmar/G. usi.	67.2	29.5	2.2	1.1

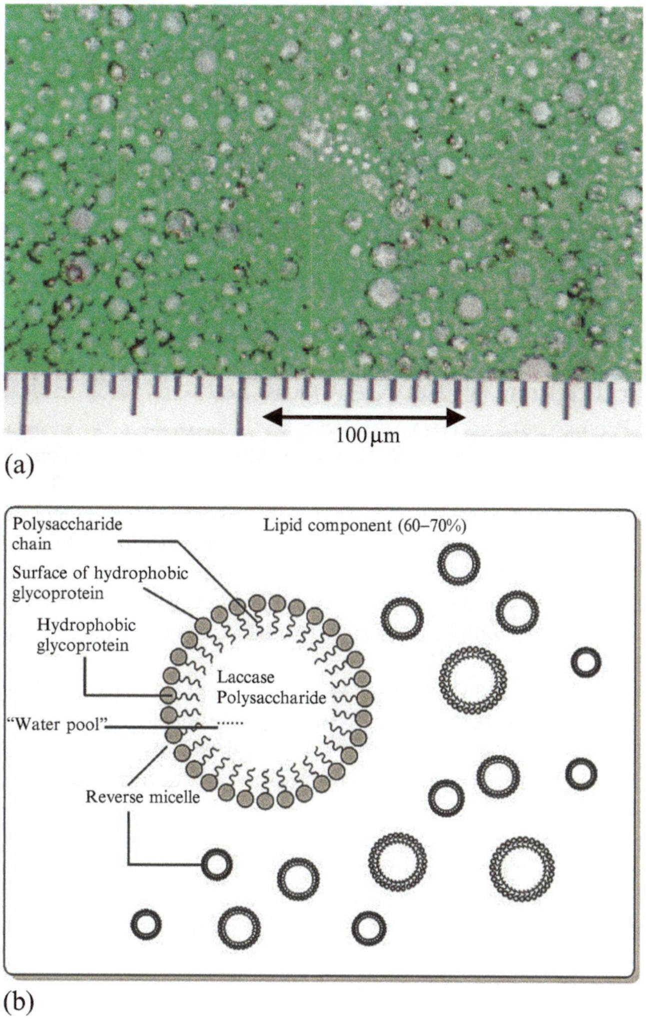

FIGURE 2.3

Photomicrograph of lacquer sap (a) and image of water drop (b).

Many water drops about 10 μm diameter can be seen dispersed in the green lipid ocean. This is evidence of the water-in-oil (w/o) emulsion of lacquer sap. The same results also were reported by Ota et al. using Sudan III as the dye [26], and in this case, the colorless water drops are dispersed in the red lipid ocean. The water-soluble components, such as laccase, polysaccharides, and stellacyanin are dissolved in the reverse water micelles (Figure 2.3B).

REFERENCES

[1] J. Kumanotani, Urushi (Oriental lacquer)—a natural aesthetic durable and future-promising coating, Prog. Org. Coat. 26 (1995) 163–195.

[2] T. Miyakoshi, K. Nagase, T. Yoshida, Progress of Lacquer Chemistry, IPC Publisher, Tokyo, Japan, 1999, p. 43.

[3] T. Terada, K. Oda, H. Oyabu, T. Asami, Lacquer—the Science and Practice, Rikou Publisher, Tokyo, Japan, 1999, p. 25.

[4] Y. Yamauchi, R. Oshima, J. Kumanotani, Configuration of the olefinic bonds in the heteroolefinic side-chains of Japanese lacquer urushiol: separation and identification of components of dimethylurushiol by means of reductive ozonolysis and high-performance liquid chromatography, J. Chromatogr. 243 (1982) 71–84.

[5] B. Reinhammar, Y. Oda, Spectroscopic and catalytic properties of *T. vernicifluum* laccase depleted in type 2 copper, J. Inorg. Biochem. 11 (1979) 115–127.

[6] J. Kumanotani, Urushi—cultural relations, Kagakutokougyo 36 (1983) 151–154.

[7] S. Harigaya, T. Honda, R. Lu, T. Miyakoshi, C. Chen, Enzymatic dehydrogenative polymerization of urushiols in fresh exudates from the lacquer tree, *T. vernicifluum* DC, J. Agric. Food Chem. 55 (2007) 2201–2208.

[8] R. Lu, S. Harigaya, T. Ishimura, K. Nagase, T. Miyakoshi, Development of a fast drying lacquer based on raw lacquer sap, Prog Org. Coat. 51 (2004) 238–243.

[9] R. Lu, T. Ishimura, K. Tsutida, T. Honda, T. Miyakoshi, Development of a fast drying hybrid lacquer in a low relative-humidity environment based on *Kurome* lacquer sap, J. Appl. Polym. Sci. 98 (2005) 1055–1061.

[10] R. Lu, D. Kanamori, T. Miyakoshi, Characterization of thitsiol dimer structure from *Melanorrhoea usitata* with laccase catalyst by NMR spectroscopy, Inter. J. Polym. Anal. Char. 16 (2011) 86–94.

[11] M. Herreraa, G. Matuschek, Fast identification of polymer additives by pyrolysis-gas chromatography/mass spectrometry, J. Anal. Appl. Pyrolysis 70 (2003) 35–42.

[12] T. Honda, R. Lu, N. Kitano, Y. Kamiya, T. Miyakoshi, Applied analysis and identification of ancient lacquer based on pyrolysis-gas chromatography/mass spectrometry, J. Appl. Polym. Sci. 118 (2010) 897–901.

[13] Xu Shanxiang, Paint, Oil & Chem. Rev. 81 (1926) 15–21.

[14] Xu. Bingxi, Chemistry and Industry 5 (1930) 1–4 (in Chinese).

[15] Xu. Zhifang, Science 16 (1932) 724–728.

[16] Jiang Lijin, Experience of Production Technology in Coating Industry, Chemical Industry Publisher, Beijing, China, 1960, pp. 29–43.

[17] R. Lu, Y. Kamiya, Y. Wan, T. Honda, T. Miyakoshi, Synthesis of *T. succedanea* lacquer film and analysis by pyrolysis-gas chromatography/mass spectrometry, J. Anal. Appl. Pyrolysis 78 (2007) 117–124.

[18] Y. Wan, R. Lu, Y. Du, T. Honda, T. Miyakoshi, Does Donglan lacquer tree belong to *T. vernicifluum* species? Inter. J. Bio. Macromol. 41 (2007) 497–503.
[19] K. Urakami, A. Higashi, K. Umemoto, M. Godo, C. Watanabe, K. Hashimoto, Compositional analysis of copoly(DL-lactic/glycolic acid) (PLGA) by pyrolysis-gas chromatography/mass spectrometry combined with one-step thermally assisted hydrolysis and methylation in the presence of tetramethylammonium hydroxide, Chem. Pharm. Bull. 49 (2001) 203–205.
[20] M. Fukushima, M. Yamamoto, T. Komai, K. Yamamoto, Studies of structural alterations of humic acids from conifer bark residue during composting by pyrolysis-gas chromatography/mass spectrometry using tetramethylammonium hydroxide (TMAH-py-GC/MS), J. Anal. Appl. Pyrolysis 86 (2009) 200–206.
[21] X. Chai, T. Shimaoka, Q. Guo, Y. Zhao, Characterization of humic and fulvic acids extracted from landfill by elemental composition, 13C CP/MAS NMR and TMAH-Py-GC/MS, Waste Manage. 28 (2008) 896–903.
[22] R. Widyorini, T. Higashihara, J. Xu, T. Watanabe, S. Kawai, Self-bonding characteristics of binderless kenaf core composites, Wood Sci. Tech. 39 (2005) 651–662.
[23] R. Martine, G. Nicolas, C. Cecile, R. Christian. New analytical methodology for investigating both volatile and nonvolatile constituents from archaeological ceramic vessels, Abstracts of Papers, 225th ACS National Meeting, New Orleans, LA, United States, March 23–27, 2003, GEOC-107.
[24] H. Araki, N. Tatarazako, K. Kishi, K. Kuroda, Evaluation of bioaccumulation potential of 3,4,5-trichloroguaiacol in a zooplankton (*Daphnia magna*) by pyrolysis-GC/MS in the presence of tetramethylammonium hydroxide (TMAH), J. Anal. Appl. Pyrolysis 55 (2000) 69–80.
[25] T. Arii, Complex thermal analysis: evolved gas analysis-mass spectrometry EGA-MS, Rigaku Janaru 39 (2008) 17–25.
[26] T. Terada, K. Oda, H. Oyabu, T. Asami, Lacquer—the Science and Practice, Rikou Publisher, Tokyo, Japan, 1999, p. 57–58.

CHAPTER

Main compositions of lacquer 3

Early in 1883, Yoshida treated Japanese lacquer sap with ethanol and divided it into two parts. Because the ethanol soluble part could be precipitated by treatment with metal salts, he mistakenly believed that it was a kind of acid, and named it "urushi acid" [1]. The ethanol insoluble part can be further divided into two water-soluble and water-insoluble portions.

G. Bertrand extracted a component from Vietnamese lacquer sap, which is similar to Japanese lacquer, in 1898. He called the ethanol soluble part "laccol" and the enzyme "laccase." During further study of the laccase, he found that it promotes oxidation of Vietnamese lacquer sap and other polyhydroxy phenols, such as hydroquinone and catechol. This established the concept of urushiol (laccol, thitsiol) as polyhydroxy phenols [2–5].

3.1 URUSHIOL

Urushiol is the main component of sap tapped from the *Toxicodendron vernicifluum* lacquer tree. It is a catechol derivative having C15 unsaturated hydrocarbon chains consisting of a mixture of monoenes, dienes, and trienes mainly at the 3-position of catechol [6,7]. Rather minor amounts of phenolic lipids having C17 carbon chains instead of C15 carbon chains were also found. Urushiol was separated by Majima using ethanol for the first time in 1883. This separation method is still used; however, in order to maintain the laccase activity, acetone is used instead of ethanol.

As mentioned above, in the primary stage of separating lacquer sap, Majima called the ethanol soluble part "urushi acid." In order to confirm the structure, he continuously treated the urushi acid with petroleum, ether, and water, after removing the resinous material; he then purified it by distillation, and a colorless viscous liquid was obtained. It has the following properties.

Boiling point: ~210 °C (0.4-0.5 mmHg)
Specific gravity ($d_4^{215\,°C}$): 0.9687
Refractive index ($n_D^{215\,°C}$): 1.52341
Molecular weight: 298-305

It can reduce ammoniacal silver nitrate solution, generate a white precipitate in reaction with lead acetate, produce a black-green color in reaction with ferric chloride or alkali, and its alkali solution is easily oxidized in air. Based on these results,

Lacquer Chemistry and Applications. http://dx.doi.org/10.1016/B978-0-12-803589-4.00003-1

it is a "phenol" but not an "acid." Majima named it "urushiol" because the lacquer is called "urushi" in Japanese.

With the progress of modern instrumental analysis technology, the chemical structure of urushiol was studied by instrument analysis. After evaporation of the acetone-soluble portion under pressure, the lipid component of lacquer is obtained as a residue. However, during the collection process, urushiol monomers are oxidized in air by laccase and tend to form dimers and oligomers. It is necessary to remove the dimers and oligomers in the structural analysis of an urushiol side chain. Our laboratory has separated and purified the urushiol monomer using an HPLC-908 system equipped with an HPLC column [Tosoh, TSK-gel G1000H6, ϕ 7.6 mm × 300 mm × 2; exclusion limit $M_r = 1000$] and UV-254H (254 nm) detector, eluted with $CHCl_3$ at a rate of 1 ml/min. The result is shown in Figure 3.1.

Peaks 1, 2, and 3 were oligomer, dimer, and monomer, respectively, according to analyzed results. Laccol and thitsiol also were analyzed using the same method and the ratios of monomers and oligomers including dimer of urushiol, laccol, and thitsiol are summarized in Table 3.1.

Urushiol has been analyzed by infrared (IR), nuclear magnetic resonance (NMR) spectroscopy, and gas chromatography/mass spectrometry (GC/MS). The IR spectrum is shown in Figure 3.2. As shown, 750-770 cm^{-1} was assigned to the characteristic absorption

Table 3.1 Ratio of monomer and oligomer of urushiol, laccol, and thitsiol

Origin/species	Monomer	Oligomer
Japan/*T. vernicifluum*	85	15
China/*T. vernicifluum*	80	20
Korea/*T. vernicifluum*	80	20
Vietnam/*T. succedanea*	74	26
Myanmar/*G. usitata*	45	55

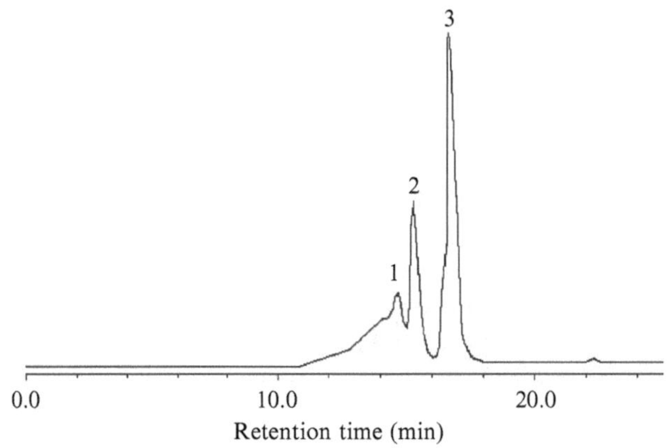

FIGURE 3.1

HPLC spectrum of urushiol.

of 1,2,3-substituted at benzene ring, 960-980 cm^{-1} for the olefin and triene, 1030 and 1080 cm^{-1} for C–O stretching vibration absorption, 1280 cm^{-1} due to the absorption band of phenol hydroxyl, 1380 cm^{-1} for symmetric bending vibration of CH_3, 1460 and 1470 cm^{-1} for ether olefin, 1600 and 1650 cm^{-1} for stretching vibration of C=C, 1745 cm^{-1} for stretching vibration of ester C=O, 2855 cm^{-1} for symmetric vibration of C–H, 2940 cm^{-1} for asymmetric vibration of C–H, 3020 cm^{-1} for C–H group, and 3500 cm^{-1} due to the absorption bands of O–H of catechol ring, as summarized in Table 3.2.

The ^1H NMR spectrum of urushiol is shown in Figure 3.3. The assignments of absorption peaks were based on the two-dimensional NMR measurement. The peak at

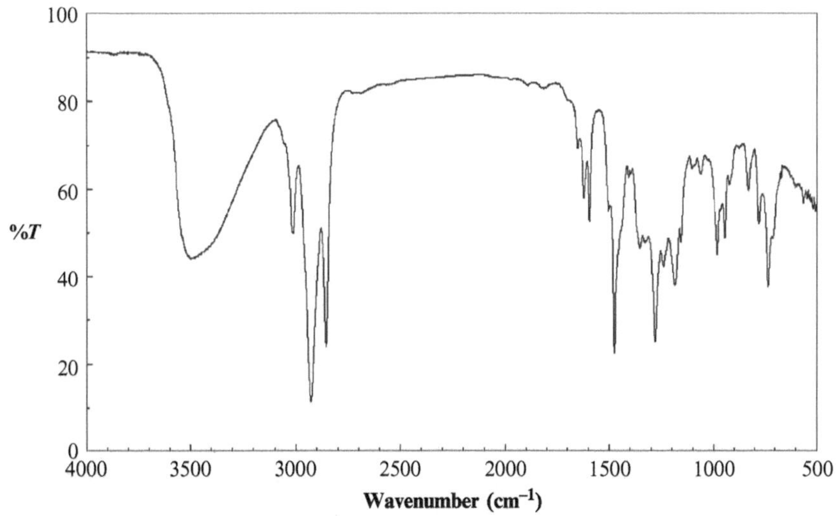

FIGURE 3.2

IR spectrum of urushiol.

Table 3.2 Results of urushiol IR assignments

Wavenumber (cm^{-1})	Groups
750-770	1,2,3-substituted at benzene ring
960-980	C=C (=CH *trans*)
1030, 1080	C–O
1280	–OH (phenol)
1380	C–H (CH_3)
1460, 1470	C–H (CH_2)
1600, 1650	C=C
1745	C=O
2855	Symmetric C–H (CH_2, CH_3)
2940	Asymmetric C–H (CH_2, CH_3)
3020	C–H (=CH *cis*)
3500	O–H

δ 0.82 ppm can be assigned to $-CH_3$; at δ 1.27 ppm to $-CH_2$; at δ 1.49 ppm to the proton in the γ-position of the side chain; at δ 1.66 ppm to the proton in the β-position of the side chain; at δ 2.00 ppm to $-CH_2-C\underline{H}_2-CH=$; at δ 2.49 ppm to the proton in the α-position of the side chain; at δ 2.81 ppm to $=CH-C\underline{H}_2-CH=$; at δ 5.42 ppm to $-CH=C\underline{H}_2-$ (*cis*); at δ 5.61 ppm to $-CH=C\underline{H}_2-$ (*trans*); at δ 5.94 ppm to $-CH=C\underline{H}-CH_2-$; at δ 6.42 ppm to $-CH=C\underline{H}-CH=$; at δ 6.60 ppm to two protons in 4- and 6-positions of the catechol ring; at δ 6.69 ppm to the proton in the 5-position of catechol ring; at δ 7.72 and 9.11 ppm to the phenolic proton of the catechol ring, respectively. The strong peak at 3.36 ppm is the solvent peak of DMSO-d_6. The results are summarized in Table 3.3.

The GC spectrum of urushiol shows 15 peaks, as seen in Figure 3.4. After analysis of the mass chromatography of each peak, the structure of urushiol was determined and is summarized in Table 3.4.

There are saturated (MW = 320), monoenes (MW = 318), dienes (MW = 316), and trienes (MW = 314) in the urushiol side chain; meanwhile, low concentrations of monoenes (MW = 346) and dienes (MW = 344) with 17-carbon side chains were also

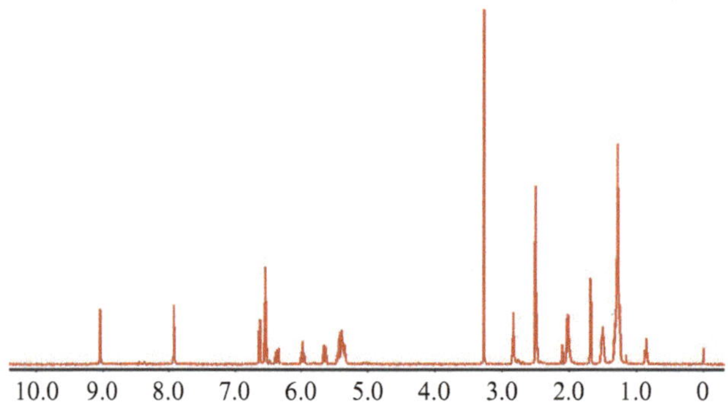

FIGURE 3.3

^{1}H NMR spectrum of urushiol.

Table 3.3 ^{1}H NMR results of urushiol

Chemical shift (ppm)	Groups
0.82	$-CH_3$
1.27	$-CH_2$
1.49	$Ar-C-C-CH_2-$
1.66	$Ar-C-CH_2-C-$
2.00	$-CH_2-C\underline{H}_2-CH=$
2.49	$Ar-CH_2-C-C-$
2.81	$=CH-C\underline{H}_2-CH=$
5.42	$-CH=C\underline{H}_2-$ (*cis*)
5.61	$-CH=C\underline{H}_2-$ (*trans*)
5.94	$-CH=C\underline{H}-CH_2-$
6.42	$-CH=C\underline{H}-CH=$

(Continued)

Table 3.3 ^{1}H NMR results of urushiol—Cont'd

Chemical shift (ppm)	Groups
6.60	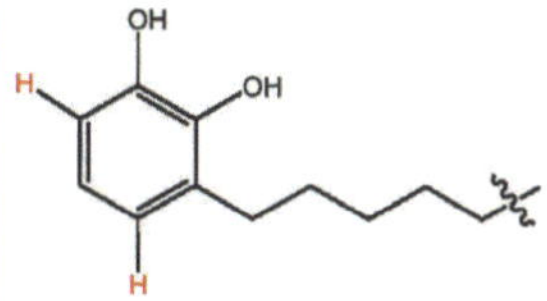
6.69	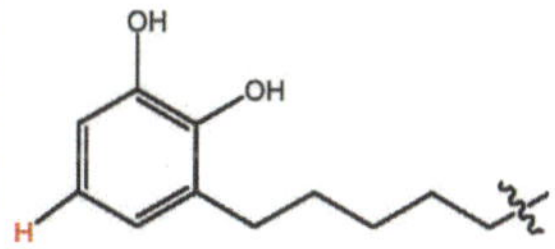
7.92	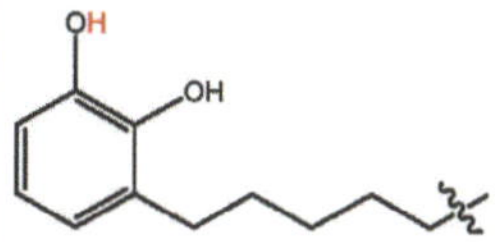
9.11	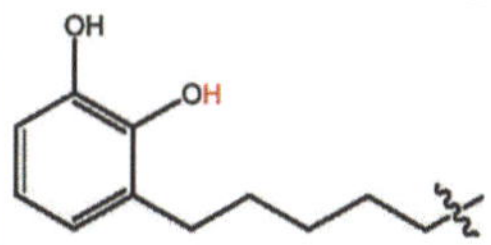

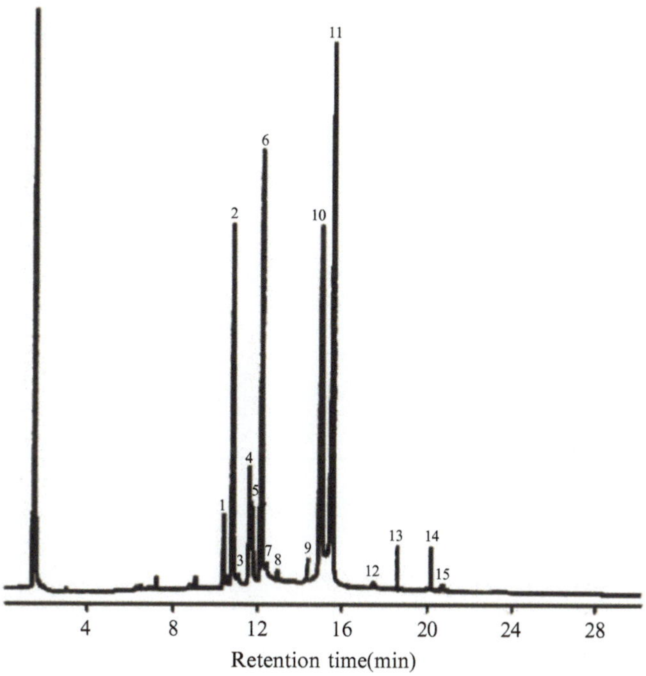

FIGURE 3.4

GC spectrum of urushiol.

Table 3.4 GC/MS results of urushiol

Peak no.	*m/z*	R.t. (min)	Area (%)	Structure
1	320	10.33	2.7	Saturated side chain
2	318	10.74	14.4	Monoene (8Z)
3	318	11.02	0.5	Pentadecenylcatechol
4	316	11.53	4.7	Diene (8Z,11R)
5	316	11.66	3.2	Diene (8Z,11Z)
6	314	12.07	17.9	Triene (8Z,11Z,13Z)
7	316	12.32	1.6	4-(Pentadecadienyl)catechol
8	316	12.73	0.2	4-(Pentadecadienyl)catechol
9	318	14.23	0.1	4-(Pentadecenyl)catechol
10	314	14.92	20.2	Triene (8Z,11R,13R)
11	314	15.47	32.9	Triene (8Z,11R,13Z)
12	346	17.43	0.4	Monoene (8Z,C17)
13	344	18.13	0.1	Diene (8Z,11Z,C17)
14	314	19.93	0.2	4-(Pentadecatrienyl)catechol
15	314	20.65	0.5	4-(Pentadecatrienyl)catechol

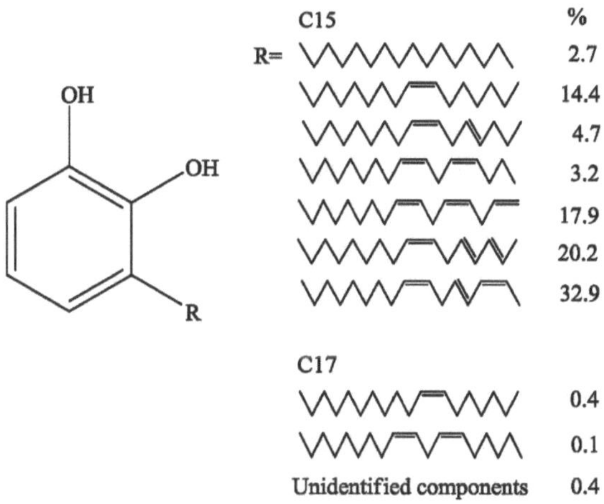

FIGURE 3.5

Chemical structure of urushiol.

detected (Table 3.4). From the above analysis, the urushiol structure was resolved as shown in Figure 3.5.

In order to reveal the enzymatic polymerization mechanism of urushiol, the dimer structures produced in the initial stage of enzymatic polymerization were examined using modern techniques including ^{1}H-^{1}H and ^{1}H-^{13}C correlation two-dimensional NMR (2D-NMR) measurements.

Raw lacquer sap was first kneaded in a mixer at a constant speed of 60 rpm for 1 h to promote the polymerization reaction. It was chromatographed by thin-layer chromatography (TLC) with silica gel as the stationary phase and $CHCl_3$ as the solvent in a preliminary test, and then was fractionated on a recycling HPLC LC-908 system (Japan Analytical Co. Ltd., Tokyo, Japan) equipped with an HPLC column [Tosoh, TSK-gel G1000H6, 7.6 mm (i.d.) × 300 mm × 3] and an RI and UV (254 nm) detector. The sample (5 mg) was dissolved in 2 ml of $CHCl_3$. The resulting solution was aspirated into a 5 ml syringe and injected into the column through a pretreatment poly(tetrafluoroethylene) cartridge containing an H-13-5 filter disc into the column, and then the column was eluted with $CHCl_3$ at a rate of 3.5 ml/min.

Figure 3.6 is the TLC profile of lacquer sap. There were six spots with R_f values of 0.79 (spot 1), 0.65 (spot 2), 0.49 (spot 3), 0.14 (spot 4), 0.08 (spot 5), and 0.01 (spot 6). Except for spot 3, the other spots were similar in size. In contrast, spot 3 was about 10 times larger than the others. Spot 1 was extracted, and the ^{1}H NMR spectrum of the resulting extract did not show the presence of phenyl protons. In addition, its IR spectrum showed the presence of a carboxylic acid ester. Therefore, it must be one of the fragrance substances present in the raw lacquer sap [8,9]. The extracts of spots 2 and 3 showed characteristic of monomers on ^{1}H and ^{13}C NMR spectra. Spots 4 and 5 could not be extracted because they were too small. Although the extract of spot 6 was also rather small, we were able to record a ^{1}H NMR spectrum, which indicated that it is an urushiol

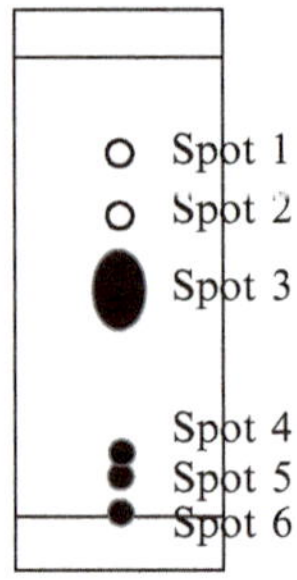

FIGURE 3.6

TLC profile of lacquer.

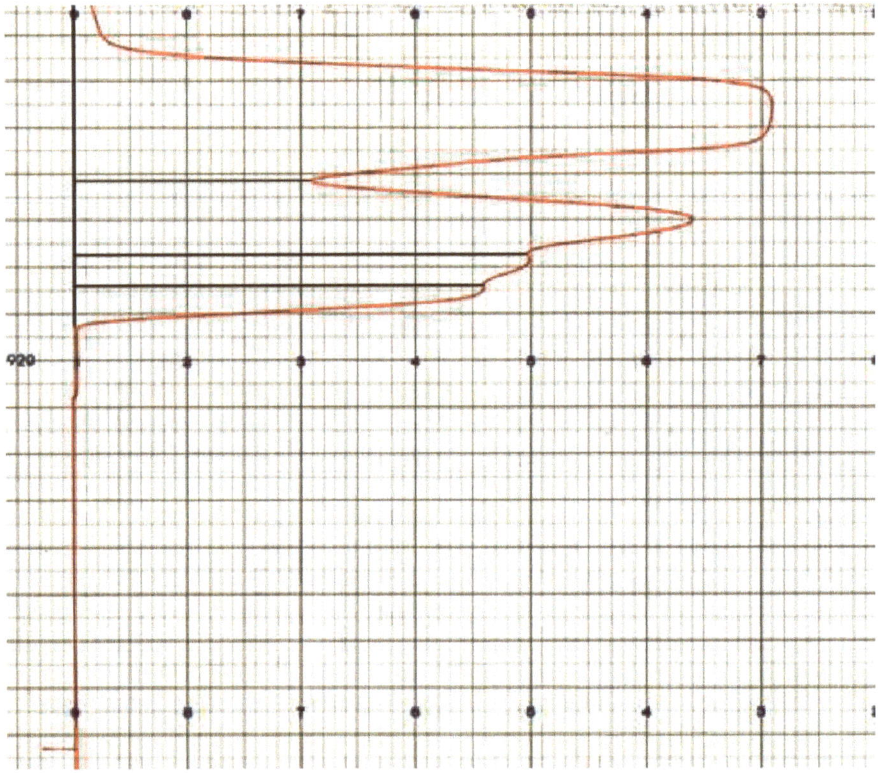

FIGURE 3.7

HPLC profile of urushiol.

polymer. Therefore, the lacquer sap was fractionated by column chromatography with recycled HPLC with chloroform as eluent to separate the dimer fractions.

Figure 3.7 is the HPLC profile. The elution volumes centered at 55, 46, and 44 ml correspond to monomer, dimer, and trimer fractions, respectively, according to standard polystyrene calibration curves; the dimer was collected.

3.1.1 CHARACTERIZATION OF URUSHIOL DIMER STRUCTURE

Figure 3.8 shows the ^{1}H NMR spectrum of the lacquer dimers. It showed signals at ranges δ_H 2.1-2.4, 2.9-4.4, 6.0-6.2, 6.37-7.00, and 7.5-9.1 ppm. Because the signals are complex and overlap, two-dimensional NMR techniques such as homonuclear shift correlation spectroscopy (COSY), heteronuclear multiple quantum coherence (HMQC), and heteronuclear multiple bond coherence (HMBC) were used to characterize the nature of these signals, in particular, signals A-M, in order to identify the chemical structure of the dimers. This resulted in identification of 8′-(3″,4″-dihydroxy-5″-alkenyl)phenyl-3-[9′*E*,-11′*E*,13′*Z*-pentadecatrienyl]catechol (**1**), 14′-(3″,4″-dihydroxy-5″-alkenyl)-phenyl-3-[8′Z,10′*E*,12′*E*-pentadecatrienyl]catechol (**2**), 2-hydroxyl-3- or -6-alkenylphenyl ethyl ether (**3**), 14′-(4″,5″-dihydroxy-6″-alkenyl)phenyl-3-[8′Z,10′*E*,12′*E*-pentadecatrienyl] catechol (**4**), 15′-(2″-hydroxy-3″- or -6″-alkenyl)phenyloxy-3-[8′Z,11′Z,13′*E*-pentadecatrienyl]catechol (**5**), 14′-(5″,6″-dihydroxy-4″-alkenyl)-phenyl-3-[8′Z,10′*E*,12′*E*-pentadecatrienyl]catechol (**6**), 1,1′,2,2′-tetrahydroxy-6,6′-dialkenyl-4,3′-biphenyl (**7**), 1,1′,2,2′-tetrahydroxy-6,6′-dialkenyl-4,4′-biphenyl (**8**), 1,1′,2,2′-tetrahydroxy-6,6′-dialkenyl-5,4′-biphenyl (**9**), and 1,2,1′-trihydroxy-6,6′-dialkenyldibenzofuran (**10**), as shown in Scheme 3.1. In the lacquer sap tapped from *T. verniciflluum*, the presence of compounds **7-10** was suggested by Du *et al.* [10] three decades ago on the basis of GC-MS analysis.

8′-(3″,4″-Dihydroxy-5″-Alkenyl)Phenyl-3-[9′E,11′E,13′Z-Pentadecatrienyl]Catechol (1)

In the ^{1}H-^{1}H COSY spectrum of urushiol dimeric fraction (Figure 3.9), a signal at δ_H/δ_H 3.03-[labeled (A)]/1.55 corresponded to ^1J coupling of H-8′ and H-7′ of compound **1**, while the signal at δ_H/δ_C (A)/47.8 in the 2D ^{1}H-^{13}C HMQC spectrum (Figure 3.10)

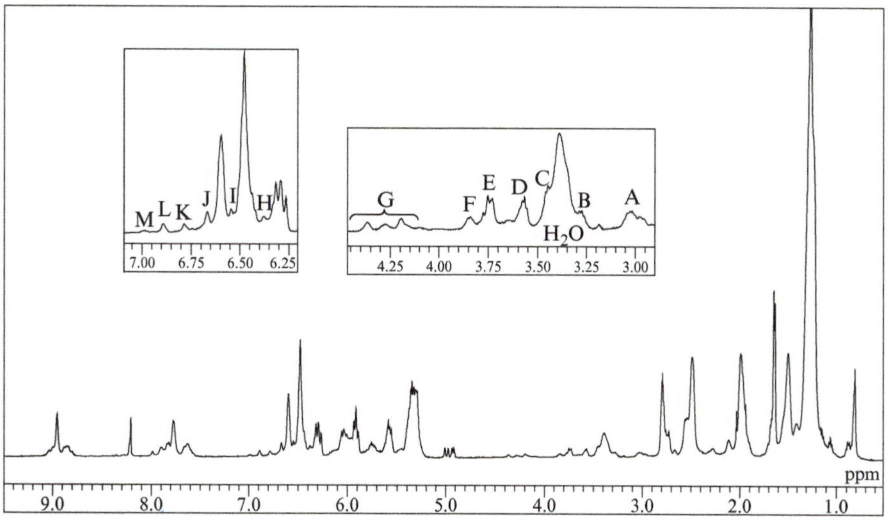

FIGURE 3.8

^{1}H NMR spectrum of lacquer dimers.

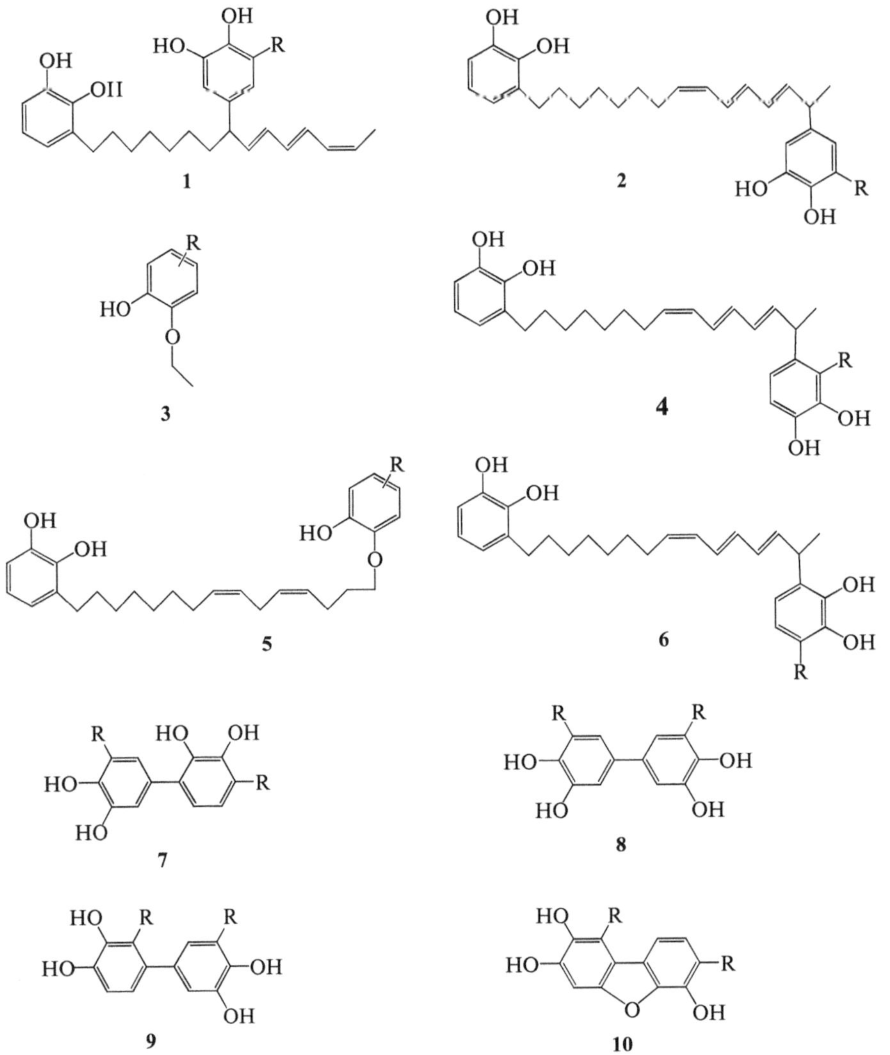

SCHEME 3.1

Structures of compounds identified in dimer fractions of *T. vernicifluum* lacquer.

corresponded to ^{1}J coupling of ^{1}H-^{13}C in the –CH– group at C-8′ in the side chain of compound **1**. Furthermore, in the 2D ^{1}H-^{13}C HMBC spectra (Figures 3.11 and 3.12), the signal at δ_H/δ_C at (A)/134.5 corresponded to ^{2}J coupling of H-8′/C-1″ of a 3″,4″-dihydroxy-5″-alkenylphenyl group, while signals at δ_H/δ_C (A)/111.6 and (A)/119.2 corresponded to ^{3}J coupling of ^{1}H-^{13}C at H-8′/C-2″and H-8′/C-6″, respectively. In addition, signals at δ_H/δ_C 6.30/47.8 and 6.44/47.8 corresponded to ^{4}J coupling of H-5″/C-8′ and H-3″/H-8′, respectively. Furthermore, signals at δ_H/δ_C (A)/35.7 and

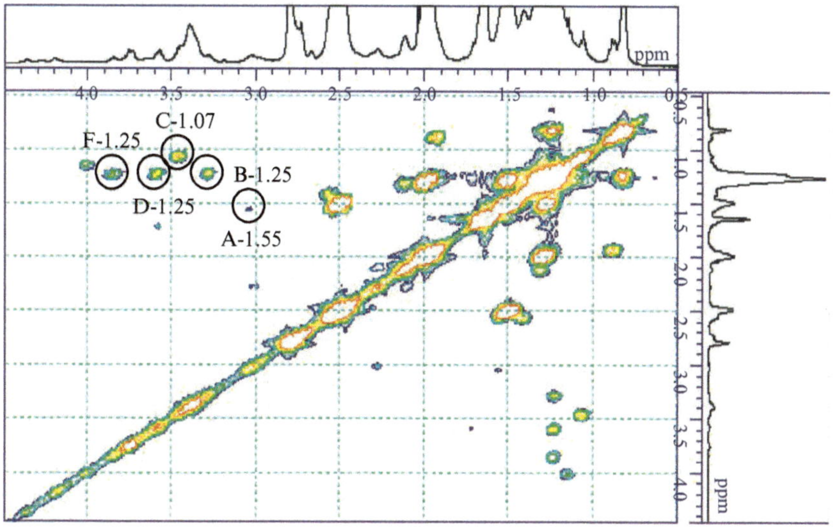

FIGURE 3.9

COSY spectrum of lacquer dimers.

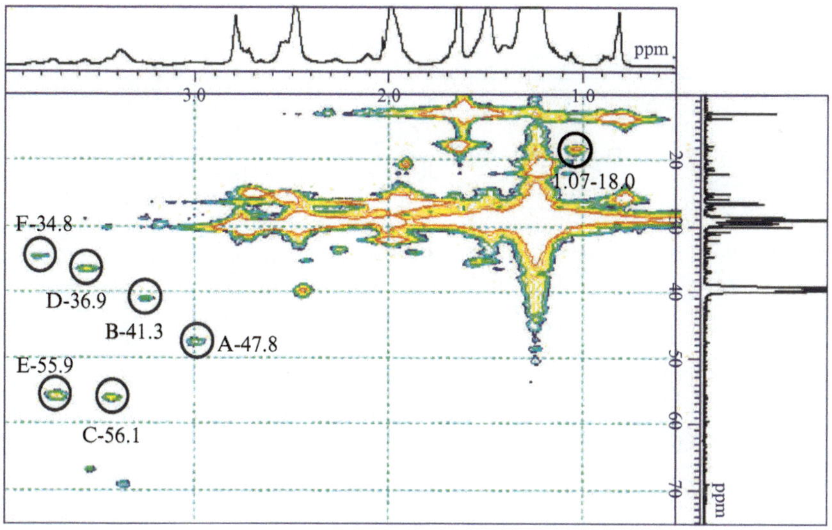

FIGURE 3.10

HMQC spectrum of lacquer dimers (^{1}H: 0.5-4.0 ppm; ^{13}C: 10.0-75.0 ppm).

(A)/139.7 corresponded to ^{2}J coupling of H-8′/C-7′ and H-8′/C-9′, respectively; the signal at δ_H/δ_C (A)/128.8 corresponded to ^{3}J coupling of H-8′/C10′; and the signal at δ_H/δ_C 5.37/47.8 corresponded to ^{2}J coupling of H-9′/C-8′. Thus, the 3″,4″-dihydroxy-5″-alkenylphenyl group should be bonded to C-8′ in the side chain of compound **1**.

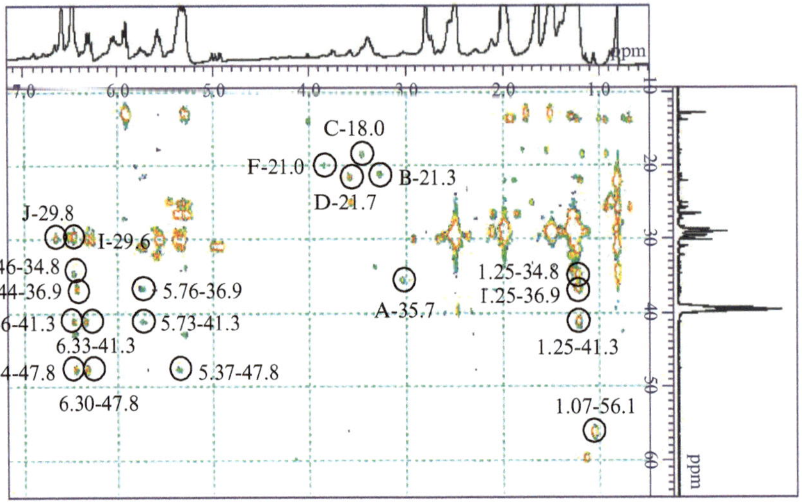

FIGURE 3.11

HMBC spectrum of lacquer dimers (^1H: 0.5-7.0 ppm; ^{13}C: 10.0-65.0 ppm).

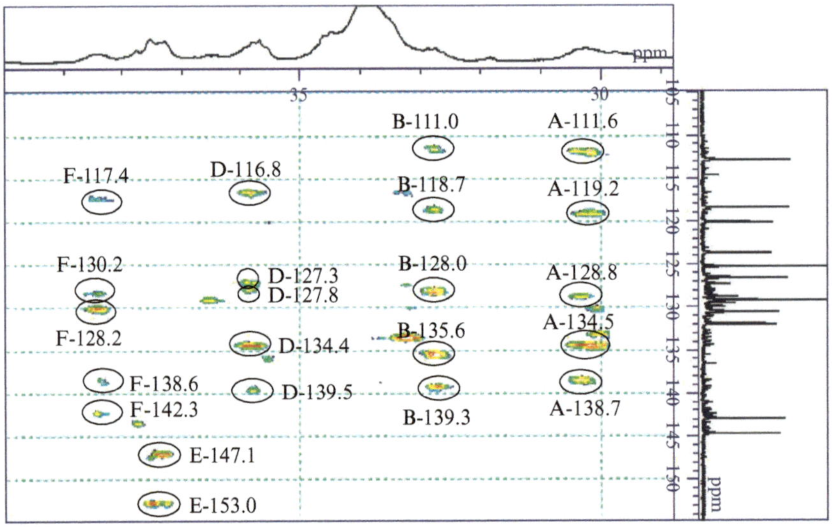

FIGURE 3.12

HMBC spectrum of lacquer dimers (^1H: 2.9-4.0 ppm; ^{13}C: 105.0-155.0 ppm).

14′-(3″,4″-Dihydroxy-5″-Alkenyl)Phenyl-3-[8′Z,10′E,12′E-Pentadecatrienyl]Catechol (2)

In the ^1H-^1H COSY spectrum (Figure 3.9), the signal at δ_H/δ_H 3.28[labeled (B)]/1.25 corresponded to ^1J coupling of H-14′ and H-15′ of compound **2**, while the signal at δ_H/δ_C (B)/41.3 in the 2D ^1H-^{13}C HMQC spectrum (Figure 3.10) corresponded to ^1J coupling of ^{1}H-^{13}C in the –CH– group at C-14′ in the side chain of compound **2**. Furthermore, in the 2D ^{1}H-^{13}C HMBC spectra (Figure 3.11), the signal at δ_H/δ_C (B)/135.6 corresponded to ^2J coupling of H-14′/C-1″ of a 3″,4″-dihydroxyl-5″-alkenylphenyl group, while the signals at δ_H/δ_C (B)/111.0 and (B)/118.7 corresponded to ^3J coupling of ^{1}H-^{13}C at H-14′/C-2″ and H-14′/C-6″, respectively. In addition, signals at δ_H/δ_C 5.74/41.3, 6.33/41.3, and 6.46/41.3 (Figure 3.11) corresponded to ^3J coupling of ^{1}H-^{13}C at H-13′/C-14′, H-14′/C-2″, and H-14′/C-6″, respectively. Thus, the 3″,4″-dihydroxy-5″-alkenylphenyl group should be bonded to C-14′ in the side chain of compound **2**.

2-Hydroxy-3- or -6-Alkenylphenyl Ethyl Ether (3)

In the ^{1}H-^{1}H COSY spectrum (Figure 3.9), the signal at δ_H/δ_H 3.44-[labeled (C)]/1.07 corresponded to ^1J coupling of H-14′ and H-15′ of an ethoxy group in compound **3**, which is confirmed by the signal at δ_H/δ_C (C)/56.1 in the 2D ^{1}H-^{13}C HMQC spectrum (Figure 3.10), corresponding to ^1J coupling of ^{1}H-^{13}C in the –CH– group at C-14′ in the side chain of compound **2**. Moreover, in the 2D ^{1}H-^{13}C HMBC spectrum (Figure 3.11), a signal at δ_H/δ_C (C)/18.0 corresponded to ^2J coupling of H-14′/C-15′, while the signal at δ_H/δ_C 1.07/56.1 corresponded to ^2J coupling of H-15′/C-14′. In addition to the absence of ^2J and ^3J couplings of both H-14′ and C-14′ with a 3,4-dihydroxy-5-alkenylphenyl group in the 2D ^{1}H-^{13}C HMBC (Figure 3.11), the deshielding of both δ_H and δ_C of –CH-14′ indicates that the C-14′ is bonded to a phenoxy group but not with a phenyl group, to form a phenoxy ether moiety. Thus, the compound should have structure **6** that is produced by coupling of a 3,4-dihydroxyl-5-alkenylphenyl group with concomitant elimination of a 3-[8′Z,10′*E*,12′Z-tridecatrienyl]catechol fragment.

14′-(4″,5″-Dihydroxy-6″-Alkenyl)Phenyl-3-[8′Z,10′E,12′E-Pentadecatrienyl]Catechol (4)

The ^{1}H-^{1}H COSY spectrum (Figure 3.9) showed a signal at δ_H/δ_H 3.57[labeled (D)]/1.25 corresponding to ^1J coupling of H-14′/H-15′, while the signal at δ_H/δ_C (D)/36.9 in the 2D ^{1}H-^{13}C HMQC spectrum (Figure 3.10) corresponded to ^1J coupling of ^{1}H-^{13}C in the –CH– group at C-14′ in the side chain of compound **4**. In the 2D ^{1}H-^{13}C HMBC spectra (Figures 3.11 and 3.12), signals at δ_H/δ_C 3.54(D)/3.57 and 1.25/41.3 corresponded to ^1J coupling of H-14′/C-15′ and H-15′/C-14′, respectively. Furthermore, the signal at δ_H/δ_C (D)/134.4 corresponded to ^2J coupling of H-14′/C-1″ of a 4″,5″-dihydroxy-6″-alkenylphenyl group, while signals at δ_H/δ_C (D)/127.8, (D)/116.8, and 6.44/36.9 corresponded to ^3J coupling of ^{1}H-^{13}C at H-14′/C-2″, H-14′/C-6″, and H-2″/C-14′, respectively. In addition, signals at δ_H/δ_C (D)/139.5 and 5.76/36.9 corresponded to ^1J coupling of H-14′/C-13′ and H-13′/C-14′, respectively. Thus, the 4″,5″-dihydroxy-6″-alkenylphenyl group should be bonded to C-14′ in the side chain of compound **2**.

15′-(2″-Hydroxy-3″- or -6″-Alkenyl)Phenyloxy-3-[8′Z,11′Z,-13′E-Pentadecatrienyl]Catechol (5)

The 2D ^{1}H-^{13}C HMQC spectrum (Figure 3.10) showed a signal at δ_H/δ_C 3.75[labeled (E)]/55.9 corresponding to ^1J coupling of H-15′/C-15′, while signals at δ_H/δ_C (E)/147.1 and (E)/153.0 in the 2D ^{1}H-^{13}C HMBC spectrum (Figure 3.12) corresponded to ^4J coupling of H-15′/C-1″ and H-15′/C-2″ of either a 2″-hydroxy-3″-

or -6″-alkenylphenoxyl group with the terminal –CH_2– group at C-15′in the side chain of compound **8**. The deshielding of both C-1″ and C-2″ indicates that the substituent at the C-15′ is a phenoxy group, not a phenyl group. This would lead to structure **5** for the compound.

14′-(5″,6″-Dihydroxy-4″-Alkenyl)Phenyl-3-[8′Z,10′E,12′E-Pentadecatrienyl]Catechol (6)

The ^{1}H-^{1}H COSY spectrum (Figure 3.9) showed signals at δ_H/δ_H 3.85[labeled (F)]/1.25 corresponding to ^1J coupling of H-14′/H-15′, while the signal at δ_H/δ_C (F)/34.8 in the 2D ^{1}H-^{13}C HMQC spectrum (Figure 3.10) corresponded to ^1J coupling of ^{1}H-^{13}C in the –CH– group at C-14′ in the side chain of compound **6**. Furthermore, in the 2D ^{1}H-^{13}C HMBC spectra (Figures 3.11 and 3.12), signals at δ_H/δ_C (F)/20.1 and 1.25/34.8 corresponded to ^2J coupling of H-14′/C-15′ and H-15′/C-14′, respectively, while signals at δ_H/δ_C (F)/128.2 correspond to ^2J coupling of a H-14′/C-1″, 5″,6″-dihydroxy-4″-alkenylphenyl group, and (F)/142.3, (F)/117.4, and 6.46/34.8 correspond to ^3J coupling of ^{1}H-^{13}C at H-14′/C-2″, H-14′/C-6″, and H-2″/C-14′, respectively. In addition, signals at δ_H/δ_C (F)/138.6 and (F)/130.2 corresponded to ^1J coupling of H-14′/C-13′ and ^2J coupling of H-14′/C-12′, respectively. Thus, the 4″,5″-dihydroxy-6″-alkenylphenyl group should be bonded to C-14′ in the side chain of compound **2**.

1,1′,2,2′-Tetrahydroxy-6,6′-Dialkenyl-4,3′-Biphenyl (7)

The 2D ^{1}H-^{13}C HMQC spectrum (Figure 3.13) showed a signal at δ_H/δ_C 6.38[labeled (H)]/116.8 corresponding to ^1J coupling of ^{1}H-^{13}C in the –CH– groups in biphenyl ring at C-3 and C-4′ of compound **10**. According to Du *et al.* [7], the biphenyl of this

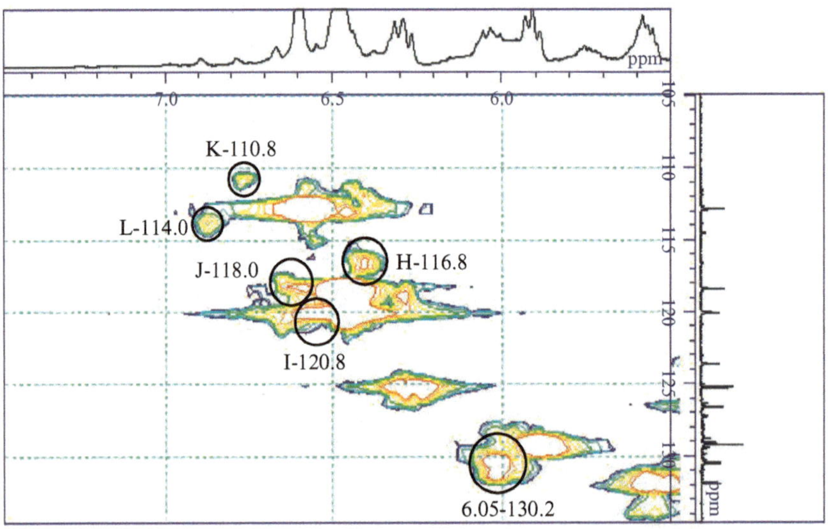

FIGURE 3.13

HMQC spectrum of lacquer dimers (^{1}H: 5.5-7.5 ppm; ^{13}C: 105-135 ppm).

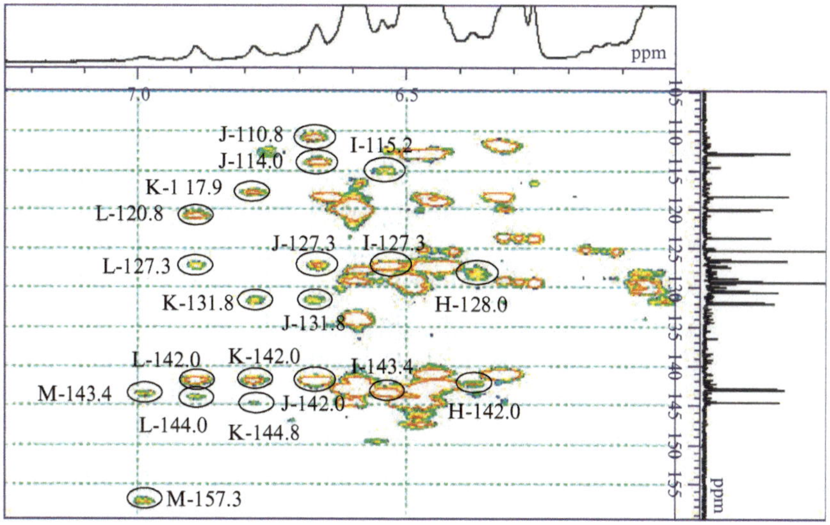

FIGURE 3.14

HMBC spectrum of lacquer dimers (^{1}H: 6.0-7.2 ppm; ^{13}C: 105.0-160.0 ppm).

type has only one broad signal for aromatic hydrogen atoms at H-3, H-5, H-4′, and H-5′ with δ_H 6.53. In the 2D ^{1}H-^{13}C HMBC spectra (Figure 3.14), signals at δ_H/δ_C (H)/128.0 and (H)/142.0 corresponded to ^2J coupling of H-3/C-4 and H-3/C-2, respectively. This would lead to structure **7** for the compound.

1,1′,2,2′-Tetrahydroxy-6,6′-Dialkenyl-4,4′-Biphenyl (8)

The 2D ^{1}H-^{13}C HMQC spectrum (Figure 3.13) showed a signal at δ_H/δ_C 6.53[labeled (I)]/120.8 corresponding to ^1J coupling of ^{1}H-^{13}C in the –CH– groups of the biphenyl ring at C-3 and C-3′ of compound **8**, because the molecule is symmetric with respect to a plane perpendicular to the center of molecule that bisects the (C-4)-(C-4′) bond, if the nature of the alkenyl groups substituted at C-6 and C-6′ is the same or not considered. In the 2D ^{1}H-^{13}C HMBC spectra (Figure 3.14), the signal at δ_H/δ_C (I)/147.1 corresponded to ^2J coupling of H-3/C-2 and H-3′/C-2′, while the signal at δ_H/δ_C (I)/115.2 corresponded to ^3J coupling of H-3/C-5 and H-3′/C-5′. Furthermore, the signal at δ_H/δ_C (I)/127.3 corresponded to ^2J coupling of H-3/C-4 and H-3′/C-4′, while the signal at δ_H/δ_C (I)/143.4 corresponded to ^3J coupling of H-3/C-1 and H-3′/C-1′. This would lead to structure **8** for the compound.

1,1′,2,2′-Tetrahydroxy-6′,6-Dialkenyl-5,4′-Biphenyl (9)

The 2D ^{1}H-^{13}C HMQC spectrum (Figure 3.13) showed signals at δ_H/δ_C 6.67[labeled (J)]/118.0, 6.78[labeled (K)]/110.8, and 6.89[labeled (L)]/114.0, corresponding to ^1J couplings of ^{1}H-^{13}C in the –CH– groups of the biphenyl ring at C-4, C-3, and C-3′ of compound **9**, respectively. In the 2D ^{1}H-^{13}C HMBC spectrum (Figure 3.14), signals at δ_H/δ_C (J)/110.8 and (J)/131.8 corresponded to ^2J coupling of H-4/C-3 and H-4/C-5, respectively, while signals at δ_H/δ_C (J)/127.3 and (J)/142.0 corresponded to ^3J coupling of H-4/C-4′ and H-4/C-2, respectively. In addition, the signal at δ_H/δ_C (J)/114.3

corresponded to ^4J coupling of H-4/C-3′. Furthermore, signals at δ_H/δ_C (K)/117.9 and (K)/142.0 corresponded to ^2J coupling of H-3/C-4 and H-3/C-2, respectively, while signals at δ_H/δ_C (K)/131.8 and (K)/144.8 corresponded to ^3J coupling of H-3/C-5 and H-3/C-1, respectively. Similarly, the signals at δ_H/δ_C (L)/127.3 and (L)/142.0 corresponded to ^2J coupling of H-3′/C-4′ and H-3′/C-2′, respectively, while signals at δ_H/δ_C (L)/120.8 and (L)/144.0 corresponded to ^2J coupling of H-3′/C-5′ and H-3′/C-1′, respectively. The deshielding of the C-1, C-2, C-1′, and C-2′ indicated that these carbons are substituted by hydroxy groups. Thus, the spectroscopic data are compatible with structure **9** for the compound.

1,1′,2-Trihydroxy-6,6′-Dialkenyldibenzofuran (10)

In both the ^{1}H-^{1}H COSY and 2D ^{1}H-^{13}C HMQC spectra (Figures 3.9 and 3.13), no signals for dibenzofuran-type structures are discernible. However, in the 2D ^{1}H-^{13}C HBQC spectra (Figure 3.14), signals at δ_H/δ_C 6.99 [labeled (M)]/143.4 and (M)/157.3 are likely to correspond to ^2J coupling of H-3/C-2 and H-3/C-4 of compound **10**, respectively. The deshielding of C-4 as compared to C-2 indicated that C-4 is a part of furan ring, while C-2 is substituted by a hydroxyl group. Thus, the compound should have structure **10**.

Signals in Area G of ^{1}H NMR Spectrum of Lacquer Dimer

In the 2D ^{1}H-^{13}C HMQC and HMBC spectra, no signals related to the signals in area G at the δ_H range of 4.10-4.45 in the ^{1}H NMR spectrum of the dimeric fraction (Figure 3.8) are discernible. The deshielding of δ_H of protons in this region seems to indicate that hydrogen atoms in −CH− groups bonded to oxygen atoms are likely giving rise to these signals.

Monomeric urushiols would undergo autoxidation in the initial stage of reactions, producing radical species involving the conjugated system produced by oxidation with the atmospheric oxygen. Oxygen would abstract a single electron from the conjugated system in the urushiol, producing corresponding radical species and superoxide (O_2^-). This would result in formation of dimeric urushiols with −HC−CH−, ether, and peroxide linkages between two monomeric urushiols to form dimer structures. However, further investigation is required to verify this assumption.

3.1.2 REACTION MECHANISM OF URUSHIOL

Compounds **1** and **2** could be produced by coupling a C-5-centered radical species of catechol radical from one of the monomeric urushiols with C-8′- and C-14′-centered radical species, with a concomitant hydrogen atom abstraction at C-10′ to form the 9′*E*,11′*E*,13′*Z* and 8′Z,10′*E*,12′*E* conjugate systems, respectively. Compounds **3** and **5** could be produced by radical addition of either 2-hydroxy-3-alkenylphenoxyl or 2-hydroxy-6-alkenylphenoxyl radical species to C-14′, followed by either a McLafferty-type rearrangement [11] resulting in elimination of a 3-[8′Z,10′*E*,12′Z-tridecatrienyl]-catechol fragment or a concomitant hydrogen atom abstraction at C-13′ to form the 8′Z,11′Z,13′*E* conjugate system, respectively. Compounds **4** and **6** could be produced by coupling C-4- and C-6-centered catechol radical species from the monomeric urushiols with a C-14′-centered radical species with a concomitant hydrogen atom abstraction at C-10′ to form the 8′Z,10′*E*,12′*E* conjugate system, respectively.

Compounds **7** and **8** could be produced by coupling a C-5-centered catechol radical species with a C-6-centered radical species and two C-4-centered catechol radical species from any of the monomeric urushiols, respectively. Compound **9** could be produced by coupling a C-4-centered catechol radical species with a C-5-centered radical species from the monomeric urushiols. Compound **10** could be produced by coupling a C-4-centered catechol radical with a C-6-centered radical species from any of the monomeric urushiols, followed by a dehydration involving the hydroxyl group at C-2′ and H-4 of the resulting intermediate catalyzed by laccase.

3.2 LACCOL

Laccol is the main component of sap tapped from *Toxicodendron succedanea* lacquer trees and has a C17 unsaturated hydrocarbon chain at the 3-position of the catechol ring. *T. succedanea* lacquer trees grow mainly in North Vietnam. The *T. succedanea* lacquer trees grown in Taiwan were transplanted from Vietnam. The chemical structure of laccol also was analyzed by the same methods as urushiol. Figure 3.15 is the IR spectrum of laccol. It shows the absorptions due to the hydroxyl group (O–H) at 3450 cm^{-1}, methylene (C–H) at 2940 cm^{-1}, aromatic skeletal at 1600 and 1470 cm^{-1}, bending vibrations of conjugated alkene at 987 (*trans*) cm^{-1}, and alkenes at 966 (*trans*) and 732 (*cis*) cm^{-1}, respectively.

The proton NMR spectrum of laccol is shown in Figure 3.16. It has a 1,2,3-substituted benzene (3-alkenylcatechol) structure. Aromatic protons of the laccol monomer (H-4, H-5, and H-6) were detected at $\delta_H = 6.49$ (dd, $J = 7.4$, 2.3 Hz), 6.51 (t, $J = 7.4$ Hz), and 6.59 (dd, $J = 7.4$, 2.3 Hz), and two protons of the phenolic hydroxyl group were observed at $\delta_H = 8.99$ (OH-1) and 7.86 (OH-2), respectively.

The carbon NMR spectrum of laccol is shown in Figure 3.17. Almost all absorption peaks were assigned. The detail analysis results of Figures 3.16 and 3.17 are summarized in Table 3.5. It was concluded that 3-(heptadeca-10*Z*,13*E*,15*E*-trienyl)

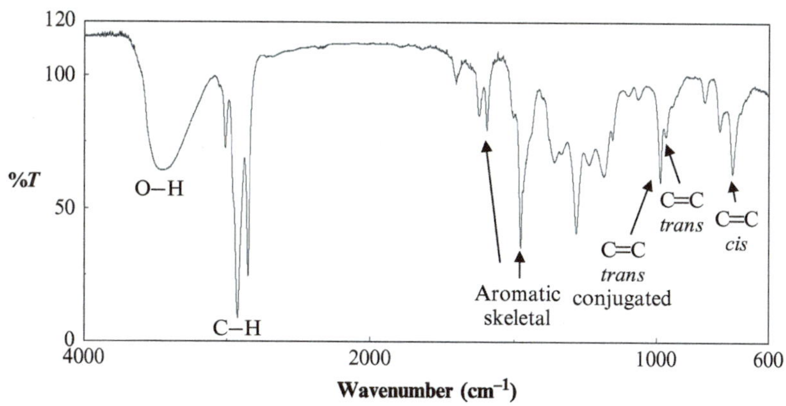

FIGURE 3.15

IR spectra of laccol monomer, dimer, and oligomer.

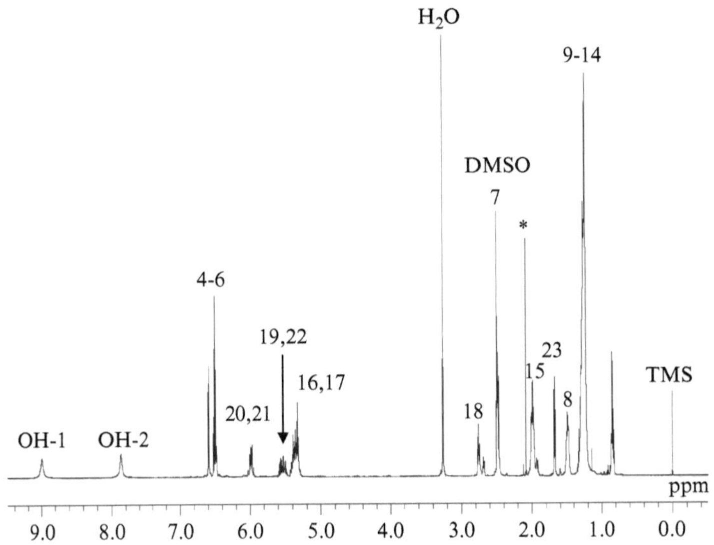

FIGURE 3.16

^{1}H NMR spectrum of laccol [solvent: DMSO-d_6, *: impurity (acetone)].

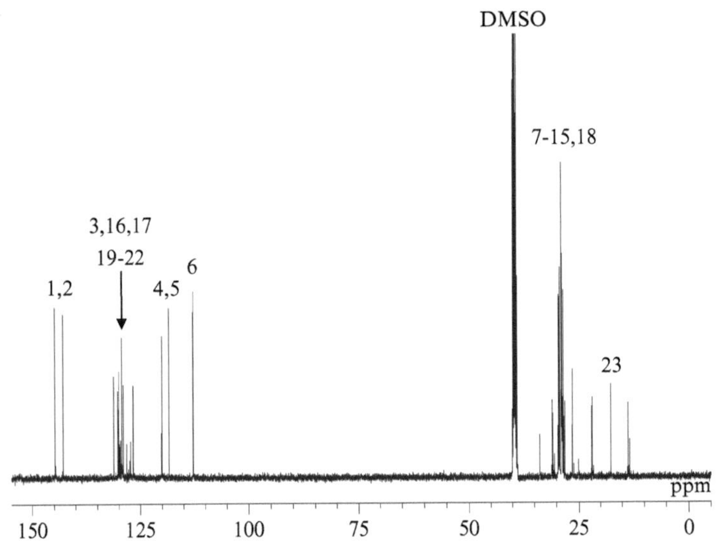

FIGURE 3.17

^{13}C NMR spectrum of laccol (solvent: DMSO-d_6).

Table 3.5 NMR data of laccol monomer

	10*Z*,13*E*,15*E*, 36.27%			10*Z*, 23.76%			10*Z*,13*E*, 14.30%		
C/H no.	δ_H	Multiplicity (*J*, Hz)[a]	δ_C	δ_H	Multiplicity (*J*, Hz)[a]	δ_C	δ_H	Multiplicity (*J*, Hz)[a]	δ_C
1	-	-	144.7	-	-	144.7	-	-	144.7
2	-	-	142.9	-	-	142.9	-	-	142.9
3`	-	-	129.5	-	-	129.5	-	-	129.5
4	6.49	dd (7.4, 2.3)	120.0	6.49	dd (7.4, 2.3)	120.0	6.49	dd (7.4, 2.3)	120.0
5	6.51	t (7.4)	118.4	6.51	t (7.4)	118.4	6.51	t (7.4)	118.4
6	6.59	dd (7.4, 2.3)	112.8	6.59	dd (7.4, 2.3)	112.8	6.59	dd (7.4, 2.3)	112.8
7 (1′)	2.48	d (7.7)	29.5[d]	2.48	d (7.7)	29.5[b]	2.48	d (7.7)	29.5[b]
8 (2′)	1.49	Quintet (7.7)	29.3[d]	1.49	Quintet (7.7)	29.3[b]	1.49	Quintet (7.7)	29.3[b]
9 (3′)	1.28	m	29.0[d]	1.28	m	29.0[b]	1.28	m	29.0[b]
10 (4′)	1.24	m	29.0[d]	1.24	m	29.0[b]	1.24	m	29.0[b]
11 (5′)	1.24	m	29.0[d]	1.24	m	29.0[b]	1.24	m	29.0[b]
12 (6′)	1.24	m	29.0[d]	1.24	m	29.0[b]	1.24	m	29.0[b]
13 (7′)	1.24	m	29.0[d]	1.24	m	29.0[b]	1.24	m	29.0[b]
14 (8′)	1.29	m	29.0[d]	1.29	m	29.0[b]	1.29	m	29.0[b]
15 (9′)	1.99	m	26.4	1.98	m	26.4	1.99	m	26.4
16 (10′)	5.35	m	126.8	5.32	m	129.1	5.35	m	126.8
17 (11′)	5.40	m	130.4	5.32	m	129.1	5.40	m	130.4
18 (12′)	2.76	t (6.6)	29.6	1.98	m	26.4	2.69	t (6.6)	29.7[b]
19 (13′)	5.50	dt (14.3, 6.6)	129.5	1.29	m	29.0[b]	5.40	m	130.4
20 (14′)	5.98	ddt (14.3, 10.3, 1.4)	131.4	1.24	m	29.0[b]	5.35	m	126.8
21 (15′)	6.00	ddq (14.3, 10.3, 1.4)	130.2	1.24	m	29.0[b]	1.99	m	26.4
22 (16′)	5.55	dq (14.3, 6.6)	126.6	1.24	m	29.0[b]	1.29	m	29.0[b]
23 (17′)	1.68	d (6.6)	17.6	0.85	t (6.8)	13.7	0.86	t (6.8)	13.7
OH-1	8.99	br s	-	8.99	br s	-	8.99	br s	-
OH-2	7.86	br s	-	7.86	br s	-	7.86	br s	-

[a] *s, singlet; d, doublet; t, triplet; dd, doublet-doublet; dt, doublet-triplet; dq, doublet-quartet; ddt, doublet-doublet-triplet; ddq, doublet-doublet-quartet; m, multiplet; br, broad signal.*
[b] *Overlapped by other signals.*

catechol, 3-(heptadeca-10*Z*-nyl)catechol, and 3-(heptadeca-10*Z*,13*E*-dienyl)catechol are the major monomeric structures of laccol.

Figure 3.18 is the GC spectra of laccols extracted from raw and refined lacquer sap. They show 12 peaks. After analysis of mass chromatography of each peak, the structure of laccol was concluded and is summarized in Table 3.6.

There are saturated alkenes (MW = 346 and MW_{C15} = 320), monoene (MW = 344), diene (MW = 342), and triene (MW = 340) in the laccol side chain;

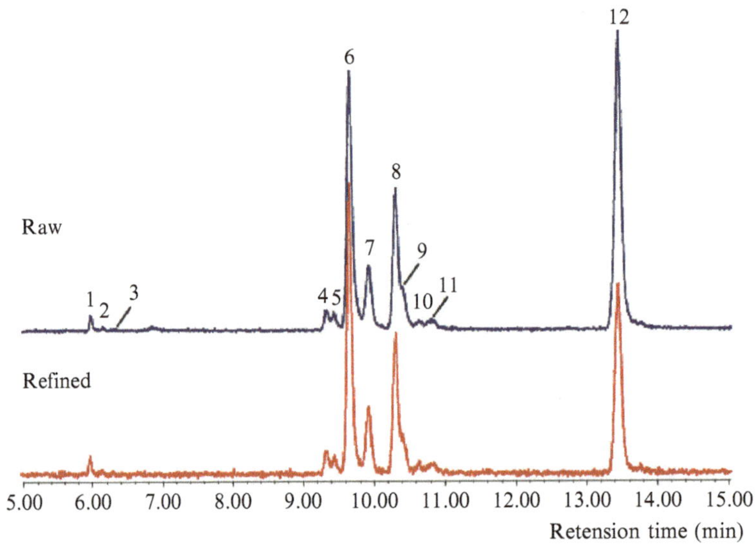

FIGURE 3.18

GC spectra of laccol extracted from raw and refined lacquer.

Table 3.6 GC/MS results of laccol

Peak no.	*m/z*	R.t. (min)	Area (%)	Structure
1	320	6.0	1.1	Saturated side chain (C15)
2	318	6.1	0.3	Monoene (8Z,C15)
3	318	6.3	0.1	Monoene (10Z,C15)
4	346	9.3	1.7	Saturated side chain
5	344	9.4	1.7	Monoene (10Z)
6	344	9.7	27.2	Monoene (12Z)
7	344	9.9	8.2	Monoene (14Z)
8	342	10.3	13.6	Diene (10Z,13R)
9	342	10.4	4.2	Diene (10Z,13Z)
10	340	10.6	1.3	Triene (10Z,13R,16)
11	342	10.8	1.9	Diene (12Z,15R)
12	340	13.5	38.7	Triene (10Z,13R,15R)

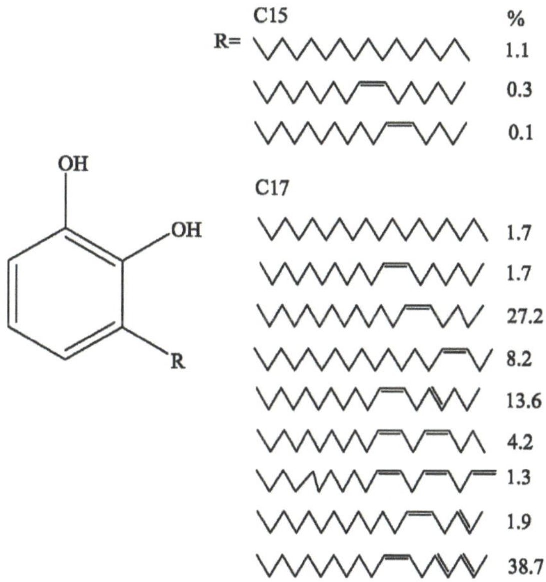

FIGURE 3.19

Chemical structure of laccol.

meanwhile, low concentrations of saturated alkenes ($MW_{C15}=320$) and monoenes ($MW_{C15}=318$) with 15-carbon side chains were also detected (Table 3.6). From the above analysis, the laccol structure was concluded as shown in Figure 3.19.

3.2.1 CHARACTERIZATION OF LACCOL DIMER STRUCTURE

The dimer structure of laccol produced in the initial stage of enzymatic polymerization also was analyzed by NMR including 2D-NMR measurements to discuss the enzymatic polymerization mechanism.

Analysis of the mechanism of enzymatic polymerization of urushiol from the preferentially produced dimer structures by NMR measurements showed that urushiol monomers are first oxidized by laccase to form semiquinone radicals, then the semiquinone radicals react by two routes: (i) semiquinone radicals react vigorously with each other to form nuclear-nuclear biphenyl urushiol (C–C) and (ii) semiquinone radicals react with side chain radicals produced by autoxidation to form nuclear-side chain structures (C–C or C–O–C) [12]. Although laccol has a catechol ring like urushiol, the side chain has 17 carbons unlike urushiol (C15), and the enzymatic polymerization of laccol should be different. In order to reveal and compare it to urushiol, the chemical structure of laccol dimers produced during enzymatic polymerization was analyzed, and the laccol polymerization mechanism is investigated. Polymerization of raw laccol lacquer was promoted by kneading in a mixer at a constant speed of 60 rpm at 50-60% relative humidity at room temperature for 30-180 min.

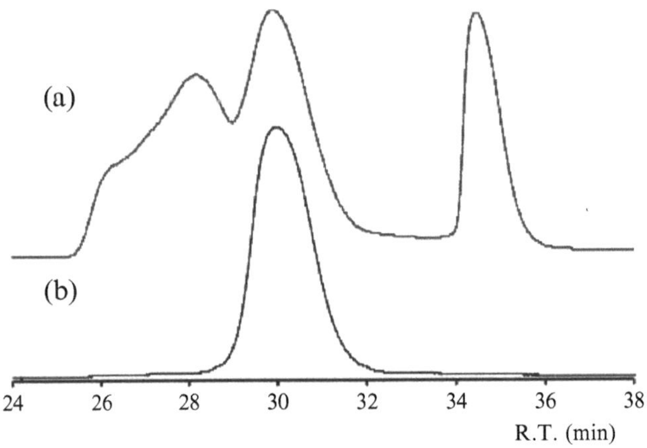

FIGURE 3.20

HPLC charts of laccol (a) and purified dimer (b).

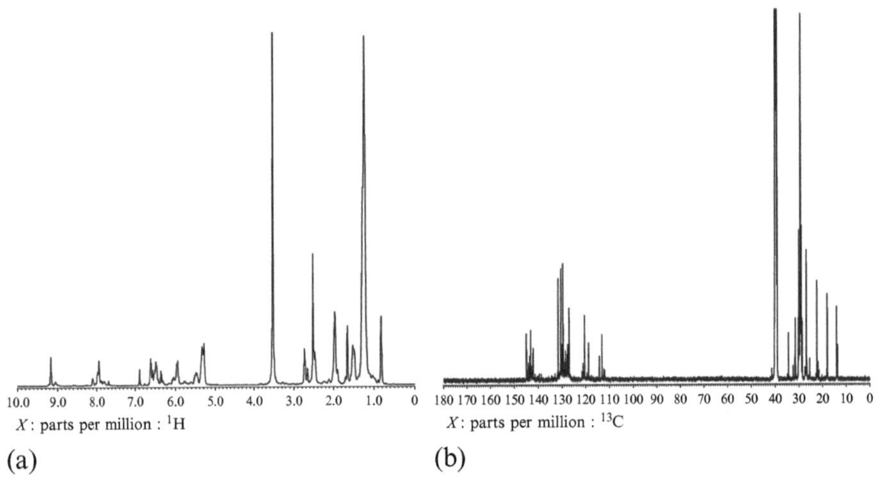

FIGURE 3.21

(a) 500 MHz ^1H NMR and (b) 125 MHz ^{13}C NMR spectra of laccol dimer.

Like the analysis method of urushiol dimer, laccol dimers also were separated by HPLC as shown in Figure 3.20.

Laccol was extracted with acetone. The HPLC chart shows three main and one acromial peaks at the retention times of 35, 30, 28, and 26 min corresponding to monomer, dimer, trimer, and oligomer, respectively (Figure 3.20a). The dimer was separated and purified using the same column, and then the chemical structure was analyzed to discuss the polymerization mechanism (Figure 3.20b).

The 1D-NMR spectra of laccol dimers are shown in Figure 3.21. Like the urushiol dimer, peaks at 3.0-4.0 ppm in the proton NMR and at 35-45 ppm in the carbon

NMR spectrum due to nuclear-side chain structures were observed [13], suggesting that laccol dimers commonly generate nuclear-nuclear and nuclear-side chain structures like urushiol. However, no peak around 7.0 ppm in the proton NMR and 160 ppm in the carbon NMR spectra due to dibenzofuran-type dimers appeared, implying that laccol possibly does not tend to form dibenzofuran-type dimers, which is different from urushiol.

The COSY, HMQC, and HMBC spectra of laccol dimers are shown in Figures 3.22–3.24, respectively. The detailed analyses of results are as follows.

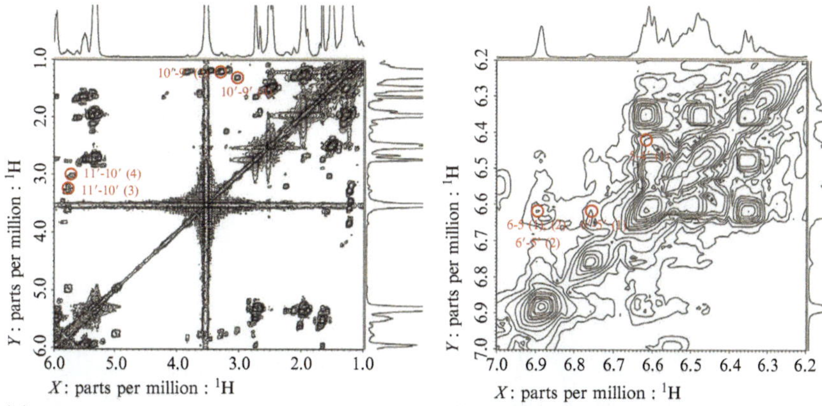

(a) (b)

FIGURE 3.22

¹H-¹H COSY spectra of laccol dimer: (a) 1.0-6.0 ppm and (b) 6.2-7.0 ppm.

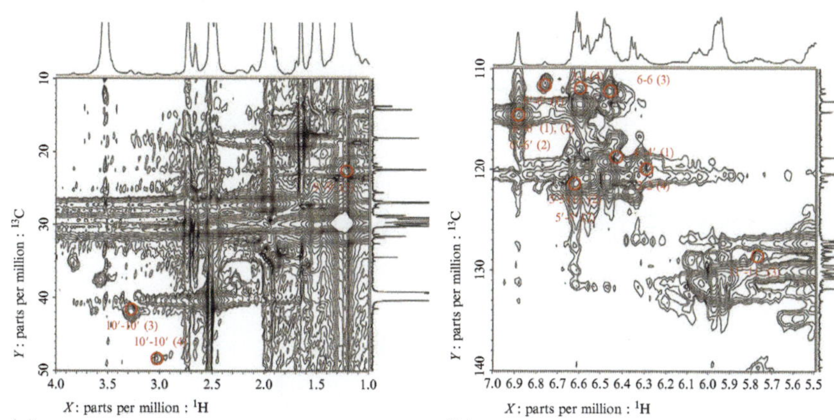

(a) (b)

FIGURE 3.23

HMQC spectra of laccol dimer: (a) ¹H: 1.0-4.0 ppm; ¹³C: 10.0-50.0 ppm and (b) ¹H: 5.5-7.0 ppm; ¹³C: 110.0-140.0 ppm.

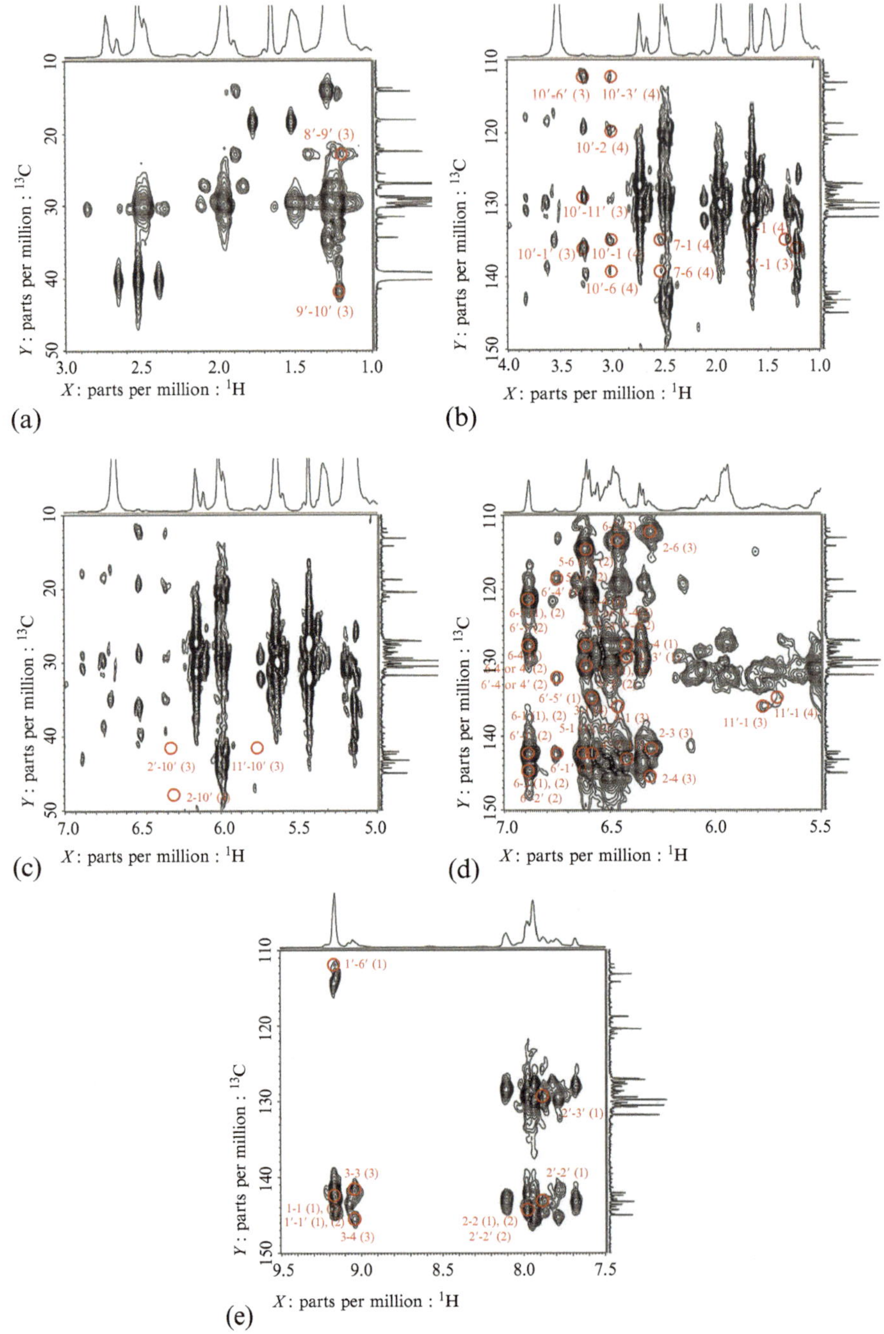

FIGURE 3.24

HMBC spectra of laccol dimer: (a) ^{1}H: 1.0-3.0 ppm; ^{13}C: 10.0-50.0 ppm, (b) ^{1}H: 1.0-4.0 ppm; ^{13}C: 110.0-150.0 ppm, (c) ^{1}H: 5.0-7.0 ppm; ^{13}C: 10.0-50.0 ppm, (d) ^{1}H: 5.5-7.0 ppm; ^{13}C: 110.0-150.0 ppm, and (e) ^{1}H: 7.5-9.5 ppm; ^{13}C: 110.0-150.0 ppm.

1,1′,2,2′-Tetrahydroxy-3,3′-Dialkenyl-4,5′-Biphenyl (1)

The ^1H-^1H COSY spectrum (Figure 3.22b) shows a signal at δ_H/δ_H 6.88/6.61 corresponding to ^1J coupling of H-6/H-5 and H-6′/H-5′ in the catechol ring. In the HMQC spectrum (Figure 3.23b), the signals at δ_H/δ_C 6.61/121.3 correspond to ^1J coupling of H-5/C-5 and H-5′/C-5′ and at δ_H/δ_C 6.88/114.7 correspond to ^1J coupling of H-6/C-6 and H-6′/C-6′. Furthermore, in the HMBC spectrum (Figure 3.24d), the signals at δ_H/δ_C 6.62/114.7 correspond to ^2J coupling of H-5/C-6 and H-5′/C-6′, at δ_H/δ_C 6.62/128.0 correspond to ^2J or ^3J coupling of H-5/C-4 or C-4′ and H-5′/C-4 or C-4′, at δ_H/δ_C 6.62/130.6 correspond to ^3J coupling of H-5/C-3 and H-5′/C-3′, at δ_H/δ_C 6.62/142.6 correspond to ^3J coupling of H-5/C-1 and H-5′/C-1′, at δ_H/δ_C 6.89/121.3 correspond to ^2J coupling of H-6/C-5 and H-6′/C-5′, at δ_H/δ_C 6.89/128.0 correspond to ^3J or ^4J coupling of H-6/C-4 or C-4′ and H-6′/C-4 or C-4′, at δ_H/δ_C 6.89/142.6 correspond to ^2J coupling of H-6/C-1 and H-6′/C-1′, and at δ_H/δ_C 6.89/144.6 correspond to ^3J coupling of H-6/C-2 and H-6′/C-2′. In the HMBC spectrum (Figure 3.24e), the signals at δ_H/δ_C 7.98/144.6 correspond to ^2J coupling of H-2 in the –OH catechol to C-2 and/or H-2′ in the –OH catechol to C-2′, at δ_H/δ_C 9.16/142.6 correspond to ^2J coupling of H-1 in the –OH catechol to C-1 and/or H-1′ in the –OH catechol to C-1′. This would lead to structure 2 for the compound.

1,1′,2,2′-Tetrahydroxy-3,3′-Dialkenyl-4,4-Biphenyl (2)

The ^1H-^1H COSY spectrum (Figure 3.22b) shows a signal at δ_H/δ_H 6.88/6.61 corresponding to ^1J coupling of H-6/H-5 and H-6′/H-5′ in the catechol ring. In the HMQC spectrum (Figure 3.23b), the signals at δ_H/δ_C 6.61/121.3 correspond to ^1J coupling of H-5/C-5 and H-5′/C-5′ and at δ_H/δ_C 6.88/114.7 correspond to ^1J coupling of H-6/C-6 and H-6′/C-6′. Furthermore, in the HMBC spectrum (Figure 3.24d), the signals at δ_H/δ_C 6.62/114.7 correspond to ^2J coupling of H-5/C-6 and H-5′/C-6′, at δ_H/δ_C 6.62/128.0 correspond to ^2J or ^3J coupling of H-5/C-4 or C-4′ and H-5′/C-4 or C-4′, at δ_H/δ_C 6.62/130.6 correspond to ^3J coupling of H-5/C-3 and H-5′/C-3′, at δ_H/δ_C 6.62/142.6 correspond to ^3J coupling of H-5/C-1 and H-5′/C-1′, at δ_H/δ_C 6.89/121.3 correspond to ^2J coupling of H-6/C-5 and H-6′/C-5′, at δ_H/δ_C 6.89/128.0 correspond to ^3J or ^4J coupling of H-6/C-4 or C-4′ and H-6′/C-4 or C-4′, at δ_H/δ_C 6.89/142.6 correspond to ^2J coupling of H-6/C-1 and H-6′/C-1′, and at δ_H/δ_C 6.89/144.6 correspond to ^3J coupling of H-6/C-2 and H-6′/C-2′. In the HMBC spectrum (Figure 3.24e), the signals at δ_H/δ_C 7.98/144.6 correspond to ^2J coupling of H-2 in the –OH catechol to C-2 and/or H-2′ in the –OH catechol to C-2′, at δ_H/δ_C 9.16/142.6 correspond to ^2J coupling of H-1 in the –OH catechol to C-1 and/or H-1′ in the –OH catechol to C-1′. This would lead to structure 2 for the compound.

10′-(3,4-Dihydroxy-5-Alkenyl)Phenyl-3-[12′E,14′E,16′ E-Heptadecatrienyl]Catechol (3)

The ^1H-^1H COSY spectrum (Figure 3.22a) shows signals at δ_H/δ_H 3.27/1.22 corresponding to ^1J coupling of H-10′/H-9′ and at δ_H/δ_H 5.78/3.27 corresponding to ^1J coupling of H-11′/H-10′ in the unsaturated side chain. In the HMQC spectrum (Figure 3.23a), the signals at δ_H/δ_C 1.21/22.7 correspond to ^1J coupling of H-9′/C-9′ and at δ_H/δ_C 3.27/41.9 correspond to ^1J coupling of H-10′/C-10′. In the other area HMQC spectrum (Figure 3.23b), the signals at δ_H/δ_C 5.78/129.0 correspond to ^1J coupling of H-11′/C-11′

and at δ_H/δ_C 6.46/112.3 correspond to ^{1}J coupling of H-6/C-6. Furthermore, in the HMBC spectrum (Figure 3.24a), the signals at δ_H/δ_C 1.20/22.7 correspond to ^{2}J coupling of H-8′/C-9′ and at δ_H/δ_C 1.21/41.9 correspond to ^{2}J coupling of H-9′/C-10′. In the HMBC spectra (Figure 3.24b), the signals at δ_H/δ_C 1.21/135.9 correspond to ^{3}J coupling of H-9′/C-1, at δ_H/δ_C 3.27/112.3 correspond to ^{3}J coupling of H-10′/C-6, at δ_H/δ_C 3.27/129.0 correspond to ^{2}J coupling of H-10′/C-11′, and at δ_H/δ_C 3.27/135.9 correspond to ^{2}J coupling of H-10′/C-1. In the HMBC spectrum (Figure 3.24c), the signals at δ_H/δ_C 5.77/41.9 correspond to ^{2}J coupling of H-11′/C-10′ and at δ_H/δ_C 6.31/41.9 correspond to ^{3}J coupling of H-2/C-10′, and in the HMBC spectra (Figure 3.24d), the signals at δ_H/δ_C 5.77/135.9 correspond to ^{3}J coupling of H-11′/C-1, at δ_H/δ_C 6.31/112.3 correspond to ^{3}J coupling of H-2/C-6, at δ_H/δ_C 6.31/141.6 correspond to ^{2}J coupling of H-2/C-3, at δ_H/δ_C 6.31/145.9 correspond to ^{3}J coupling of H-2/C-4, at δ_H/δ_C 6.46/113.7 correspond to ^{3}J coupling of H-6/C-2 and at δ_H/δ_C 6.46/135.9 correspond to ^{2}J coupling of H-6/C-1. Meanwhile, in the HMBC spectrum (Figure 3.24e), the signals at δ_H/δ_C 9.04/141.6 correspond to ^{2}J coupling of H-3 in the –OH catechol to C-3 and at δ_H/δ_C 9.04/145.9 correspond to ^{3}J coupling of H-3 in the –OH catechol to C-4. Therefore, the 3,4-dihydroxy-5-alkenylphenyl group should be bonded to C-10′ in the structurally altered side chain, as shown in Scheme 3.2.

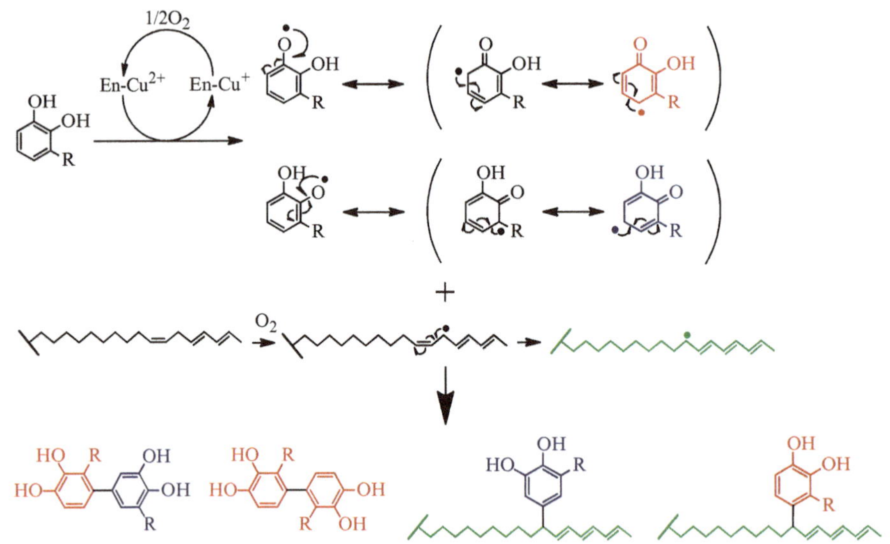

Compound 1 Compound 2 Compound 3 Compound 4

SCHEME 3.2

Enzymatic polymerization mechanism of laccol: nuclear-nuclear C–C bond and nuclear-side chain C–C bond formation.

10′-(4,5-Dihydroxy-6-Alkenyl)Phenyl-3-[12′E,14′E,16′ E-Heptadecatrienyl]Catechol (4)

The ^{1}H-^{1}H COSY spectrum (Figure 3.22a) shows signals at δ_H/δ_H 3.03/1.31 corresponding to ^1J coupling of H-10′/H-9′ and at δ_H/δ_H 5.73/3.03 corresponding to ^1J coupling of H-11′/H-10′ in the unsaturated side chain. In the HMQC spectrum (Figure 3.23a), the signal at δ_H/δ_C 3.02/48.2 corresponds to ^1J coupling of H-10′/C-10′, and in the HMQC spectrum (Figure 3.23b), the signals at δ_H/δ_C 6.29/120.0 correspond to ^1J coupling of H-2/C-2 and at δ_H/δ_C 6.58/112.0 correspond to ^1J coupling of H-3/C-3. Furthermore, in the HMBC spectrum (Figure 3.24b), the signals at δ_H/δ_C 1.31/134.9 correspond to ^3J coupling of H-9′/C-1, at δ_H/δ_C 2.53/134.9 correspond to ^3J coupling of H-7 in the –CH– side chain to C-1, at δ_H/δ_C 2.53/139.2 correspond to ^2J coupling of H-7 in the –CH– side chain to C-6, at δ_H/δ_C 3.02/112.3 correspond to ^4J coupling of H-10′/C-3, at δ_H/δ_C 3.02/119.7 correspond to ^3J coupling of H-10′/C-2, at δ_H/δ_C 3.02/134.9 correspond to ^2J coupling of H-10′/C-1 and at δ_H/δ_C 3.02/139.2 correspond to ^3J coupling of H-10′/C-6. In the HMBC spectrum (Figure 3.24c), the signal at δ_H/δ_C 6.29/48.2 corresponds to ^3J coupling of H-2/C-10′, and in the HMBC spectrum (Figure 3.24D), the signals at δ_H/δ_C 5.70/134.9 correspond to ^3J coupling of H-11′/C-1, at δ_H/δ_C 6.58/134.9 correspond to ^3J coupling of H-3/C-1 and at δ_H/δ_C 6.58/142.9 correspond to ^2J coupling of H-3/C-4. Thus, the 3,4-dihydroxy-5-alkenylphenyl group should be bonded to C-10′ in the structurally altered side chain as shown in Scheme 3.2.

3.2.2 REACTION MECHANISM OF LACCOL

Based on the polymerization mechanism of urushiol and the results of this research, we concluded that the enzymatic polymerization mechanism of laccol occurs as shown in Scheme 3.2. First, laccol is oxidized by laccase to form a radical structure, and then laccol radicals are transferred to give semiquinone radicals. The semiquinone radicals undergo reactions in two ways, like urushiol: (i) attacking the laccol nucleus to give biphenyl laccol or (ii) attacking the side-chain methylene group inserted between conjugated diolefines and monoolefines to give a conjugated triene that may subsequently produce a nuclear-side chain structure. Focusing attention on the binding site of laccol dimers, new binding sites formed at C-4 and C-5 were observed. These results showed that if the semiquinone radical pulls a hydrogen out of the hydroxyl group at position 1 to form another radical, a radical at position 4 is the most stable structure, and if the semiquinone radical pulls a hydrogen out of the hydroxyl group at position 2 to form another radical, a radical at position 5 is the most stable structure. Focusing on the coupling form, there was no C–O–C bonding observed in the laccol dimers. These results showed that laccol easily forms C–C bonds during laccase catalytic polymerization.

3.3 THITSIOL

Thitsiol is the main component of sap tapped from the *Gluta usitata* lacquer tree. It is a mixture of 4-substituted and/or 3-substituted catechol derivatives with $n=17$ carbon chains. Figure 3.25 is IR spectra of laccol monomer, dimer, and oligomer. It showed several main characteristic absorption peaks such as a hydroxyl group at $3500\,cm^{-1}$, methylene at $2930\,cm^{-1}$, phenolic hydroxyl group at $1100\,cm^{-1}$, and ether at $1080\,cm^{-1}$, respectively, like urushiol and laccol.

The ^{1}H NMR and ^{13}C NMR of thitsiol monomer are shown in Figure 3.26. After analysis of two-dimensional NMR such as DQF-COSY, HMQC, and HMBC, almost all peaks were assigned and the structure of thitsiol monomer was concluded.

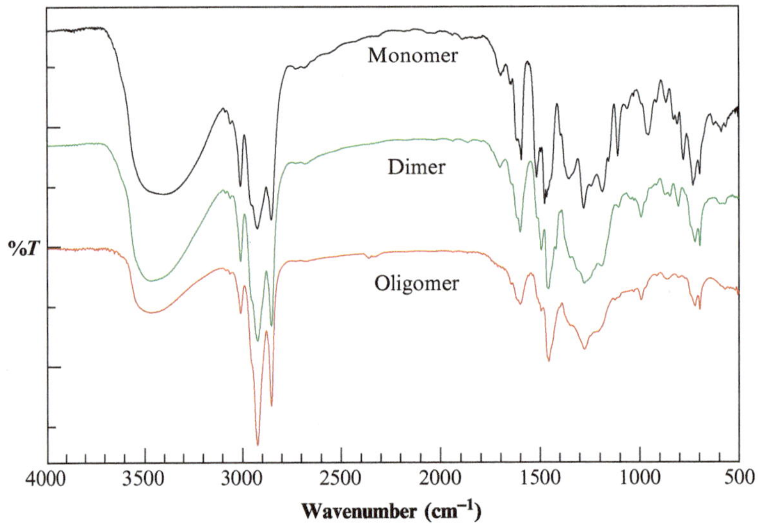

FIGURE 3.25

IR spectra of thitsiol monomer, dimer, and oligomer.

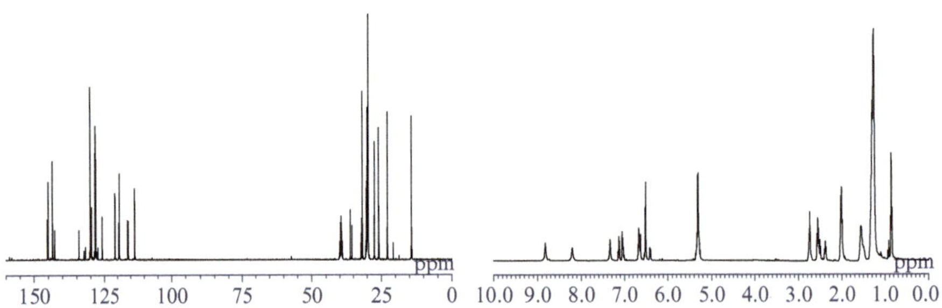

FIGURE 3.26

^{1}H (a) and ^{13}C NMR (b) spectra of thitsiol monomer.

Structure **1**:

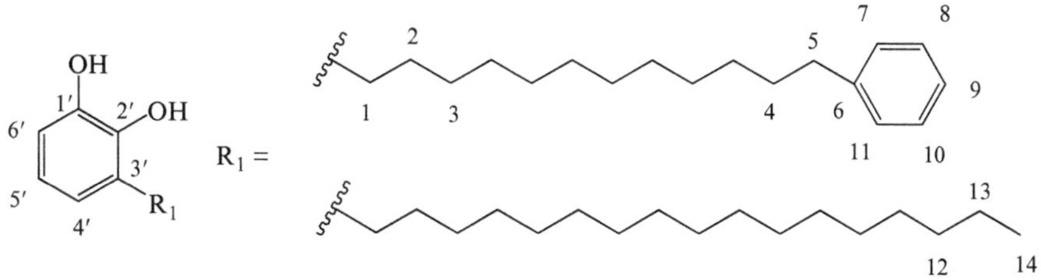

[^{1}H NMR (400 MHz, DMSO-d_6/TMS)]: δ (ppm) = 0.75-0.80 [t, H-14], 1.15-1.30 [m, H-3-4, H-12, 13], 1.47-1.50 [quin, H-2], 2.55[t, H-1, 5], 6.45-6.55 [m, H-4′, 6′], 6.65-6.70 [dd, H-5′], 7.02-7.05 [m, H-9], 7.04-7.08 [m, H-7, 11], 7.10-7.15 [t, H-8, 10], 7.35 [s, OH-2′], 8.80 [s, OH-1′].

[^{13}C NMR (100 MHz, DMSO-d_6)]: δ (ppm) = 14.0 [C-14], 22.3 [C-13], 29.1-29.7 [C-3-4], 29.8 [C-2], 29.9 [C-20], 31.0 [C-12], 36.0 [C-5], 113.2 [C-5′], 118.7 [C-6′], 120.8 [C-4′], 125.5 [C-9], 127.8 [C-8, 10], 128.1 [C-7, 11], 129.6 [C-3′], 142.4 [C-6], 143.6 [C-2′], 145.2 [C-1′].

Structure **2**:

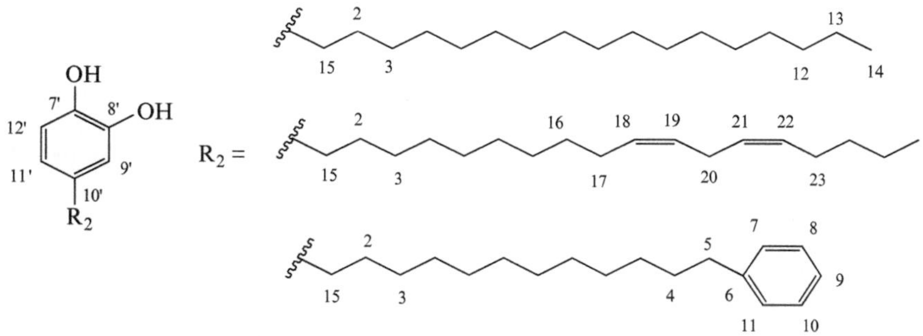

[^{1}H NMR (400 MHz, DMSO-d_6/TMS)]: δ (ppm) = 0.75-0.80 [t, H-14], 1.15-1.30 [m, H-3-4, 12, 13, 16],1.47-1.50 [quin, H-2], 1.89-2.00 [q, H-17, 23], 2.38 [t, H-15], 2.55 [t, H-5], 2.70 [t, H-20], 5.30 [m, H-18, 19, 21, 22], 6.40 [dd, H-12′], 6.61 [s, H-9′], 6.67 [dd, H-11′], 7.02-7.05 [m, H-9], 7.04-7.08 [m, H-7, 11], 7.10-7.15 [t, H-8, 10], 8.23 and 8.80 [s, OH × 2].

[^{13}C NMR (100 MHz, DMSO-d_6)]: δ (ppm) = 14.0 [C-14], 22.3 [C-13], 25.5 [C-20], 27.0 [C-17, 23], 29.1-29.7 [C-3-4, 16], 29.8 [C-2], 31.0 [C-12], 35.4 [C-15], 36.0 [C-5], 115.8 [C-11′], 116.0 [C-9′], 119.5 [C-12′], 125.5 [C-9], 127.8 [C-8, 10], 128.1 [C-7, 11], 127.1 [C-22], 128.0 [C-19], 129.8 [C-18], 129.9 [C-21] 134.0 [C-10′], 142.4 [C-6], 143.4 [C-8′], 145.7 [C-7′].

Structure **3**:

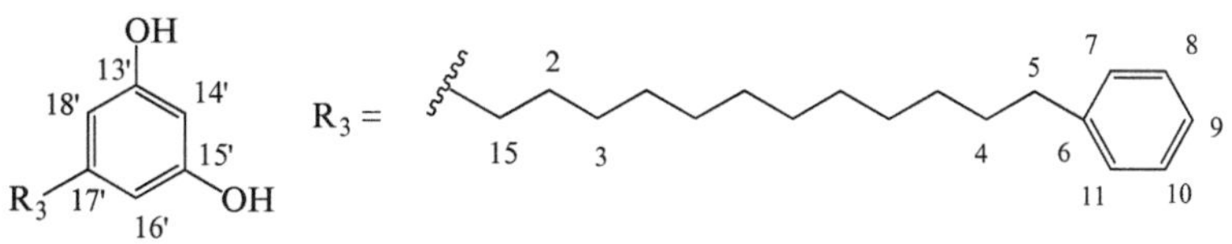

[^{1}H NMR (400 MHz, DMSO-d_6/TMS)]: δ (ppm) = 1.47-1.50 [quin, H-2], 2.38 [t, H-15], 2.55 [t, H-5] 6.12 [s, H-16′, 18′] 6.17 [s, H-14′], 7.02-7.05 [m, H-9], 7.04-7.08 [m, H-7, 11], 7.10-7.15 [t, H-8, 10], 8.80 [s, OH × 2].

[^{13}C NMR (100 MHz, DMSO-d_6)]: δ (ppm) = 29.8 [C-2], 35.4 [C-15], 36.0 [C-5], 100.9 [C-14′], 107.3 [C-16′, 18′], 125.5 [C-9], 127.8 [C-8, 10], 128.1 [C-7, 11], 142.4 [C-6], 144.2 [C-17′], 159.0 [C-13′, 15′].

Structure **4**:

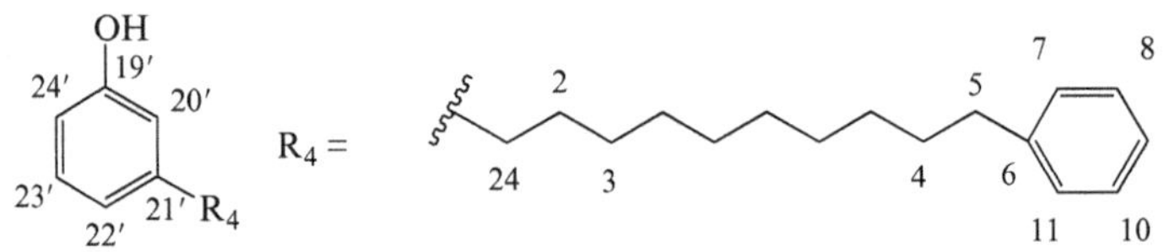

[^{1}H NMR (400 MHz, DMSO-d_6/TMS)]: δ (ppm) = 1.47-1.50 [quin, H-2], 2.50 [t, H-24], 2.55 [t, H-5], 6.50 [m, H-23′], 6.65 [m, H-21′], 7.01 [t, H-22′], 7.02-7.05 [m, H-9], 7.04-7.08 [m, H-7, 11], 7.10-7.15 [t, H-8, 10], 8.90 [s, OH].

[^{13}C NMR (100 MHz, DMSO-d_6)]: δ (ppm) = 29.8 [C-2], 32.0 [C-24], 113.1 [C-23′], 119.8 [C-21′], 125.5 [C-9], 127.8 [C-8, 10], 128.1 [C-7, 11], 129.8 [C-22′], 142.4 [C-6], 144.3 [C-20′], 158.1 [C-24′].

Figure 3.27 is the GC spectra of thitsiol. It shows 19 peaks. Each peak mass in chromatography was analyzed, and the results are summarized in Table 3.7.

There are saturated 3-, and 4-heptadecenylcatechol (MW = 346), monoenes of 3- and 4-heptadecadienylcatechol (MW = 344), and phenyldodecylbenzenediol (MW = 354) in the thitsiol side chain; meanwhile, 3- and 4-pentadecylcatechol (MW = 320) and pentadecenylcatechol were also detected (Table 3.6). From the above analysis, the thitsiol structure was concluded as shown in Figure 3.28.

In the three kind lacquers, we have found that the ratio of dimers and trimers to monomers is higher in thitsiol, and that the color of the lacquer sap is black even when raw. After polymerization, thitsiol lacquer films are black even if dried in slow-drying conditions. However, in the case of urushiol or laccol

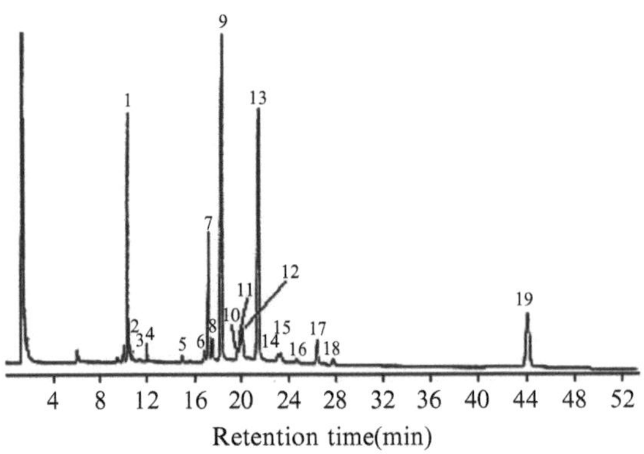

FIGURE 3.27

GC spectrum of thitsiol.

Table 3.7 GC/MS results of thitsiol

Peak no.	*m/z*	R.t. (min)	Area (%)	Structure
1	320	10.33	9.3	3-Pentadecylcatechol
2	328	10.49	1.8	3-(10Z,13Z-Heptadecadienyl)phenol
3	318	10.73	0.3	3-(8Z-pentadecenyl)catechol
4	320	12.01	0.8	4-Pentadecylcatechol
5	310	14.90	0.4	3-Phenyldecylphenol
6	348	16.75	0.8	3-Heptadecylcatechol
7	346	17.11	8.8	3-Heptadecenylcatechol
8	346	17.46	1.5	3-Heptadecenylcatechol
9	344	18.22	35.2	3-Heptadecadienylcatechol
10	344	19.64	0.9	3-Heptadecadienylcatechol
11	346	19.86	2.0	4-Heptadecenylcatechol
12	346	20.06	2.8	4-Heptadecenylcatechol
13	344	21.35	21.4	4-Heptadecadienylcatechol
14	344	23.03	0.8	4-Heptadecadienylcatechol
15	344	23.29	0.9	4-Heptadecadienylcatechol
16	338	24.60	0.4	3-Phenyldodecylphenol
17	326	26.36	2.4	Phenyldecylbenzenediol
18	344	27.73	0.7	4-Heptadecadienylcatechol
19	354	44.11	8.8	Phenyldodecylbenzenediol

lacquer, the color of the film is dark brown when dried quickly in high relative humidity and buff in a slow, low RH drying condition. We speculate that the deep color reflects the nuclear-nuclear (C–C) structure, whereas the light color is due to nuclear-side chain (C–O–C) structure, but this has not been confirmed yet.

3.3.1 CHARACTERIZATION OF THITSIOL DIMER STRUCTURE

In order to reveal the enzymatic polymerization mechanism of thitsiol, the dimer structures produced in the initial stage of enzymatic polymerization were analyzed by NMR including 2D-NMR measurements, like the analysis of urushiol and laccol lacquer, and then, the color of the lacquer film and the thitsiol polymerization mechanism are discussed based on the result of dimer structures.

Figure 3.29 is the proton (a) and carbon (b) NMR spectra of thitsiol dimers. Compared with urushiol, the absence of peaks due to nuclear-side chain dimer structures at 3.5-4.0 ppm in the proton NMR spectrum (Figure 3.29a), suggested that thitsiol dimers preferentially generate the nuclear-nuclear structure. On the other hand, the lack of a peak due to dibenzofuran dimers around 160 ppm in the carbon NMR spectrum (Figure 3.29b) also implied that thitsiol possibly does not generate dibenzofuran-type dimers.

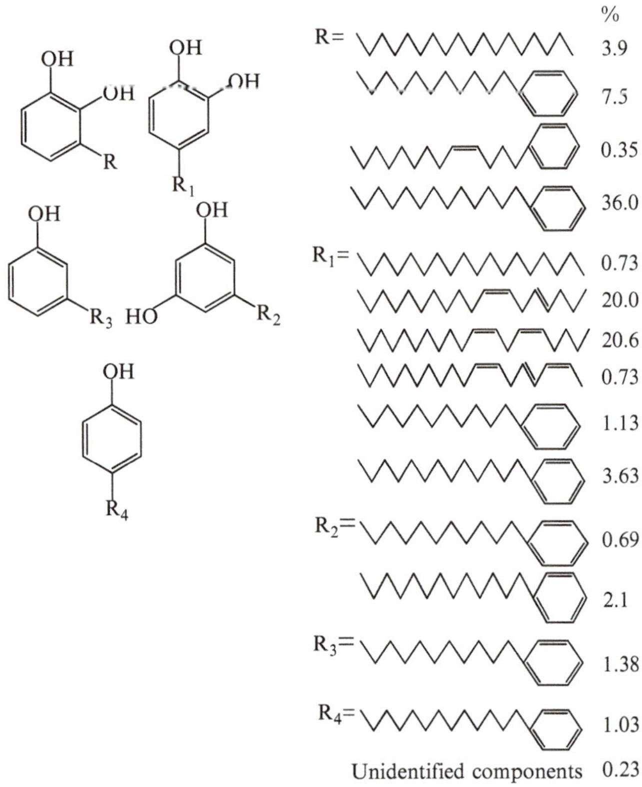

FIGURE 3.28

Chemical structure of thitsiol.

The DQF-COSY spectrum of the thitsiol dimeric fraction is shown in Figure 3.30, and the HMQC and HMBC spectra are shown in Figure 3.31. The detailed analysis follows.

1,1′,2,2′-Tetrahydroxy-3,3′-Dialkenyl-5,5′-Biphenyl (1)

The 2D ^{1}H-^{1}H COSY spectrum (Figure 3.30) showed a signal at δ_H/δ_H 6.63/6.86 corresponding to ^2J coupling of H-4/H-6 and H-4′/H-6′ in the catechol ring. The signals at δ_H/δ_C 6.63/120.7 in the HMQC spectra (Figure 3.31b) corresponded to ^1J coupling of H-4/C-4 and H-4′/C-4′, while those at δ_H/δ_C 6.86/113.8 corresponded to H-6/C-6 and H-6′/C-6′, respectively. In addition, signals at δ_H/δ_C 6.63/113.8 corresponded to H-4/C-6 and H-4′/C-6′ and at δ_H/δ_C 6.86/120.7 corresponded to H-6/C-4 and H-6′/C-4′, respectively, in the HMBC spectrum (Figure 3.31d). In addition, signals at δ_H/δ_C 6.63/29.7 (Figure 3.31c) corresponded to H-4/C-7 and H-4′/C-7′, at δ_H/δ_C 6.63/127.0 corresponded to H-4/C-5 and H-4′/C-5′, at δ_H/δ_C 6.63/141.8 corresponded to H-4/C-2 and H-4′/C-2′, at δ_H/δ_C 6.86/127.0 corresponded to H-6/C-5 and H-6′/C-5′, those at δ_H/δ_C 6.86/141.8 corresponded to H-6/C-1 and H-6′/C-1′, respectively. Thus, the spectroscopic data are compatible with compound **1** for the dimer.

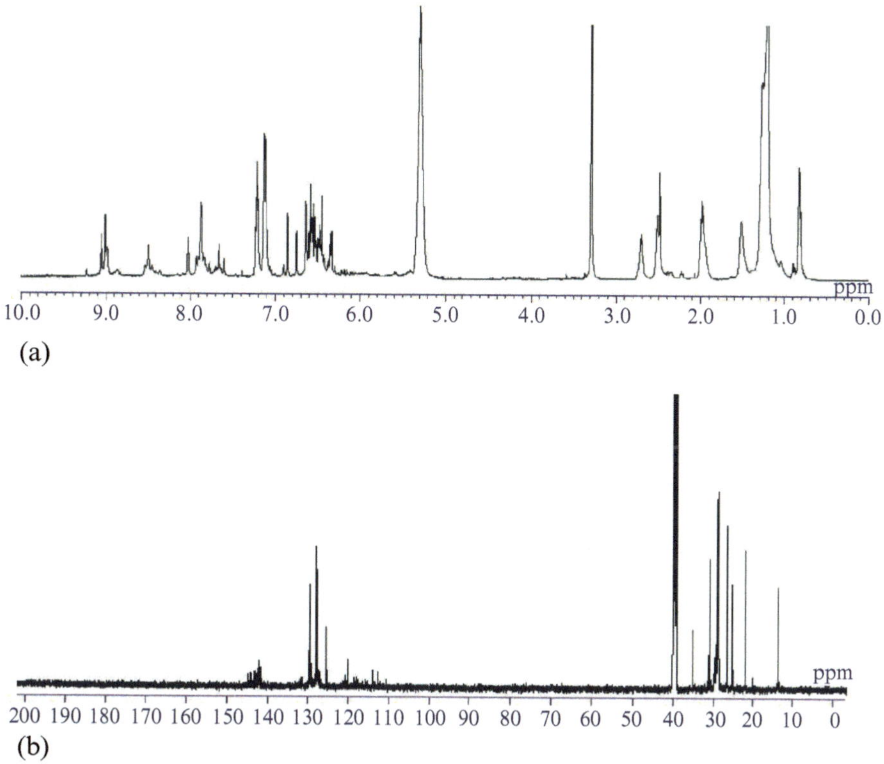

FIGURE 3.29

^1H NMR (a) and ^{13}C NMR (b) spectra of thitsiol dimers.

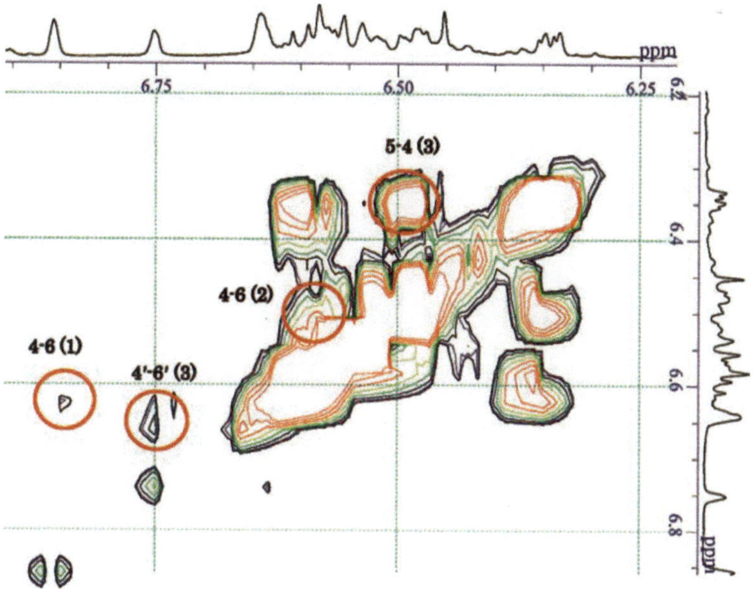

FIGURE 3.30

^1H-^1H DQF-COSY spectrum of thitsiol dimers (6.2-7.0 ppm).

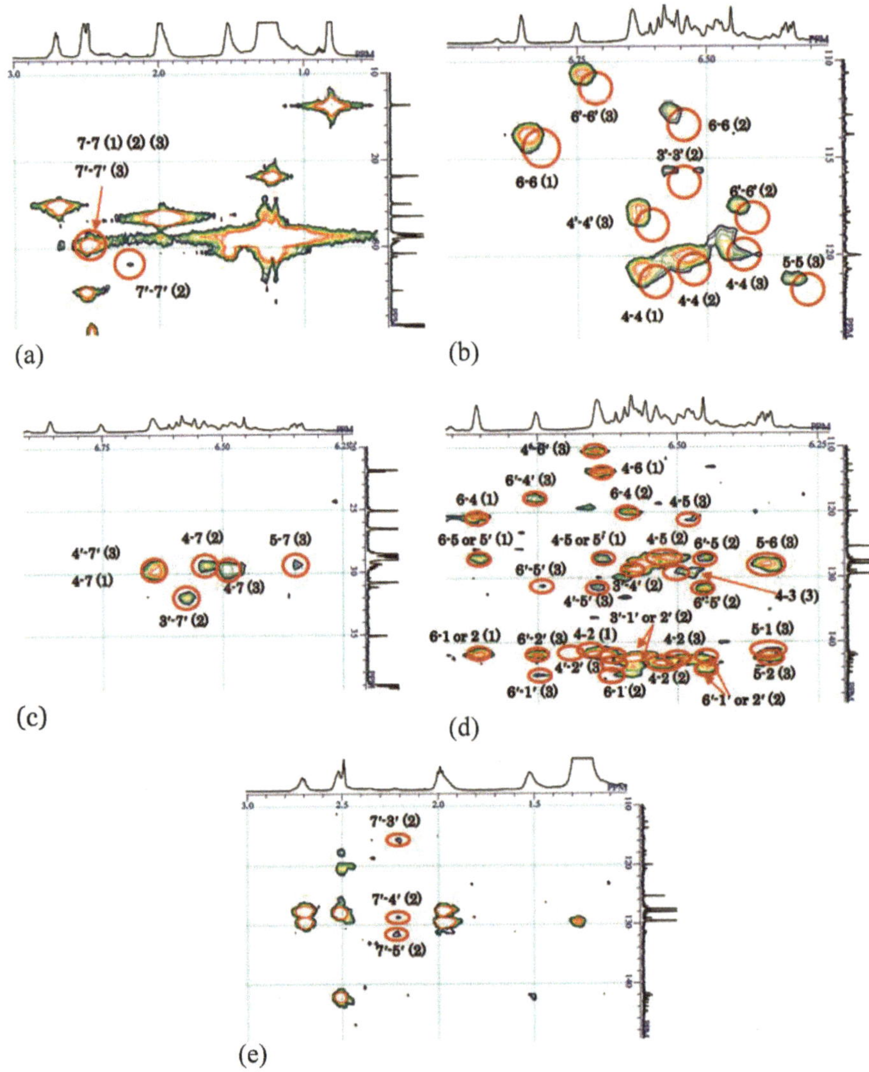

FIGURE 3.31

HMQC spectra of dimeric fraction: (a) ^{1}H: 0.5-3.0 ppm; ^{13}C: 10.0-40.0 ppm, and (b) ^{1}H: 6.2-7.0 ppm; ^{13}C: 110.0-125.0 ppm. HMBC spectra of dimeric fraction: (c) ^{1}H: 6.2-7.0 ppm; ^{13}C: 20.0-40.0 ppm, (d) ^{1}H: 6.2-7.0 ppm; ^{13}C: 110.0-150.0 ppm, and (e) ^{1}H: 1.0-3.0 ppm; ^{13}C: 110.0-150.0 ppm.

1,1′,2,2′-Tetrahydroxy-3,3′-Dialkenyl-6,5′-Biphenyl (2)

The 2D ^{1}H-^{1}H COSY spectrum (Figure 3.30) showed a signal at δ_H/δ_H 6.34/6.47 corresponding to ^1J coupling of H-5/H-4 and δ_H/δ_H 6.64/6.75 corresponding to ^2J coupling of H-4′/H-6′ in catechol ring, respectively. In the HMQC spectra (Figure 3.31b), signals at δ_H/δ_C 6.34/121.2 corresponded to ^1J coupling of H-5/C-5, at δ_H/δ_C 6.47/118.5 corresponded to ^1J coupling of H-4/C-4, at δ_H/δ_C 6.64/117.8 corresponded to ^1J coupling of H-4′/C-4′, and at δ_H/δ_C 6.75/110.4 corresponded to ^1J coupling of H-6′/C-6′, respectively. Furthermore, in the HMBC spectra (Figure 3.31c), the signal at δ_H/δ_C 6.34/29.5 corresponded to ^4J coupling of H-5 to the –CH– group at C-7 in the side chain, and signals at δ_H/δ_C 6.47/29.5 corresponded to ^3J coupling of H-4 to the –CH– group at C-7 in the side chain, at δ_H/δ_C 6.64/29.7 corresponded to ^3J coupling of H-4′ to the –CH– group at C-7′ in the side chain of compound **2**, respectively. In addition, signals at δ_H/δ_C 6.34/128.3 (Figure 3.31d and e) corresponded to ^2J coupling of H-5/C-6, at δ_H/δ_C 6.34/141.6 corresponded to ^3J coupling of H-5/C-1, at δ_H/δ_C 6.34/142.5 corresponded to ^4J coupling of H-5/C-2, at δ_H/δ_C 6.47/121.2 corresponded to ^2J coupling of H-4/C-5, at δ_H/δ_C 6.47/129.4 corresponded to ^2J coupling of H-4/C-3, at δ_H/δ_C 6.47/142.5 corresponded to ^3J coupling of H-4/C-2, at δ_H/δ_C 6.64/110.4 corresponded to ^3J coupling of H-4′/C-6′, at δ_H/δ_C 6.64/131.4 corresponded to ^2J coupling of H-4′/C-5′, at δ_H/δ_C 6.64/141.7 corresponded to ^3J coupling of H-4′/C-2′, at δ_H/δ_C 6.75/117.8 corresponded to ^3J coupling of H-6′/C-4′, at δ_H/δ_C 6.75/131.4 corresponded to ^2J coupling of H-6′/C-5′, at δ_H/δ_C 6.75/141.7 corresponded to ^3J coupling of H-6′/C-2′, and at δ_H/δ_C 6.75/144.8 corresponded to ^2J coupling of H-6′/C-1′, respectively. This would lead to compound **2** structure.

1,1′,2,2′-Tetrahydroxy-3,4′-Dialkenyl-5,5′-Biphenyl (3)

The 2D ^{1}H-^{1}H COSY spectrum (Figure 3.30) shows a signal at δ_H/δ_H 6.54/6.58 corresponding to ^3J coupling of H-4/H-6 in the catechol ring. In HMQC spectra (Figure 3.31B), signals at δ_H/δ_C 6.45/117.6 corresponded to ^1J coupling of H-6′/C-6′, at δ_H/δ_C 6.54/120.0 corresponded to ^1J coupling of H-4/C-4, at δ_H/δ_C 6.57/115.5 corresponded to ^1J coupling of H-3′/C-3′, and at δ_H/δ_C 6.58/112.5 corresponded to ^1J coupling of H-6/C-6, respectively. In addition, there was a signal at δ_H/δ_C 2.22/31.9 (Figure 3.31a) corresponding to ^1J coupling of H-7′/C-7′ in the side chain. In HMBC spectra (Figure 3.31d and e), signals were seen at δ_H/δ_C 6.45/126.8 corresponding to ^3J coupling of H-6′/C-5, at δ_H/δ_C 6.45/131.4 corresponding to ^2J coupling of H-6′/C-5′, at δ_H/δ_C 6.45/142.5 corresponding to ^2J coupling of H-6′/C-1′, at δ_H/δ_C 6.45/144.0 corresponding to ^3J coupling of H-6′/C-2′, and at δ_H/δ_C 6.54/29.6 corresponding to ^3J coupling of H-4 to the –CH– group at C-7 in the side chain, while signals were seen at δ_H/δ_C 6.54/126.8 corresponding to ^2J coupling of H-4/C-5, at δ_H/δ_C 6.54/143.0 corresponding to ^3J coupling of H-4/C-2, and at δ_H/δ_C 6.57/31.9 (Figure 3.31c) corresponding to ^3J coupling of H-3′ to the –CH– group at C-7′ in the side chain, while signals were seen at δ_H/δ_C 6.57/120.9 corresponding to ^2J coupling of H-3′/C-4′, at δ_H/δ_C 6.57/142.5 corresponding to ^3J coupling of H-3′/C-1′, at δ_H/δ_C 6.57/144.0 corresponding to ^2J coupling of H-3′/C-2′, at δ_H/δ_C 6.58/120.0 corresponding to ^3J coupling of H-6/C-4, and at δ_H/δ_C 6.58/144.8 corresponding to ^2J coupling of H-6/C-1.

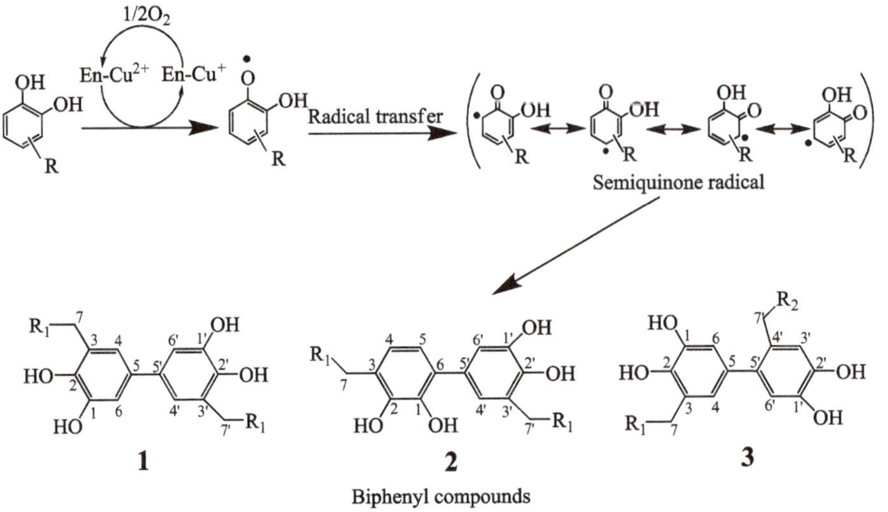

SCHEME 3.3

Nuclear-nuclear structures of thitsiol dimers produced from laccase catalyst.

In addition, signals seen at δ_H/δ_C 2.22/115.5 corresponded to ^3J coupling of H-7′ in the –CH– side chain to C-3′, at δ_H/δ_C 2.22/129.0 corresponded to ^2J coupling of H-7′ in the –CH– side chain to C-4′, and at δ_H/δ_C 2.22/131.3 corresponded to ^3J coupling of H-7′ in the –CH– side chain to C-5′. The above analysis results would lead to compound **3** structures.

From analyses of the NMR spectra, it can be concluded that combined nuclear-nuclear (C–C) couplings were produced preferentially by the early enzymatic polymerization, and that at least three kinds of thitsiol dimers were generated (Scheme 3.3).

3.3.2 REACTION MECHANISM OF THITSIOL

Based on the polymerization mechanism of urushiol and laccol, and the results of thitsiol dimers, it can be concluded that the reaction mechanism of thitsiol catalyzed by laccase is as shown in Scheme 3.3.

First, laccase oxidizes thitsiol monomers to form a radical structure, and then, radicals are transferred to give semiquinone radicals. In urushiol, the semiquinone radicals would undergo reactions in two ways: (i) attacking the urushiol nucleus to give biphenyl urushiol, with some of the nuclei being converted into dibenzofuran compounds through their further laccase-catalyzed oxidation or (ii) attacking the side-chain methylene group inserted between conjugated diolefines and monoolefines to give heptatriene cations that may subsequently produce a nuclear-side chain structure. However, because the side chain in thitsiol is bigger than that in urushiol and there are very few olefin structures in side chains (C-17 with benzene ring), no route (ii) reaction occurred; only route (i) occurred to form a nuclear-nuclear structure.

Urushiol Laccol Thitsiol

FIGURE 3.32

Photos of urushiol, laccol, and thitsiol.

In general, biphenyl phenols absorb most of the wavelengths of visible light, while the single phenols absorb less, because both the commercial thitsiol lacquer sap and its film are black and the proportions of oligomers including dimers, trimers, and tetramers are very high compared with urushiol lacquer. In order to reveal the reason for this, we examined the activity of each laccase, but the result showed that the laccase component in urushiol lacquer has higher activity. Furthermore, the thitsiol lacquer sap just exuded from a *G. usitata* lacquer tree is white, and the color darkens after contact with the air for several minutes, and then becomes black. Compared with the behaviors of urushiol and laccol lacquer, we speculated that the change of color of thitsiol lacquer may be due to the nuclear-nuclear (C-C) biphenyl phenol type structure.

In summary, urushiol is a mixture of 3-substituted catechol derivatives with $n=15$ carbon chains with 0-3 olefins, contaminated with other minor phenolic lipids such as those having $n=17$ carbon chains instead of an $n=15$ carbon chain, while laccol is a mixture of 3-substituted catechol derivatives with $n=17$ carbon chains with 0-3 olefins, and thitsiol is a mixture of 4-substituted and/or 3-substituted catechol derivatives with $n=17$ carbon chains. The photo of urushiol, laccol, and thitsiol is shown in Figure 3.32.

REFERENCES

[1] H. Yoshida, Chemistry of lacquer (urushi), Part I, J. Chem. Soc. 43 (1883) 472–486.

[2] G. Bertrand, Chimie industrielle—Sur le latex de l'arbre à laque, Comptes Rendus 118 (1894) 1215–1218.

[3] G. Bertrand, Chimie organique—Sur la laccase et sur le pouvoir oxydant de cette diastase, Comptes Rendus 120 (1895) 266–269.

[4] G. Bertrand, Chimie organique—Sur les rapports qui existent entre la constitution chimique des composes organiques et leur oxydabilitè sous l'influence de la laccase, Comptes Rendus 122 (1896) 1132–1134.

[5] G. Bertrand, Chimie organique—Sur I'intervention du manganèse dans les oxydations provoquées par la laccase, Comptes Rendus 124 (1897) 1032–1035.

[6] J. Kumanotani, Urushi (Oriental lacquer)—a natural aesthetic durable and future-promising coating, Prog. Org. Coat. 26 (1995) 163–195.

[7] Y.M. Du, R. Oshima, J. Kumanotani, Reversed-phase liquid chromatographic separation and identification of constituents of urushiol in the sap of the lac tree, *Rhus vernicifera*, J. Chromatogr. 284 (1984) 463–473.
[8] T. Miyakohsi, K. Nahashe, K. Yoshida, Progress in Urushi Science, IPC Publisher, Tokyo, Japan, 2000 (in Japanese).
[9] K. Nahashe, T. Miyakoshi, A guide to urushi chemistry 2. *Toso Toryo* 1998, No. 574, 30-41, Chem. Abstr. 128 (1998) 231625 (in Japanese).
[10] Y.M. Du, R. Oshima, H. Iwatsuki, J. Kumanotani, High-resolution gas—liquid chromatographic analysis of urushiol of the lac tree, *Rhus vernicifera*, without derivatization, J. Chromatogr. 295 (1984) 179–186.
[11] H. Budzikiewicz, C. Djerassi, D.H. Williams, Chapter 1, hydrocarbons, Mass Spectrometry of Organic Compounds, Holden-Day Inc., San Francisco, Cambridge, London, and Amsterdam, 1967 pp. 45–93.
[12] S. Harigaya, T. Honda, R. Lu, T. Miyakoshi, L. Chen, Enzymatic dehydrogenative polymerization of urushiols in fresh exudates from the lacquer tree, *Rhus vernicifera* DC, J. Agric. Food Chem. 55 (2007) 2201–2208.
[13] R. Lu, D. Kanamori, T. Miyakoshi, Characterization of thitsiol dimer structures from *Melanorrhoea usitata* with laccase catalyst by NMR spectroscopy, Int. J. Polym. Anal. Charact. 16 (2011) 86–94.

CHAPTER

Lacquer polysaccharide

4

A lacquer polysaccharide is a component that is soluble in water as reverse micelles, and functions as a dispersing agent. It is a flocculent acidic polysaccharide with a white color after freeze-drying that is separated from the water soluble portion as shown in Figure 4.1.

4.1 STRUCTURAL ANALYSIS

In 1963, Oda *et al.* reported the sugar analysis of Japanese lacquer gum by partial acid hydrolysis. The detailed separation method is thus: 50 mg crude lacquer gum was dissolved in 5 ml 1 N H_2SO_4, and refluxed for 6 h. After cooling to room temperature, it was neutralized with barium hydroxide [$Ba(OH)_2$] to pH 4. The $Ba(OH)_2$ precipitate was removed by centrifugation, and the supernatant was treated with Amberlite IR-120 to remove ions. The deionized solution was dried by evaporation, then dissolved in 10 ml 0.02 N sodium hydroxide (NaOH), and analyzed using a Dowex-1 column (φ 0.8 × 10 cm). D-galactose, D-xylose, L-arabinose, L-rhamnose, D-galacturonic acid, and D-glucuronic acid were detected. Furthermore, in 1984, Oshima and Kumanotani [1] described the partial chemical structure of Chinese lacquer polysaccharide by methylation analysis and Smith degradation. The results showed that lacquer polysaccharides can be separated into two fractions having average molecular weights around 84,000 and 27,000 by aqueous-phase gel permeation chromatography (GPC). Each fraction contained D-galactose (65 mol%), 4-*O*-methyl-D-glucuronic acid (24 mol%), D-glucuronic acid (3 mol%), L-arabinose (4 mol%), and L-rhamnose (3 mol%). The linkage of these sugars was ascertained by methylation analysis; that is, β-(1 → 3)-linked galactopyrannan main chains having complex side chains. 4-*O*-methyl-D-glucuronic acid, D-glucuronic acid, L-arabinose, and L-rhamnose were present at the terminal of side chains. Lacquer polysaccharides showed obvious polyelectrolyte properties in aqueous solution. Zhang *et al.* [2] classified the Chinese lacquer polysaccharide, and investigated the molecular weight (M_w, M_n) and intrinsic viscosity [η] of each fraction by light scattering, osmotic pressure, and viscosity measurements. The results showed that in a 0.1 M NaCl solution at 30 °C, the Mark-Houwink equation is:

$$[\eta] = 3.59 \times 10^{-2} M_w^{\;0.50}$$

Lacquer Chemistry and Applications. http://dx.doi.org/10.1016/B978-0-12-803589-4.00004-3

FIGURE 4.1

Photo of lacquer polysaccharide.

They concluded that the Chinese lacquer polysaccharide in aqueous solution exists as random coils with a compact chain conformation [3,4]. After reduction of carboxyl groups, methylation, periodate oxidation, and Smith degradation, Du *et al.* [5] reported the chemical structure of Chinese lacquer polysaccharide by GC/MS measurements.

Because lacquer trees are sensitive to environmental changes and the locations of sap production are becoming limited, changes in the structures of both components and hardened lacquer are occurring. In order to investigate the relationship between chemical and physical properties and structure, a convenient and efficient method for structural analysis of Chinese lacquer polysaccharide by means of high-resolution nuclear magnetic resonance (NMR) spectroscopy involving several two-dimensional measurements was reported recently [6].

4.1.1 PURIFICATION OF LACQUER POLYSACCHARIDE

Chinese lacquer polysaccharide was obtained from an acetone-extracted residue of the sap of a lacquer tree at Maoba in Hubei province of China by the following procedures. The sap (1.0 kg) in acetone (3.5 L) was stirred at 25 °C for 8 h, and then the mixture was filtered to give a precipitate, which was washed with acetone several times and dried to afford a gray-white powder (100 g) in 10% yield. The powder was dissolved in deionized water (1 L) and stirred at 25 °C for 5 h. The aqueous solution was concentrated to 200 ml with a rotary evaporator under reduced pressure below 40 °C, and an insoluble precipitate was removed by filtration. Ammonium sulfate (80 g) was added to the clear filtrate solution. The mixture was stirred at 25 °C for 1 h, and then the insoluble precipitate was removed again. The filtrate was dialyzed for 3 days against deionized water. Ethanol (95%, 1 L) was added to the dialyzate to yield a white precipitate, which was filtered and washed with acetone several times to give 17.0 g of crude lacquer polysaccharide in 1.7% yield after drying under reduced pressure. The crude polysaccharide was purified by passing through a Sephadex G-100 column (3 cm × 25 cm) eluted with 0.1 N NaCl solution to give 10.0 g of pure lacquer polysaccharide (salt form) in 1.0% yield after dialyzing by deinonized water, followed by precipitation with 95% ethanol, and then drying under reduced pressure.

4.1.2 IR AND GPC MEASUREMENTS

Infrared spectra were determined on a Shimadzu FT-IR 8300 spectrometer by a KBr pellet method. Molecular weights of the Chinese lacquer polysaccharide were estimated at 40 °C by means of aqueous-phase GPC (column; Tosoh TSK-gel G2500PW$_{XL}$, G3000PW$_{XL}$, and G4000PW$_{XL}$, 7.6 × 300 mm × 3; eluent, 66.7 mmol/L of phosphate buffer, pH = 6.86) using pullulan as a standard. The Chinese lacquer polysaccharide is a mixture of two fractions having number-average molecular weights of 8.6×10^4 and 2.3×10^4 in the proportion of 25 and 75 mol%. The molecular weight distribution was $M_w/M_n = 1.08$ and 1.11, respectively. The polysaccharide was used directly without separation because it was impossible to separate. Thus, the number-average molecular weight of the Chinese lacquer polysaccharide was calculated to be $M_n = 4.1 \times 10^4$ taking into account the proportion of the mole fractions.

Figure 4.2 is the IR spectrum of Chinese lacquer polysaccharide. Absorption at 3450 cm^{-1} due to stretching vibration of O–H, at 2940 cm^{-1} due to stretching vibration of C–H, at 1740 cm^{-1} due to stretching vibration of COO–, at 1460 cm^{-1} due to symmetric stretching vibration of COO– or deformation planar vibration of C–OH, at 1070 cm^{-1} due to nonplanar vibration of C–C and C–O, at 885 cm^{-1} due to β-glycosidic bonds, and at 835 cm^{-1} due to α-glycosidic bonds were detected.

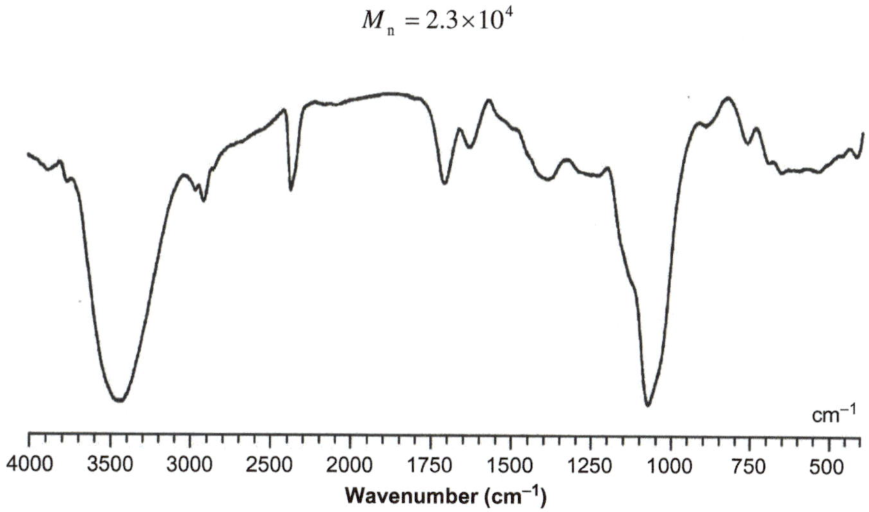

FIGURE 4.2

IR spectrum of Chinese lacquer polysaccharide.

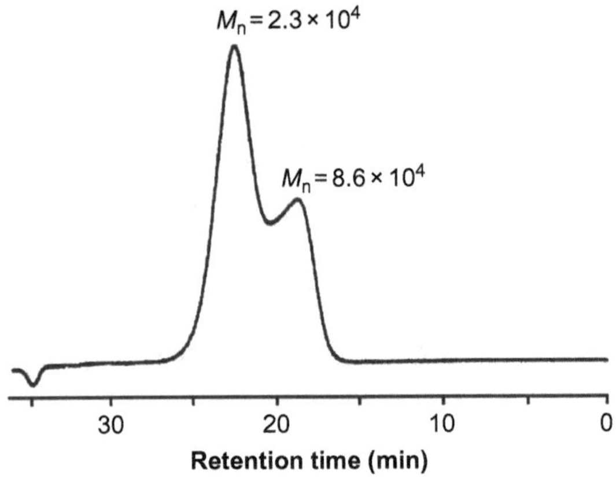

FIGURE 4.3

GPC profile of Chinese lacquer polysaccharide.

Figure 4.3 is GPC profile of Chinese lacquer polysaccharide showing two number-average molecular weight fractions around 8.6×10^4 and 2.3×10^4 in the proportion of 25% and 75%, respectively.

However, other Asian lacquer polysaccharides showed different ratios of fractions in the GPC profiles. In particular, polysaccharides from Taiwan (c), Vietnam (d), and Cambodia (f) had almost one molecular weight fraction, as shown in Figure 4.4.

Therefore, it can be assumed that polysaccharides in the sap of lacquer tree originally had one fraction with a high molecular weight. After collection of the sap and during storage in contact with air, the polysaccharides were degraded gradually to lower molecular weights. To verify this hypothesis, we collected the sap of lacquer tree in the Aizu lacquer field of Fukushima Prefecture, Japan, directly and then immediately added acetone to isolate the polysaccharide. After the sap had been kept for 3 and 21 days, the polysaccharides were isolated. In addition, the GPC profiles of polysaccharides extracted from the sap kept for 7 years in Aizu shows polysaccharides isolated (a) immediately (within 1 min), (b) after 3 days, (c) after 21 days, and (d) after 7 years (Figure 4.5).

In Figure 4.5a, the polysaccharide was revealed to have one fraction with a molecular weight of $n = 6.7 \times 10^4$. After 3 days, the polysaccharide had separated into two fractions with $n = 6.7 \times 10^4$ and $n = 2.3 \times 10^4$ in the proportion of 30:70, respectively (Figure 4.5b). The proportion of fractions had changed further to 75:25 21 days after collection (Figure 4.5c). After 7 years in Figure 4.5d, the proportion was no longer changed. Therefore, it was found that the lacquer polysaccharide of the lacquer tree originally had one molecular weight fraction. The molecular weight decreased gradually with the elapse of time after collection. After more than 21 days, the proportion of molecular weights did not change, but was maintained at least for 7 years without changing. In addition, it was found that no degradation occurred in the isolated lacquer polysaccharides for at least several years. After collection, the sap was fermented in the air, and carbon dioxide bubbled out changed the pH of the sap.

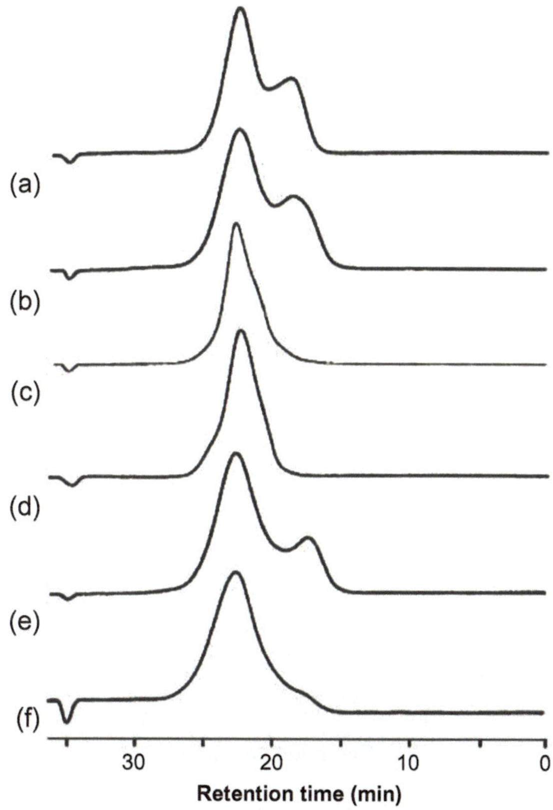

FIGURE 4.4

GPC profiles of lacquer polysaccharides in Asian countries. (a) China, (b) Japan, (c) Taiwan, (d) Vietnam, (e) Myanmar, and (f) Cambodia.

4.1.3 NMR MEASUREMENTS

About 60 mg Chinese lacquer polysaccharide was freeze-dried several times in 99.96% D_2O to prepare a sample for NMR measurement. All NMR spectra were recorded at 40 °C in D_2O on a JEOL α-400 spectrometer by using a phase-sensitive mode and a field gradient probe. Sodium 4,4-dimethyl-4-silapentane-1-sulfonate (DSS) was used as an internal standard at δ 0 ppm for ^1H and δ 0.015 ppm for ^{13}C spectra. JEOL standard software was applied to the two-dimensional spectra.

A double-quantum filtered ^1H-^1H correlated spectroscopy (DQF-COSY) was carried out with a 512 × 1024 data matrix, and eight transients were acquired for each t_1 value. The spectral width was 2173.44 Hz in both dimensions. The acquisition time and the pulse delay were 0.47 s and 1.50 s, respectively. Before Fourier transition using sine squared, the matrix was zero filled to a 1024 × 1024 matrix.

Total correlated spectroscopy (TOCSY) was performed after mixing times of 80, 120, 150, and 200 ms, and other conditions of measurement were the same as those of DQF-COSY.

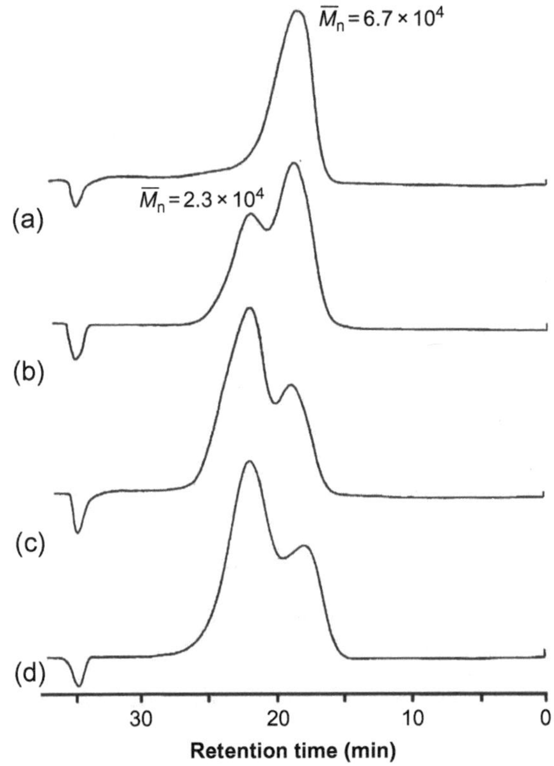

FIGURE 4.5

GPC profiles of molecular weight changes of lacquer polysaccharide from Aizu, Fukushima Prefecture, Japan, (a) immediately after collection of sap, (b) after 3 days, (c) after 21 days, and (d) after 7 years.

A heteronuclear multiple quantum coherence (HMQC) was carried out with a 1024×512 data matrix, and eight transients were acquired for each t_1 value. The proton and carbon spectral widths were 2027.99 and 19731.65 Hz, respectively.

After zero filling to a 1024×1024 data matrix, the Fourier transition was carried out. ^{1}H detected multiple bond connectivity (HMBC) was obtained with a delay time of 100 m, and other conditions were nearly the same as those of HMQC.

Chinese lacquer polysaccharides gave complex ^{1}H and ^{13}C NMR spectra, as shown in Figure 4.6, in which the anomeric signals were separated in both ^{1}H and ^{13}C spectra. Taking into account the chemical shifts of mono- and oligosaccharides and the results of methylation analysis, anomeric carbon signals were assigned as follows: a signal at δ 111.50 ppm due to terminal α-L-arabinofuranose, at δ 103.25 ppm for terminal L-rhamnose, at δ 106.04 ppm as overlapped signals for 1,3- and 1,3,6-linked D-galactopyranoses, at δ 105.85 ppm for 1,6-linked D-galactopyranose, at 105.20 ppm for terminal 4-*O*-methyl-glucuronic acid, at δ 106.82 ppm for terminal D-glucuronic acid, and at δ 102.80 ppm for terminal D-galactopyranose.

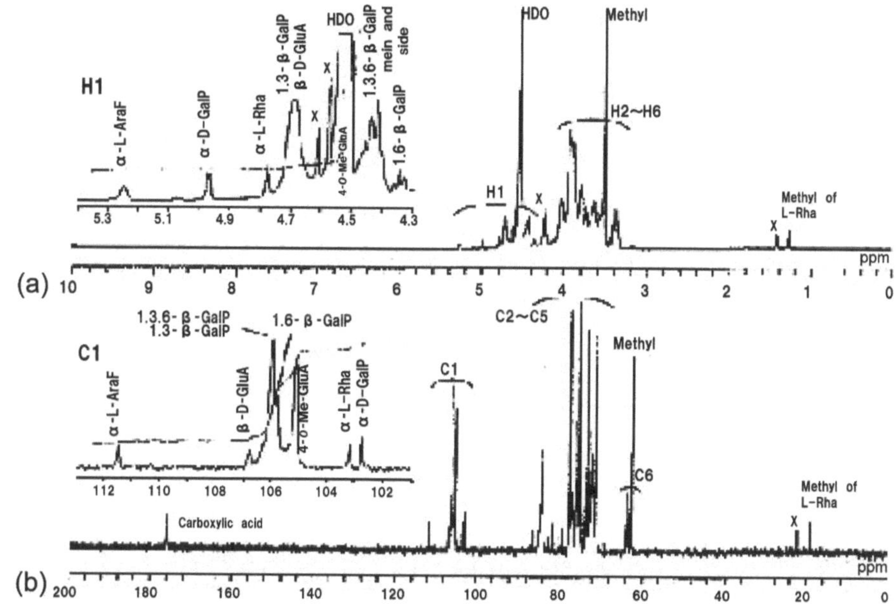

FIGURE 4.6

400 MHz ^1H (a) and 100 MHz ^{13}C NMR (b) spectra of Chinese lacquer polysaccharides and expanded spectra of anomeric regions (D_2O as solvent, 40 °C).

The integration data of the anomeric carbon signals in the ^{13}C NMR spectrum (Figure 4.6b) gave the average proportions of each sugar unit. Thus, Table 4.1 summarizes the proportions of the component sugar residues calculated by the NMR and reported chemical methods. The proportions of sugar residues by means of the NMR measurements were D-galactose (62.0 mol%), 4-*O*-methyl-D-glucuronic acid (28.0 mol%), D-glucuronic acid (3.2 mol%), L-arabinose (2.8 mol%), and L-rhamnose (4.0 mol%). The proportions were in good agreement with those of the chemical analyses, indicating that the NMR analysis is a convenient and useful method to confirm the structures of complex lacquer polysaccharides.

Further, anomeric proton signals were assigned by the correlation signals to the corresponding ^{13}C signals in the HMQC spectrum, as shown in Figure 4.7. The H1 signals at δ 4.33, δ 4.43, and δ 4.69 ppm were assigned to the main components of the lacquer polysaccharide, D-galactopyranose residues, having 1,6-di-*O*-, 1,3,6-tri-*O*-, and 1,3-di-*O*-substituted structures in the main and side chains. The anomeric proton signal of a terminal 4-*O*-methyl-glucuronic acid was detected as an overlapped signal with HDO at δ 4.51 ppm. A small peak of the lowest magnetic field at δ 5.25 ppm was found to be a H1 signal of a terminal α-L-arabinofuranose. Terminal D-galactopyranose and L-rhamnose appeared at δ 4.97 and δ 4.78 ppm, respectively.

The anomeric configuration of the sugar units were determined by the coupling constants ($J_{1,2}$) of each anomeric proton in the DQF-COSY (Figure 4.8). The coupling constants of anomeric protons at δ 4.33, δ 4.43, and δ 4.69 ppm were 6.7, 7.7,

Table 4.1 Monosaccharide residues in Chinese lacquer polysaccharide

Sugar	Content (mol%)			
	Reported[a]			
	Fr. 1	Fr. 2	Ave.	NMR[b]
D-Galactose	65.1	67.0	66.1	62.0
4-O-Methyl-D-glucuronic acid	24.5	23.6	24.1	28.0
D-Glucuronic acid	3.3	2.6	3.0	3.2
L-Arabinose	4.8	5.2	5.0	2.8
L-Rhamnose	2.0	4.1	3.1	4.0

[a]*From Carbohydr. Res. 127 (1984) 43, Ref. [1].*
[b]*Calculated from the integration data of anomeric region in proton and carbon spectra.*

and 7.2 Hz, respectively, suggesting that the main components of D-galactopyranose residues in the Chinese lacquer polysaccharide had a β-linkage. Furthermore, the $J_{1,2}$ of terminal D-galactopyranose (H1, at δ 4.97 ppm) was 3.6 Hz, and the H1 signals of L-rhamnose and L-arabinofuranose at δ 4.78 and δ 5.25 ppm showed no splitting signals, demonstrating that these three terminal sugar units had an α-linkage.

In addition, the 4-*O*-methyl signal of the terminal 4-*O*-methyl-β-D-glucuronic acid also appeared as a single peak at δ 3.62 and δ 62.5 ppm in the ^{1}H and ^{13}C spectra, respectively, and was easily assigned. The signal at δ 176.0 ppm in the carbon spectrum should be assigned to the carboxyl carbon of 4-*O*-methyl-D-glucuronic and D-glucuronic acids.

The DQF-COSY spectrum was measured to determine the H2 proton signals as represented in Figure 4.8. Although the cross peaks between H1 and H2 signals were observed clearly except for the L-arabinofuranose unit, the H2 signals were found to overlap other signals in the proton spectrum. The overlapped H1 signals due to 1,3-linked β-galactopyranose and terminal D-glucuronic acid units at δ 4.69 ppm were separated into two peaks correlated to the corresponding H2 signals at δ 3.64 and δ 3.80 ppm, respectively. The H3 signal of 1,3-linked β-galactopyranose was identified at δ 4.02 ppm by the cross peak of H2 and H3 signals. The two cross peaks on the H1 track of 1,6-linked β-galactopyranose unit appeared as the H1-H2 and H1-H3 cross peaks, indicating that the H2 and H3 signals were absorbed at δ 3.72 and δ 4.04 ppm, respectively. The connectivities from the H1-H5 signals in the terminal 4-*O*-methyl-glucuronic acid were observed as clear cross peaks, and the proton signals at δ 3.35, δ 3.59, δ 3.40, and δ 3.88 ppm could be assigned to H2, H3, and H4 (overlapped), and H5 protons, respectively. Further assignments of all proton signals were impossible because the DQF-COSY spectrum showed complex correlation peaks.

Assignments of proton signals other than those of the H1 and H2 signals were carried out by a TOCSY spectrum, which was measured by changing the mixing

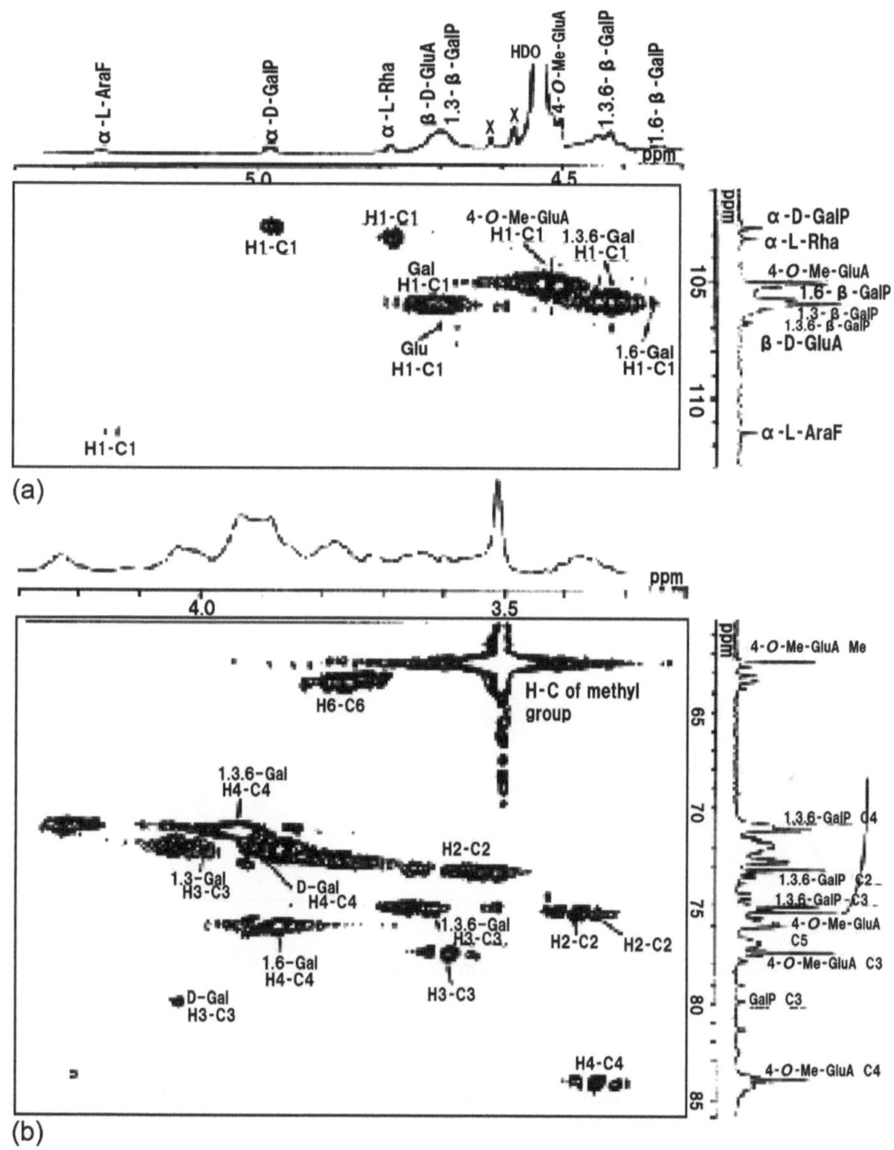

FIGURE 4.7

HMQC spectra of (a) anomeric proton and carbon regions and (b) the other regions of Chinese lacquer polysaccharide (D_2O as solvent, 40°C).

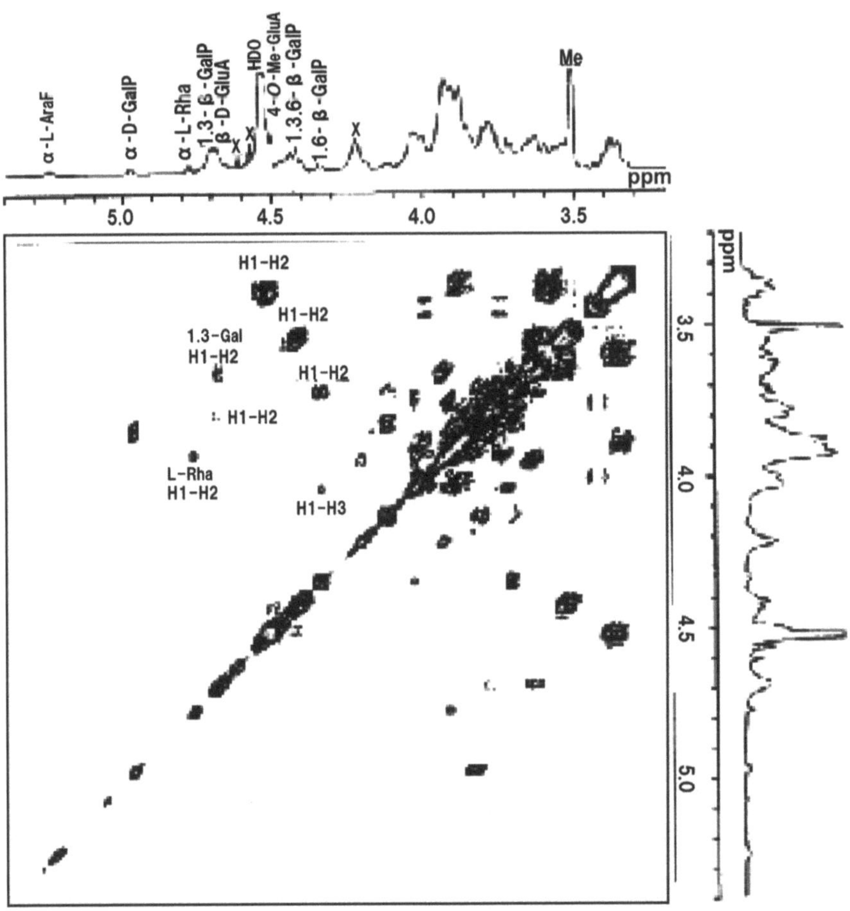

FIGURE 4.8

DQF-COSY spectrum of Chinese lacquer polysaccharide (D_2O as solvent, 40°C).

times from 80 to 200 ms as shown in Figure 4.9 (mixing time, 200 ms). It was found that the cross peaks between H1 and H2-H4 or H5 were well resolved and appeared successively on each H1 track or spin system of sugar units with increasing mixing times, and most proton signals were assigned as overlapped peaks. However, no cross peaks between the H1 and H6 signals appeared. The proton signals due to overlapped 1,3-di-*O*-substituted galactopyranose and D-glucuronic acid at δ 4.69 ppm could be determined by the connectivity from the H1 and H6 methyl signals in the TOCSY spectrum, respectively. The signals due to terminal L-arabinose were found to appear successively in the TOCSY spectrum at δ 4.19, δ 3.94, δ 3.78, and δ 3.73 ppm for H2, H3, H4, and H5 protons, respectively.

In the higher magnetic field of the proton spectrum, two doublet signals appeared at δ 1.25 and δ 1.40 ppm, which were correlated to the carbon signals at 19.0 and 21.9 ppm, respectively, in the HMQC spectrum.

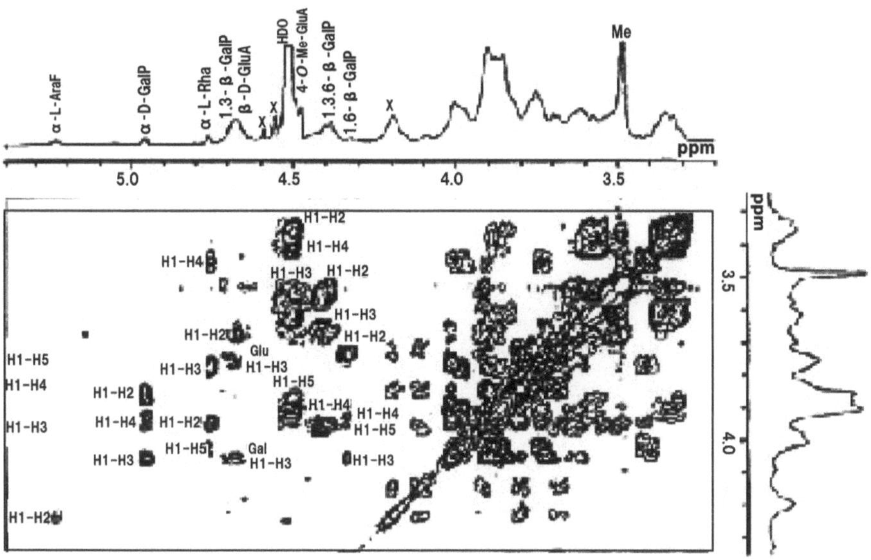

FIGURE 4.9

TOCSY spectrum of anomeric region of Chinese lacquer polysaccharide obtained with a mixing time of 200 ms (D_2O as solvent, 40 °C).

In the TOCSY spectrum shown in Figure 4.10, the signal at δ 1.25 ppm was clearly correlated to the five signals at δ 4.78, δ 3.94, δ 3.79, δ 3.46, and δ 4.02 ppm due to H1-H5 signals, respectively, but the signal at δ 1.40 ppm was correlated to a signal at δ 4.35 ppm. In addition, in the HMBC spectrum as described later, the signal at δ 1.25 ppm was correlated to the two carbon signals at 71.6 and 73.2 ppm, which could be assigned to the C5 and C4 signals of L-rhamnose. The proton signal at δ 1.40 ppm was correlated to the small carbon signal at 69.2 ppm. These results suggested that the C6 methyl group of L-rhamnose could be assigned at δ 1.25 and 19.0 ppm for proton and carbon signals, respectively, and the signals at δ 1.40 and 21.9 ppm should be impurity peaks.

Because most of proton signals were assigned, the carbon signals were identified by both HMQC and HMBC spectra as demonstrated in Figures 4.7 and 4.11. The HMBC spectrum showed clear correlation signals in comparison with that of the HMQC spectrum. The C6 signals of the sugar units appeared between 62 and 64 ppm, and the corresponding H6 signals appeared between δ 3.7 and δ 3.9 ppm in the HMQC spectrum. However, the assignments of each C6 and H6 signal were impossible because of complex peak overlapping. Therefore, we attempted the assignment of carbon signals of 4-*O*-methyl-β-D-glucuronic acid residue, because this unit was present at the terminal of branches and provided strong carbon absorption compared to those of other sugar units. In the HMQC spectrum, the methyl proton at δ 3.51 ppm correlated strongly to a carbon signal at 62.35 ppm due to the corresponding methyl carbon signal. An intraresidual long-range coupling in the HMBC

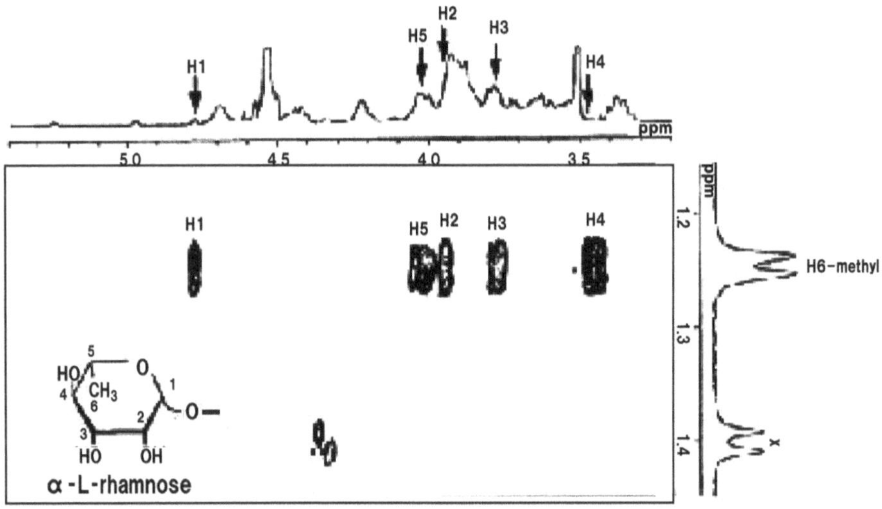

FIGURE 4.10

TOCSY spectrum of L-rhamnose region of Chinese lacquer polysaccharide obtained with a mixing time of 200 ms (D_2O as solvent, 40 °C).

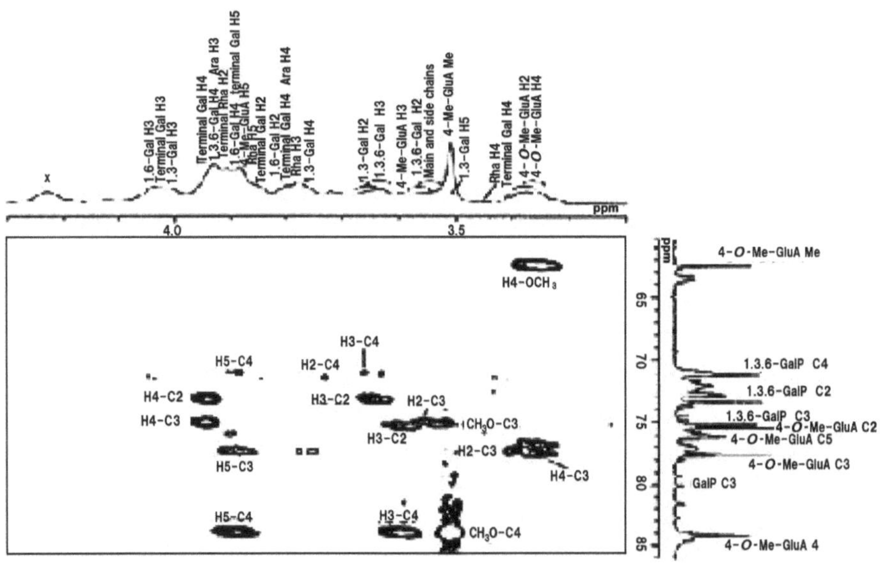

FIGURE 4.11

HMBC spectrum of Chinese lacquer polysaccharide (D_2O as solvent, 40 °C).

spectrum, the carbon signal at 62.35 ppm was found to correlate to the H4 signal at δ 3.40 ppm, which was correlated to the C4 signal at 83.9 ppm in the HMQC spectrum. Furthermore, the other carbon signals of C2, C3, and C5 appeared as sharp or strong absorption at 75.3, 77.3, and 76.0 ppm, respectively.

We continuously assigned the carbon signals having strong and sharp absorptions other than those of 4-*O*-methyl-β-D-glucuronic acid. In the HMQC spectrum, the carbon signal at 75.05 ppm was correlated to the H3 proton signal at δ 3.64 ppm of 1,3,6-tri-*O*-substituted galactopyranose. The cross peaks of carbon signal at 75.05 and both H2 and H4 proton signals at δ 3.55 and δ 3.94 ppm also appeared in the HMBC spectrum, indicating that the carbon signal at 75.05 ppm should be assigned the C3 signal of 1,3,6-tri-*O*-substituted galactopyranose. The signal at 71.0 ppm having a strong intensity was assigned to the C4 signal of 1,3,6-tri-*O*-substituted galactopyranose by the correlation signals in the HMBC and HMQC spectra.

As mentioned above, the proton and carbon signals of sugar units were mostly assigned by combinations of DQF-COSY, TOCSY, HMQC, and HMBC spectra. The anomeric proton and carbon signals were completely assigned as represented in Figure 4.5 and the chemical shifts are summarized in Table 4.2.

It was concluded that the Chinese lacquer polysaccharide had an 1,3-β-galactopyranan main chain with 1,6- and 1,3-β-linked galactopyranosidic side chains. 4-*O*-Methyl-β-D-glucuronic acid, β-D-glucuronic acid, α-L-arabinose, and α-L-rhamnose were present at the terminals of the side chains, as shown in Figure 4.12.

4.2 BIOLOGICAL ACTIVITIES

Lacquer has been used in Asian countries for thousands of years as not only a coating material, but also a traditional medicine based on experience. Although the precise constituents of the sap and the mechanism of the polymerization of the lacquers have been revealed, the biological activities are still unclear. In general, neutral branched 1,3-β-glucans have a strong antitumor activity. Lentinan and schizophyllan, which are isolated from *Lentinus edodess* and *Schizophyllum commune* mushrooms, have a 1,3-β-D-glucopyranosidic main chain with 1,6-β-D-glucopyranosidic branches, and have been used as an antitumor drugs for stomach cancer in Japan since 1985 [7]. The activity originated from the immune adjuvant activity.

Sulfonated polysaccharides having potent anti-HIV activity have been successfully synthesized in our laboratory, and the relationship between the structure and activity of the polysaccharides has been investigated [8]. Curdlan sulfonates completely inhibited HIV infection at a concentration as low as 3.3 mg/ml, which is equal to the EC_{50} of 0.1 mg/ml [9]. In 1994, the phase I/II test to the patients indicated that curdlan sulfonate was found to increase the number of CD4-positive T4 lymphocytes in the blood [10]. Recently, 3-amino-3-*O*-deoxy-(1→6)-α-D-allopyranans has been synthesized by the ring-opening polymerization and copolymerization of benzylated 1,6-anhydro-3-azido-3-*O*-deoxy-α-D-allopyranose [11]. Sulfonated amino polysaccharides were revealed to exhibit potent anti-HIV activity and low cytotoxicity [12].

Table 4.2 Proton and carbon chemical shift (ppm) of Chinese lacquer polysaccharide

Sugar unit[a]	H1	H2	H3	H4	H5	H6	C1	C2	C3	C4	C5	C6
1,3,6-Gal	4.43	3.55	3.64	3.94			106.04	73.2	75.05	71.0		
1,3-Gal	4.69	3.64	4.02	3.76	3.52		106.04	72.0	72.5	75.0		
1,6-Gal	4.33	3.72	4.04	3.91	3.93		105.85	72.6	72.5	76.0		
4-*O*-Me-GlcA	4.51	3.35	3.59	3.40	3.88		105.20	75.3	77.3	83.9	76.0	176.0
Terminal Gal	4.97	3.86	4.02	3.94			102.80	72.5	79.9	72.5		
Terminal GlcA	4.69	3.80	3.77	3.44			106.82	72.5	72.6	75.1		176.0
Terminal L-Ara	5.25	4.19	3.94	3.78	3.73		111.50	72.0	76.0	72.5		
Terminal L-Rha	4.78	3.94	3.79	3.46	4.02	1.25	103.25	71.5	72.8	73.2	71.6	19.0

[a] *1,3,6-Gal, 1,3,6-tri-O-substituted D-galactopyranose; 1,3-Gal, 1,3-di-O-substituted D-galactopyranose; 1,6-Gal, 1,6-di-O-substituted D-galactopyranose; 4-O-Me-GlcA, 4-O-methyl substituted D-glucuronic acid; Terminal Gal, terminal D-galactopyranose; Terminal GlcA, terminal D-glucuronic acid; Terminal L-Ara, terminal L-arabinofuranose; Terminal L-Rha, terminal L-rhamnopyranose.*

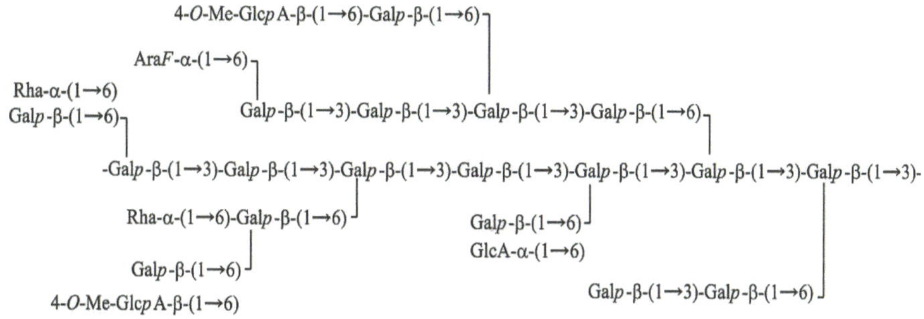

FIGURE 4.12

Partial chemical structure of Chinese lacquers polysaccharide.

In addition, it has been found that the anti-HIV and blood anticoagulant activities decreased with decreasing numbers of amino-allose units in the sulfonated polysaccharide backbone, suggesting that the amino or sulfonamide substituents in the polysaccharides play an important role in the biological activities.

Blood anticoagulation is another important biological activity of sulfonated polysaccharides. Heparin is a natural mucopolysaccharide having strong anticoagulant activity. Curdlan sulfonate had anticoagulant activity below 10 unit/mg in comparison with that of a standard dextran sulfonate (22.9 unit/mg). Although lacquer has long been used as a traditional medicine in Asian countries, there are few reports on the mechanism of biological activities. Chinese lacquer polysaccharide was found to improve the growth of leukocytes in mice by Zhang *et al.* [13]. In this section, we report in detail the specific biological activities of Chinese lacquer polysaccharide such as blood coagulation, antitumor, anti-HIV, and anticoagulant activities.

4.2.1 REDUCTION OF LACQUER POLYSACCHARIDE

Lacquer polysaccharide was reduced according to the Taylor and Conrad method. 1-Ethyl-3-(3-dimethylaminopropyl)-carbodiimide hydrochloride (EDC, 212 mg, 1.1 mmol) was added to a clear solution of lacquer polysaccharide (84 mg) dissolved in 10 ml deionized water. The solution was kept at pH = 4.75 by 0.1 N HCl. After stirring for 2 h at room temperature, 25 ml of an aqueous solution of 2 M $NaBH_4$ was added dropwise and the pH of solution was kept at 7.0 by 4 N HCl. The mixture was stirred at room temperature for one further hour, acidified with 4 N HCl, and then dialyzed against deionized water for 24 h. After evaporation and freeze-drying, 64 mg of the reduced lacquer polysaccharide was obtained.

4.2.2 SULFONATION OF LACQUER POLYSACCHARIDE

Method 1: Sulfonation by Piperidine N-Sulfonic Acid

Esterified EDC polysaccharide (94 mg) was dissolved in 20 ml dry DMSO and then the mixture was stirred at 85 °C for 1 h with piperidine *N*-sulfonic acid (99 mg, 0.6 mmol). The reaction mixture was cooled to room temperature in an ice bath, neutralized with 10% NaOH solution, and then dialyzed for 30 h with deionized water. The dialyzate was concentrated to 30 ml and then freeze-drying to give a sulfonated polysaccharide in 120 mg yield.

The reduced lacquer polysaccharide (94 mg) after drying was also sulfonated with piperidine *N*-sulfonic acid (99 mg, 0.6 mmol). The reaction mixture was neutralized with saturated $NaHCO_3$ solution, and then dialyzed to yield 168 mg sulfonated reduced polysaccharides after freeze-drying.

Method 2: Sulfonation by DMF-SO_3 Complex in DMF

Lacquer polysaccharide (200 mg) was suspended in 30 ml dry DMF under nitrogen atmosphere. After stirring at 70 °C for 10 min, DMF-SO_3 complex (1.53 g, 10 mmol) was added to the DMF solution. The mixture was stirred at 70 °C for 40 min, and then neutralized with 2 N NaOH solution below 5 °C. The neutralized solution was poured into ethanol (500 ml) to give a precipitate, which was collected by centrifugation. The precipitate was dissolved in 50 ml of deionized water and then dialyzed for 24 h to give a sulfonated polysaccharide (261 mg) after freeze-drying from water.

4.2.3 BLOOD COAGULATION ACTIVITY

The total coagulation time of blood was measured by using bovine plasma. The polysaccharide sample solutions in saline such as naturally occurring Chinese lacquer polysaccharide, sodium alginate, hyaluronic acid, and sulfonated lacquer polysaccharide were prepared by the dilution from 0.16 to 0.016 mg/ml, respectively. The standard dextran sulfonate (H-039) with anticoagulant activity of 22.7 unit/ml was used as a reference. H-039 (0.16 mg/ml, 0.05 ml) in 0.75 ml of saline and bovine plasma (0.8 ml) was added to 0.2 ml of 2% $CaCl_2$ solution. The complete coagulation time was 7 min and 35 s. In the test tube (15 ml), each sample solution (0.8 ml) and bovine plasma (1 ml) were stirred homogeneously and then 2% $CaCl_2$ solution (0.2 ml) in deionized water was added. The coagulation time was compared at 37 °C with the blank solution without polysaccharides (4 min and 25 s) and to the standard dextran solution.

Blood anticoagulant activity was measured according to the slightly modified method of the United States Pharmacopoeia using bovine plasma. The standard dextran sulfate (H-039) with an anticoagulant activity of 22.7 unit/mg was used as a reference.

In general, sulfonated polysaccharides have blood anticoagulant activity. Because the lacquer polysaccharide is an acidic polysaccharide, the coagulation time of the bovine plasma was measured by comparison with that of a standard dextran sulfonate (H-039) as shown in Figure 4.13. The bovine plasma was coagulated completely in 4 min 25 s after the addition of 2% calcium chloride solution. When 0.05 ml of dextran sulfonate (H-039) solution (0.16 mg/ml of saline) was added to the bovine plasma, the coagulation time was prolonged to 6 min 30 s. Interestingly, we found that the lacquer polysaccharide at the same concentration accelerated the blood coagulation by more than 2 min on the blank test. Acidic polysaccharides such as sulfonated lacquer polysaccharide, sodium hyaluronate, and sodium alginate had weak anticoagulant

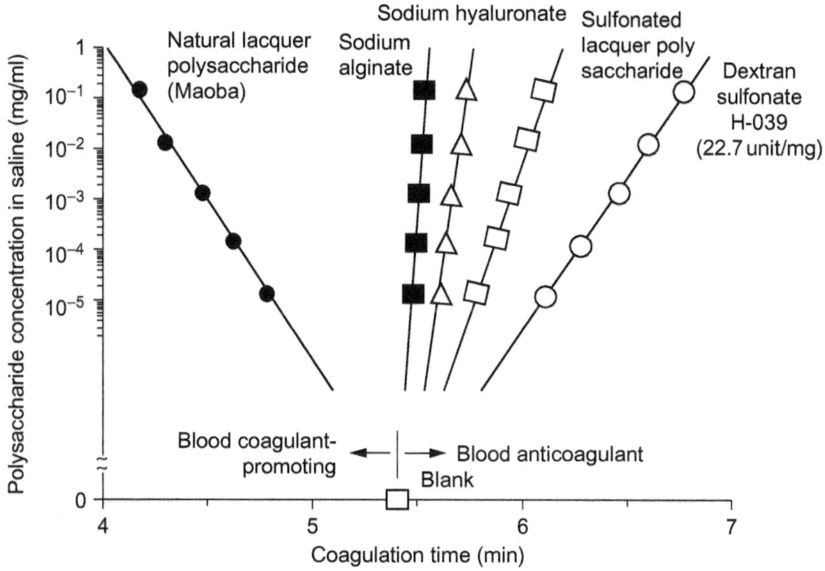

FIGURE 4.13

Coagulation time of bovine plasma at 37 °C in the presence of acidic polysaccharides. The concentrations of the sample solution were 10 times diluted with saline from 0.016 mg/ml to 0.016 μg/ml. Sample: 0.8 ml, bovine plasma; 1 ml, 2% $CaCl_2$; 0.2 ml. Dextran sulfonate: 0.05 ml, saline; 0.75 ml, 2% $CaCl_2$; 0.2 ml.

activity; that is, the coagulation time was delayed slightly. The coagulation time of these polysaccharides increased or decreased with decreasing concentration to reach 5 min and 25 s for the blank. The natural lacquer polysaccharide at the concentration as low as 0.016 mg/ml still had blood coagulation-promoting activity. If acidic polysaccharides having a sulfonate or carboxylic group were present in the plasma and the acidic group formed a complex with calcium ions, the concentration of calcium ions in the plasma decreased. As a result, the coagulation time of the plasma was delayed. However, the natural lacquer polysaccharide promoted the blood coagulation, suggesting that the branched structure having acidic groups activated some coagulant factors in the plasma.

Table 4.2 also shows the results of blood anticoagulant activity of sulfonated lacquer polysaccharides. The blood anticoagulant activity was determined by comparison with that of a standard dextran sulfonate (22.9 unit/mg) according to a modified method of the United States Pharmacopoeia. Before sulfonation, the lacquer polysaccharide had the blood coagulant-promoting activity described above. After sulfonation, the lacquer polysaccharides showed weak anticoagulant activity. The activity increased slightly with increasing sulfur concentration (Nos. 1-6). These activities were lower than that of dextran sulfonate, 22.7 unit/mg. Although the branched structure increased the blood anticoagulant activity, the lower activity might originate because the specific branched structures promoted blood coagulation.

4.2.4 ANTITUMOR ACTIVITY

Antitumor activity of the natural lacquer and reduced and sulfonated polysaccharides was assayed in vivo in rats. Lentinan, a potent stomach cancer drug, was used as a standard. Five normal healthy 8-week-old rats were used in each group. Sarcoma 180 ascite cells (5×10^6) were inoculated subcutaneously into the groin. The test polysaccharides were administered once a day by intraperitoneal injection (*i.p.*) or oral administration (*p.o.*) for 7 days from day 7 after the tumor cell inoculation. On days 7, 9, 11, 14, 16, 18, 21, 28, and 35 after inoculation, the rats were sacrificed, and the tumors were extirpated and weighed. The control was not administered polysaccharide samples. The results are exhibited in Figure 4.14.

Because branched 1,3-β-glucans have potent antitumor activity, the antitumor activity of the lacquer polysaccharides was assayed by using Salcoma 180 tumor, as shown in Figure 4.13. Thirty-five days after transplantation, the weights of tumors in rats not administered polysaccharides had increased to 20 g. When lentinan (1 mg/kg) was administered to the rats by intraperitoneal injection (*i.p.*), the tumor had disappeared 35 days after the transplantation. However, oral administration (*p.o.*) of lentinan showed no antitumor effects. After administration of the natural lacquer polysaccharide at 50 and 5 mg/kg *p.o.*, and 1 mg/kg *i.p.*, the tumor weights had decreased to 10, 11, and 13 g, respectively. It is interesting that oral administration of the lacquer polysaccharide was effective on the tumor, even though lentinan did not inhibit the tumor growth. The antitumor activity was decreased with administration of the reduced and sulfonated lacquer polysaccharides, as shown in Figure 4.12b and c. These results indicate that the glucuronic acid in the terminal of the branches may play an important role in the antitumor activity of the lacquer polysaccharide.

4.2.5 ANTI-HIV ACTIVITY

Anti-HIV activity of the sulfonated polysaccharides was assayed by the MTT method. MT-4 cells were added to each well of 96-well microplates at 2.5×10^4 cells/well, and then HIV-infected MT-4 cells were added at the multiplicity of 0.01. The HIV- and mock-infected MT-4 cells were incubated with sulfonated polysaccharides at various concentrations for 5 days at 37 °C in a CO_2 incubator. The viability of both HIV-1 and mock-infected cells were measured spectrophotometrically by the reduction of 3-(4,5-dimethylthiazol-2-yl)-2,5-diphenyltetrazolium (MTT). Anti-HIV activity was defined by the EC_{50}, the 50% effective concentration of the test compounds. The 50% cytoxic concentration (CC_{50}) of the test compounds on MT-4 cells was also determined. The results are summarized in Table 4.3.

The anti-HIV mechanism of sulfonated polysaccharides was proposed to be a conformational change in the gp120 due to the occurrence of the electrostatic interaction of negatively charged sulfonate groups in the polysaccharides and the positively charged amino acid residues in the envelope glycoprotein gp120 of HIV. The interaction of sulfonated polysaccharides inhibited the binding of HIV to the CD4 positive T-lymphocytes. NMR studies revealed that curdlan sulfonate interacted strongly with polylysine, a model gp120 compound, to cause chemical shifts

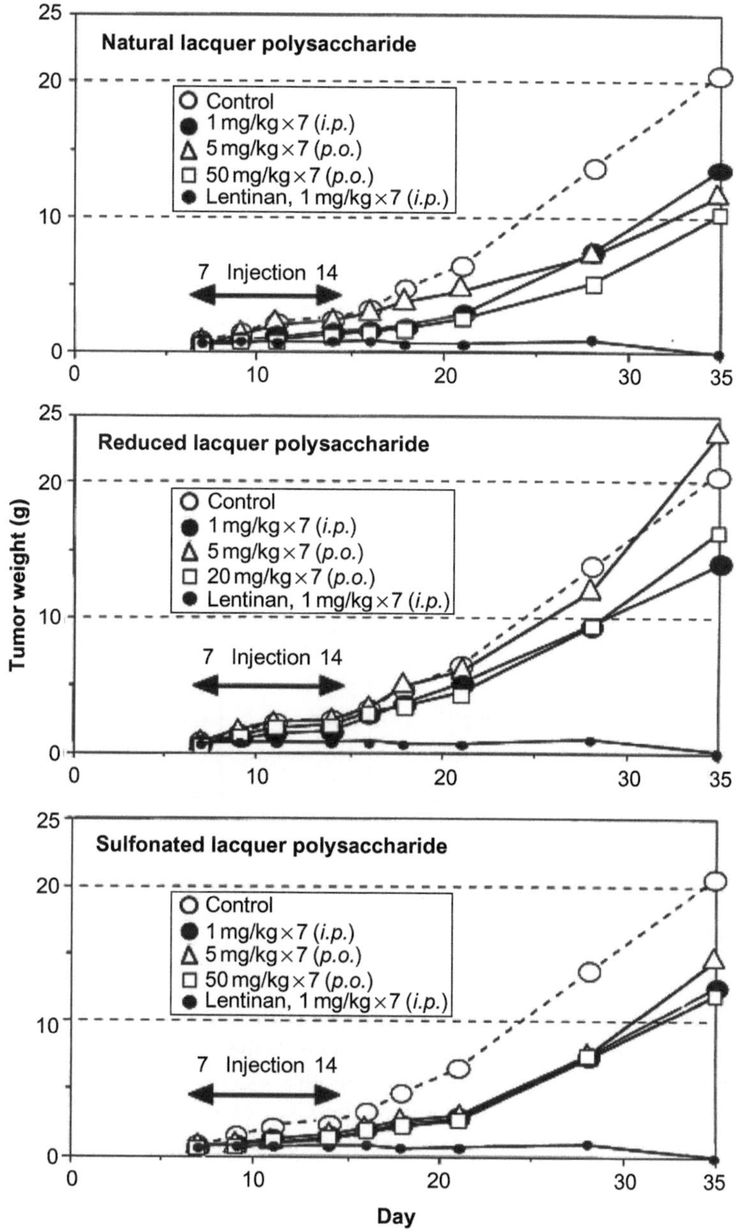

FIGURE 4.14

Antitumor activities of lacquer polysaccharides *in vivo*. The polysaccharides were administrated by intraperitoneal injection (*i.p.*) or oral administration (*p.o.*) once a day on days 7-14 after transplantation of sarcoma 180 tumor cells. Lentinan was used as a standard.

Table 4.3 Anti-HIV activities of sulfonated Chinese lacquer polysaccharides

	Sulfonation of lacquer polysaccharides				
Nos.	**M_n ($\times10^3$)**	**S content (%)**	**DS**	**EC_{50}[a] (μg/ml)**	**CC_{50}[b] (μg/ml)**
Natural	86.0	0		>1000	>1000
1	10.7	14.68	1.5	0.6	767
2	10.9	14.25	1.5	0.4	>1000
3	10.6	13.42	1.3	0.8	879
4[c]	42.7	10.63	1.0	0.5	554
5[c]	16.4	13.38	1.1	0.6	457
6[c]	14.9	15.48	1.6	0.6	566
7	4.4	3.89	0.3	>1000	>1000
8	5.2	8.56	0.7	37	>1000
9	6.5	11.58	1.1	109	>1000
10	5.0	13.33	1.3	32	492
11	3.4	14.12	1.3	155	>1000
12	3.1	16.69	2.0	48	>1000
13	2.9	16.32	2.0	35	>1000
14	3.9	12.82	1.3	143	>1000
15	3.3	14.77	2.0	29	>1000
16	4.5	10.25	0.9	655	>1000
DS[d]	8.5	18.40	1.8	0.84	>1000
CS[e]	79.0	14.10	1.6	0.13	>1000

[a]*50% effective concentration.*
[b]*50% cytotoxic concentration.*
[c]*Reduced lacquer polysaccharide.*
[d]*Standard dextran sulfonate, H-039.*
[e]*Standard curdlan sulfonate.*

in polylysine, suggesting that the anti-HIV activity originated from ionic interactions. Thus, the anti-HIV activity of sulfonated lacquer polysaccharides was evaluated by the EC_{50} of sulfonated polysaccharides to inhibit the infection of MT-4 cells by HIV. The cytotoxicity produced the CC_{50} of sulfonated polysaccharides. Sulfonated lacquer polysaccharides having molecular weights more than 10×10^3 had high anti-HIV activity (Table 4.3, Nos. 1-6), although the natural lacquer polysaccharide exhibited no anti-HIV activity. Because the reduced polysaccharides after sulfonation had the high activity of 0.5 and 0.6mg/ml, no glucuronic acids in the polysaccharides were involved in the anti-HIV activity. The sulfonated lacquer polysaccharides in numbers 7-16 of Table 4.3 had low anti-HIV activity, indicating that the molecular weight and degree of sulfonation are important for high anti-HIV activity. Cytotoxicity was relatively low without high-molecular-weight sulfonated polysaccharides.

4.3 PROPERTIES IN AQUEOUS SOLUTION

Lacquer polysaccharides are assumed to play an important role in the durability of a lacquer coating by protecting it from oxygen; however, their roles in the enzymatic polymerization of urushiol in the sap are still unclear. In order to reveal the properties in aqueous solution, we measured the viscosity behaviors of the Chinese lacquer polysaccharide in deionized water and NaCl solutions in detail to elucidate the structural behaviors in aqueous solutions. The dependence of viscosity on pH was also examined. The structure and roles of the polysaccharides in the sap of lacquer trees are proposed from the results of studies of viscosity behaviors.

4.3.1 VISCOSITY MEASUREMENT

Viscosity measurements were performed at 30±0.05°C in a constant temperature water bath with an Ubbelode capillary glass viscometer which has a flow time of 216.60±0.05s in water at 30°C. The stock solution was made by dissolving lacquer polysaccharide at the concentration of 0.96g/dl in freshly deionized water, and then filtering it through a core glass funnel (4G2). Dilutions were made by adding the appropriate amounts of deionized water to yield 13 lower concentrations. For each solution, a 10ml sample solution was loaded into the viscometer and kept for 30min in the temperature-controlled water bath to minimize drainage errors. The measurement was repeated more than 10 times, and the flow time of each solution was determined by the average of more than eight time-readings within a difference of 0.3s. Kinetic energy corrections were not made. The intrinsic viscosity $[\eta]$ in deionized water was estimated by Fuoss plots. Namely, C/η_{sp} and $C^{1/2}$, where η_{sp} is a specific viscosity and C is the concentration of polysaccharides, were calculated from the relationship between reduced viscosity and concentration of polysaccharides, and then $[\eta]$ was estimated by extrapolation to infinite dilution (zero concentration of polysaccharide) of the linear line obtained.

In NaCl solutions ranging between 1.0×10^{-4}N and 5N, the viscosity was measured in the manner described above. The intrinsic viscosity in the 0.5N NaCl solutions was calculated by extrapolation of the reduced viscosity to zero polysaccharide concentration.

The pH dependence of the reduced viscosity was measured in the phosphate buffers at pH 5.0, 6.0, 7.0, and 8.0.

4.3.2 BEHAVIOR OF VISCOSITY IN DEIONIZED WATER AND DILUTE NACL SOLUTIONS

As described in Section 4.1, the structure of lacquer polysaccharides was studied by the sugar, methylation, and NMR analyses, which suggested that the Chinese lacquer polysaccharide had a β-1,3-galactopyranosidic main chain with complex side chains having 4-*O*-methyl-glucuronic acid in the terminals. Furthermore, the Chinese lacquer polysaccharide was known to be a mixture of two molecular weight fractions. The number-average molecular weights of the polysaccharide used were about 8.6×10^4 and 2.3×10^4 in the proportion of 25 and 75 mol%, and the molecular weight distribution was 1.08 and 1.11, respectively. This polysaccharide was used directly without separation. It was found that the Chinese lacquer polysaccharide was stable in deionized water and in NaCl solutions for more than 3 months because no degradation occurred and no precipitation appeared during storage. In general, the acidity constant, pK_a, of carboxylic acid is about around 3.5 orders, indicating that the carboxyl group is weakly acidic. Thus, we considered that the lacquer polysaccharide is a weak polyanion and that the carboxylic acids have only a slight tendency to release protons in solution.

Figure 4.15 shows the variation of viscosity of the Chinese lacquer polysaccharide in deionized water and in NaCl solutions, respectively, with decreasing the concentrations of the polysaccharide.

Because the viscosity behavior is affected only by the ionic strength of the carboxylic acids in the polysaccharide, the reduced viscosity in deionized water increased rapidly with the decreasing concentration of polysaccharide, suggesting that the polysaccharide chain was swollen by both penetration of water molecules and electrostatic repulsion of the anionic charges due to the glucuronic acid in the side chains.

In NaCl solutions, the viscosity was lower than in deionized water because the shield effect around polyanions originated from the glucuronic acid. In NaCl solutions at 1.0×10^{-4} and 2.0×10^{-4} N, a maximum point of reduced viscosity appeared.

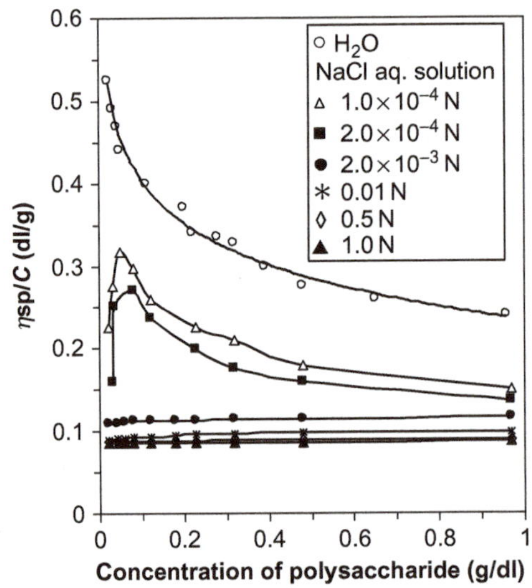

FIGURE 4.15

Vicosity of Chinese lacquer polysaccharide in deionized water and in NaCl solution at 30 ± 0.05 °C. The concentrations in NaCl solutions were: (Δ) 1.0×10^{-4}, (■) 2.0×10^{-4}, (●) 2.0×10^{-3}, (*) 0.01 N, (◊) 0.5 N, and (▲) 1.0 N.

At this point, the stoichiometric concentration of the carboxylic acids in the polysaccharide should be in competition with the ion strength due to the NaCl solutions. In NaCl solutions between 2.0×10^{-3} and 1 N, the viscosity-concentration curves were similar to those of uncharged polymers. The ionic strength of the NaCl solutions was greater than that of the polysaccharide, indicating that the viscosity behavior was determined by the ionic strength of NaCl solutions. In 0.5 N NaCl solution, the lacquer polysaccharide had the lowest reduction in viscosity and the viscosity-concentration curve was a nearly horizontal line, suggesting that the constant of Huggins, k', was close to 0. Thus, it was considered that the lacquer polysaccharide might have a dense random coil structure in 0.5 N NaCl solutions.

These results suggest that the lacquer polysaccharide was presented as a random coil in solutions and had a more swollen structure in deionized water than in NaCl solutions. It has been considered that the lacquer polysaccharide does not have a fully extended and linear structure in deionized water, unlike a strongly acidic polymer such as polystyrene sulfonic acid, because the lacquer polysaccharide is a weak polyelectrolyte and has a complex branched structure.

4.3.3 INTRINSIC VISCOSITY IN DEIONIZED WATER AND IN 0.5 N NACL SOLUTIONS

Taking into account the effects of solvents, the intrinsic viscosity in 0.5 N NaCl solution, which produced the lowest viscosity, was determined by extrapolating $C = 0$ into $[\eta] = 0.085$ dl/g (Figure 4.15).

Because the curve of the reduced viscosity increased with decreasing polysaccharide concentration in deionized water, it was impossible to calculate the intrinsic viscosity by the extrapolation of η_{sp}/C to the zero concentration. Fuoss and Cathers proposed that a curve like Figure 4.14 could be described by the following equation: $(\eta_{sp}/C)^{-1} = 1/[\eta] + BC^{1/2}$, where C and B are the concentration of polyelectrolytes and a constant, respectively. Thus, a linear line was plotted for the reciprocal of the reduced viscosity $(\eta_{sp}/C)^{-1}$ versus the square root of the concentration C in order to calculate the intrinsic viscosity $[\eta]$, as shown in Figure 4.16. By extrapolation to the concentration at $C = 0$, the intrinsic viscosity of the Chinese lacquer polysaccharide in deionized water was calculated to be $[\eta] = 0.65$ dl/g.

Comparison of these intrinsic viscosities revealed that the lacquer polysaccharide chain swelled approximately eight times in deionized water compared to that in 0.5 N NaCl solutions.

Assumption of Radius of Gyration of Lacquer Polysaccharide in 0.5 N NaCl Solution

The hydrodynamic models of polymers in dilute solutions were proposed by Einstein, Flory, and Yamakawa. It was considered the lacquer polysaccharides exist as a dense random coil in 0.5 N NaCl solution because it has the lowest viscosity. Flory reported a viscometric theory of polyelectrolytes in which salt solutions were treated similarly by straightforward extension of the theory for uncharged polymers. Thus, on the

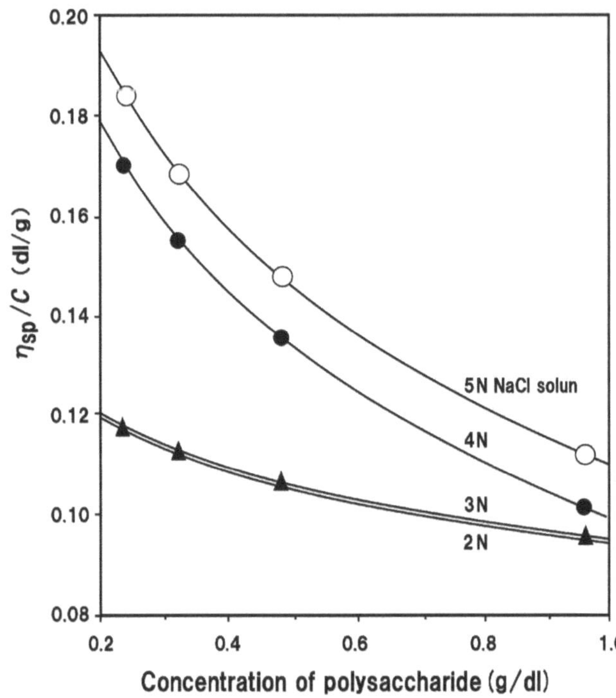

FIGURE 4.16

Relationship between reduced viscosity and concentration of Chinese lacquer polysaccharide in concentrated NaCl solutions at 30±0.05°C. The concentrations of NaCl solutions are: (▲) 2N; (△) 3N; (●) 4N; and (○) 5N.

assumption that a molecule of the lacquer polysaccharide is regarded as a hydrodynamically equivalent sphere in 0.5N NaCl solutions, the mean-square end-to-end distance $\langle R^2 \rangle$ was calculated by the following Flory equation:

$$[\eta] = \Phi\left(\frac{\langle R^2 \rangle^{3/2}}{M} M^{1/2}\right) \tag{4.1}$$

Here Φ is a constant, 2.6×10^{23}, and M is molecular weight. Introduction of $[\eta]=8.5$ l/kg and $M=41000$, which is the number-average molecular weight of the lacquer polysaccharide calculated by GPC, into (4.1) gave $\langle R^2 \rangle = 1.22\times10^{-2}\,\text{m}$. As is well known, the mean-square radius of gyration $\langle S^2 \rangle$ is equal to 1/6 $\langle R^2 \rangle$. Thus, $\langle S^2 \rangle = 0.20\times10^{-2}\,\text{m}$. The radius of gyration $r = \langle S^2 \rangle^{1/2}$ could be calculated to be 4.5×10^{-7} m.

In deionized water, the Flory equation was inapplicable to the lacquer polysaccharide because the polysaccharide structure was swollen by the penetration of water molecules and the electrostatic repulsion of the carboxyl groups, and existed in a free-draining state.

Viscosity in NaCl Solutions Above 0.5 N Concentration

It was found that the reduced viscosity decreased rapidly with increasing NaCl concentration up to 0.5 N and then increased gradually with further increases in the NaCl concentration, as shown in Figures 4.15 and 4.16. In general, for linear polyelectrolytes, the viscosity decreases sharply with increasing ionic strength of the solution. However, the lacquer polysaccharide had a minimum point of reduced viscosity in solutions with increasing ionic strength.

We assumed that these interesting features are due to the highly branched structure of the lacquer polysaccharide; that is, the counteraction Na^+ gradually penetrated the branches with increasing ionic strength of the NaCl solutions to expand the volume of polysaccharides.

Roles of Polysaccharides in Sap of Lacquer Trees

The polymerization of lacquer proceeds by polymerization of an emulsion of urushiol by the enzyme laccase, and subsequent oxygen-oxidative polymerization to give a lacquer film having a beautiful and a durable surface. Although the mechanism of polymerization of urushiol was elucidated, the roles of polysaccharides during polymerization of the sap and formation of the durable film were still unclear. It is known only that the lacquer polysaccharides in the sap work as a stabilizer and a dispersing agent for the enzyme and urushiol. In SEM photographs, the film is composed of densely packed grains with a diameter of 0.1 mm. Previous work proposed that the polysaccharides and glycoproteins composed the surface of the grains, and the polymerized urushiol was contained inside the grains. Recently, phenolic hydroxyl groups were observed on the surface of a lacquer film by means of XSCA, and no hydroxyl groups due to polysaccharides were detected. The lacquer sap contained less than 3% polysaccharides and had a pH around 7. Kumanotani [14] reported that the lacquer film was composed of densely packed grains with the diameter of 0.1 µm, and Niimura found only phenolic hydroxyl groups on the surface of the lacquer film and detected no hydroxyl groups due to polysaccharides [15]. Because the sap of a lacquer tree contains many cations such as Mg^{2+}, Na^+, Ca^{2+}, and Cu^{2+}, lacquer polysaccharides having complex branches may have a swollen structure in the sap. The results of this work suggested that the Chinese lacquer polysaccharides having complex branches were present as "swollen ball" structures with water in the sap. Thus, we propose the roles of lacquer polysaccharides in the sap and the polymerization mechanism as illustrated in Figure 4.17. The

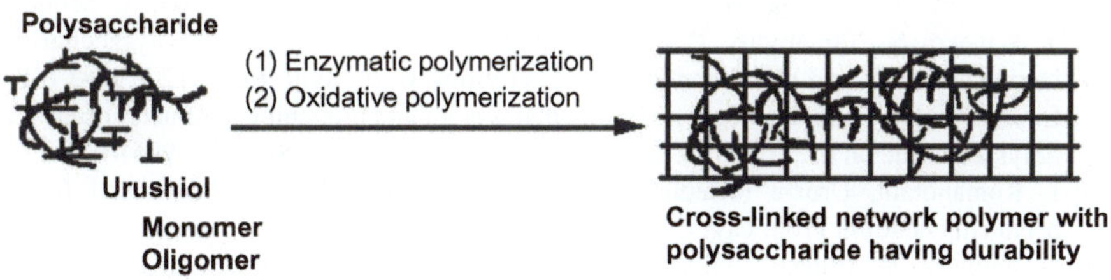

FIGURE 4.17

Proposed polymerization mechanism of urushiol in the presence of polysaccharides.

urushiol and enzyme are dispersed in the sap as an emulsion and might penetrate the swollen polysaccharides, where they are polymerized enzymatically and oxidative to afford the durable polymers with polysaccharides in the network.

REFERENCES

[1] R. Oshima, J. Kumanotani, Structural studies of plant gum from sap of the lac tree, *Rhus vernicifera*, Carbohydr. Res. 127 (1984) 43–57.

[2] X.-P. Qiu, L.-N. Zhang, Y.-M. Du, B.-G. Qian, Gel permeation chromatographic analysis of lacquer polysaccharide, Chinese J. Polym. Sci. 10 (1992) 137–143.

[3] X.-P. Qiu, L.-N. Zhang, Y.-M. Du, B.-G. Qian, Compact coil of lacquer polysaccharide: further evidence, Chinese Chem. Lett. 2 (1991) 575–576.

[4] L.-N. Zhang, X.-P. Qiu, J.-J. Xu, Y.-M. Du, B.-G. Qian, Studies on fractionation and molecular weights of Chinese lacquer polysaccharide, Carbohydr. Polym. 19 (1992) 161–166.

[5] Y.-M. Du, Z.-W. Gong, H.-P. Li, Studies on separation and structure of lacquer polysaccharide, Acta Polym. Sin. 3 (1994) 301–306 (in Chinese).

[6] R. Lu, K. Hattori, Z.Y. Xia, T. Yoshida, J.H. Yang, L.N. Zhang, Y.M. Du, T. Miyakoshi, T. Uryu, Structural analysis of polysaccharides in Chinese lacquer by NMR spectroscopy, Sen'i gakkaishi 55 (1999) 47–56.

[7] Y. Kaneko, O. Yoshida, R. Nakagawa, T. Yoshida, M. Date, S. Ogiwara, S. Shioya, Y. Matsuzawa, N. Nagashima, Y. Irie, Y. Mimura, H. Shinkai, N. Yashuda, K. Matsuzaki, T. Uryu, N. Yamamoto, Inhibition of HIV-1 infectivity with curdlan sulfate *in vitro*, Biochem. Pharmacol. 39 (1990) 793–797.

[8] T. Yoshida, T. Akasaka, Y. Choi, K. Hattori, B. Yu, T. Mimura, Y. Kaneko, H. Nakashima, E. Aragaki, M. Premanathan, N. Yamamoto, T. Uryu, Synthesis of polymethacrylate derivatives having sulfated maltoheptaose side chains with anti-HIV activities, J. Polym. Sci. A Polym. Chem. 37 (1999) 789–800.

[9] T. Yoshida, K. Hatanaka, T. Uryu, Y. Kaneko, E. Suzuki, H. Miyano, T. Mimura, O. Yoshida, N. Yamamoto, Synthesis and structural analysis of curdlan sulfate with a potent inhibitory effect in vitro of AIDS virus infection, Macromolecules 23 (1990) 3717–3722.

[10] M. Gordon, M. Guralnik, Y. Kaneko, T. Mimura, M. Baker, W. Lang, A phase I study of curdlan sulfate—an HIV inhibitor. Tolerance, pharmacokinetics and effects on coagulation and on CD4 lymphocytes, J. Med. 25 (1994) 163–180.

[11] K. Hattori, T. Yoshida, T. Uryu, Ring-opening polymerization and copolymerization of a new benzylated 1,6-anhydro-3-azido-3-deoxy-β-D-allopyranose and synthesis of amino-polysaccharides with 1,6-α-allopyranosidic structure, Macromol. Chem. Phys. 198 (1997) 29–39.

[12] K. Hattori, T. Yoshida, H. Nakashima, M. Premanathan, R. Aragaki, T. Mimura, Y. Kaneko, N. Yamamoto, T. Uryu, Synthesis of sulfonated amino-polysaccharides having anti-HIV and blood anticoagulant activities, Carbohydr. Res. 312 (1998) 1–8.

[13] L.N. Zhang, Q. Ding, Influences of molecular weight and branching parameter of lacquer polysaccharide on the growth of leucocytes, Chinese J. Polym. Sci. 13 (1995) 107–111.

[14] J. Kumanotani, Urushi (oriental lacquer)—a natural aesthetic durable and future-promising coating, Prog. Org. Coat. 26 (1995) 163–195.

[15] N. Niimura, Y. Iijima, T. Miyakoshi, Hardening process and surface structure of lacquer films studied by X-ray photoelectron spectroscopy, Surf. Interf. Anal. 24 (1996) 237–242.

CHAPTER

Glycoprotein

5

A glycoprotein is a complex carbohydrate composed of oligosaccharides and polypeptides with covalent binding. Many bioactive ingredients, such as enzymes, hormones, carrier proteins, toxins, and lectins are glycoproteins. Glycoproteins often exist in the extracellular fluid and interstitial cells in soluble or insoluble forms. In lacquer sap, a glycoprotein was found in the acetone powder that is not only insoluble in acetone or any other organic solvent, but also insoluble in water. Its concentration is about 1-5% of total lacquer, and the molecular weight is around $0.78\text{-}4.7 \times 10^4$, as shown in Figure 5.1.

Elemental analysis of the component insoluble in alcohol and water in lacquer sap by Yoshida [1] revealed 63.44% C, 7.41% H, 4.01% N, and 1.20% ash in the substance, and they named it the "nitrogenous substance." However, Hirano [2] reported 64.48% C, 8.21% H, 6.89% N, and 3.55% ash, and Matsui [3] reported 61.18% C, 7.18% H, 3.07% N, and 2.09% ash in the "nitrogenous substance." Because lacquer is a natural product, the different values are due to the different lacquer tree species or different periods of collection of lacquer sap.

Kumanotani [4,5] used IR and ^{13}C-CP/MAS to analyze the structure of lacquer glycoproteins. Figure 5.2 shows the IR spectrum of lacquer glycoprotein. It showed O–H or H–N–H at 3400 cm^{-1}, C–H at 2950 cm^{-1}, C=O at 1650 cm^{-1}, and C–O at 1050 cm^{-1}. Figure 5.3 is the ^{13}C-CP/MAS spectrum; it showed peptide carbonyl at 170 ppm, aromatic peptide carbons at 128 ppm, arginine carbons at 100 ppm, anomeric carbons at 70 ppm, peptide aliphatic carbons at 53 ppm, and carbohydrate carbons at 28 ppm, respectively. Kumanotani [6] speculated that the glycoprotein acts as a dispersant in the original lacquer sap; it has about 10% carbohydrates composed of six kinds of monosaccharides and 90% protein composed of 15 amino acids, as summarized in Table 5.1.

Because the glycoprotein is almost completely insoluble in water or organic solvents, there have been few investigations of the glycoprotein that contains in lacquer sap, and its structure and chemical properties are not yet clearly identified.

Recently, we studied the structure of the Chinese lacquer glycoprotein that was extracted with 1% sodium dodecylsulfate (SDS) in 0.05 mol pH 7.0 phosphate buffer [7]. The detailed experimental procedure is as follows. A certain amount of glycoprotein was added to ten times 1% SDS solution and refluxed at 100 °C overnight, then centrifuged at least three times. The supernatant was combined and further purified by a Sephadex G-200 column with UV monitoring at A_{280}. The chromatographic profile is shown in Figure 5.4, and peaks 1 and 2 were collected.

Lacquer Chemistry and Applications. http://dx.doi.org/10.1016/B978-0-12-803589-4.00005-5

FIGURE 5.1

Photo of lacquer glycoprotein.

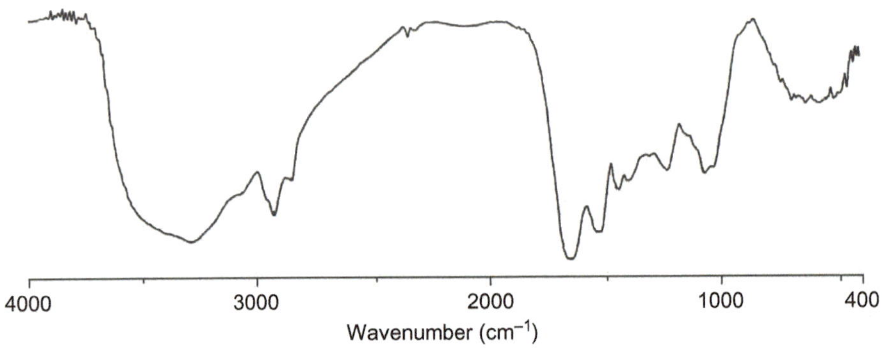

FIGURE 5.2

IR spectrum of lacquer glycoprotein.

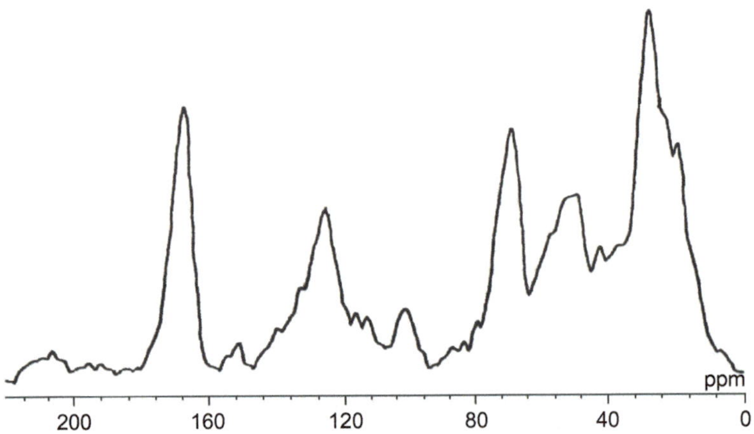

FIGURE 5.3

^{13}C-CP/MAS NMR spectrum of lacquer glycoprotein.

Table 5.1 Amino acid and sugar compositions of nitrogenous substance (by Kumanotani)

Amino acid	%	Amino acid	%	Sugar	%
Lysine	6.2	Glycine	13.0	Fucose	5.4
Histidine	1.9	Alanine	8.7	Arabinose	11.6
Arginine	1.2	Valine	8.7	Mannose	42.5
Aspartate	13.6	Isoleucine	5.0	Glucose	3.5
Threonine	6.2	Leucine	6.2	Galactose	27.5
Serine	6.8	Tyrosine	3.7	Glucosamine	9.5
Glutamine	8.1	Phenylalanine	4.3		
Proline	6.2				

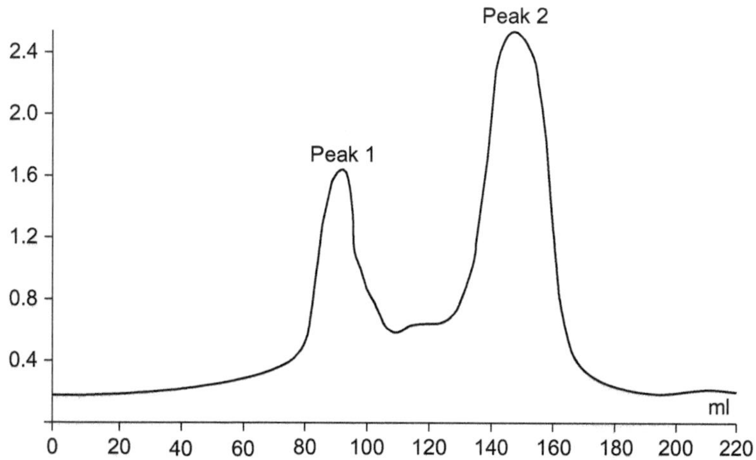

FIGURE 5.4

Sephadex G-200 chromatogram of lacquer glycoprotein.

Peaks 1 and 2 from the Sephadex G-200 column were further purified on a Sephadex G-25 with C_7H_{16}:$C_5H_{11}OH = 4:1$ as eluent. Peak 1 showed five peaks, and peak 2 showed only one peak after being separated in the Sephadex G-25 column (Figure 5.5). The purified lacquer glycoproteins obtained from the Sephadex G-25 column were named LGP-1 (Lacquer Glycoprotein) and LPG-2 corresponding to the original peaks 1 and 2. Because LPG-1 showed five peaks (Figure 5.5), suggesting the structure is very complex, we selected LGP-2 to analyze its chemical structure. The molecular weight of LGP-2 is about 2.78×10^4 measured by GPC according to the pullulan standard. Hydrazinolysis and β-elimination analysis of LGP-2 indicated asparagine linkages, that is, N-linkages, between carbohydrate and peptide.

The IR spectrum of LGP-2 shown −OH stretching vibration at 3400 cm^{-1}, $-NH_2$ stretching vibration at 3280 cm^{-1}, −OH variable angle vibrations at 1112 and 1078 cm^{-1} and at 1632 and 1542 cm^{-1} due to amino acid absorption (Figure 5.6).

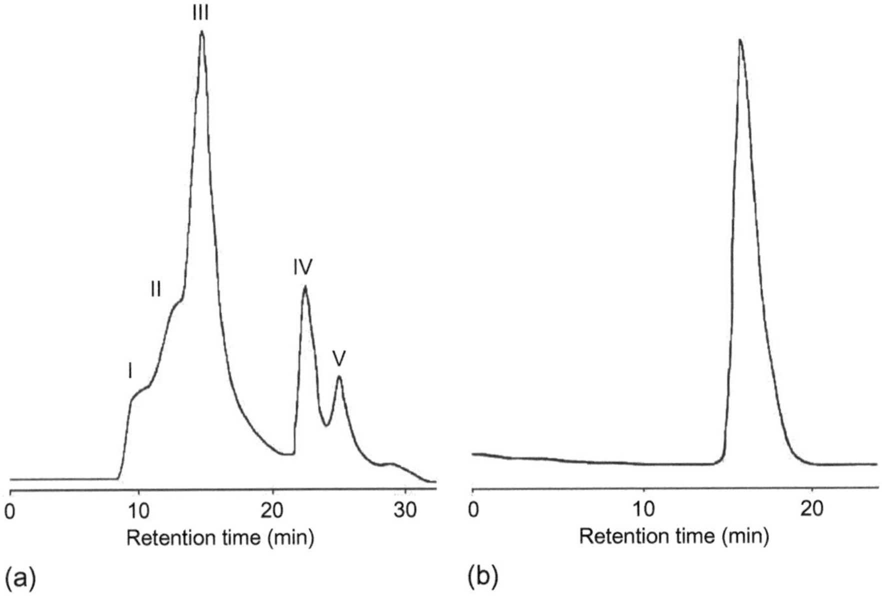

FIGURE 5.5

Chromatographic profiles of LPG-1 (a) and LPG-2 (b).

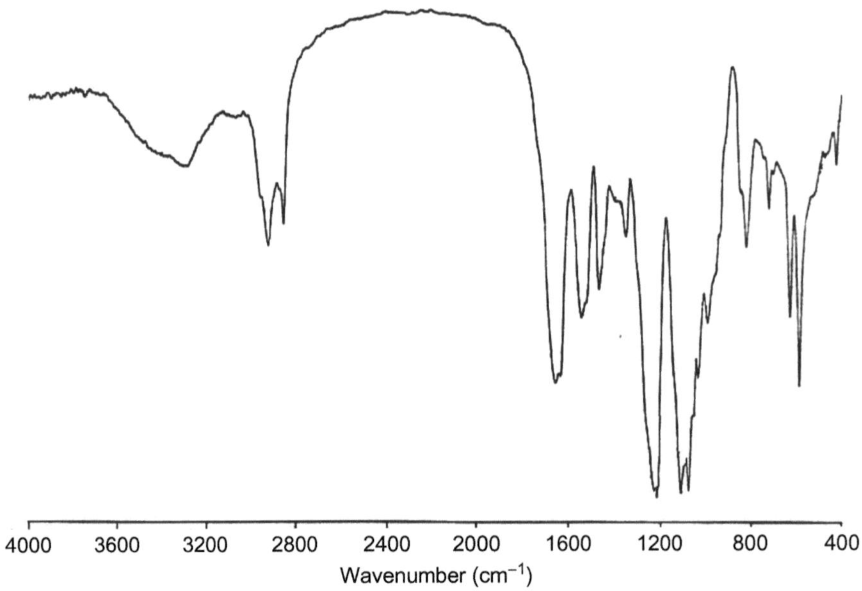

FIGURE 5.6

IR spectrum of LGP-2.

Table 5.2 Amino acid and sugar compositions of LGP-2

Amino acid	%	Amino acid	%	Sugar	%
Aspartate	12.86	Threonine	5.96	Fucose	12.8
Serine	5.50	Glutamine	10.56	Arabinose	23.7
Glycine	6.05	Alanine	4.94	Mannose	31.2
Cystine	0.91	Valine	6.72	Galactose	9.7
Methionine	0.79	Isoleucine	5.50	Glucose	16.0
Leucine	6.90	Tyrosine	9.05	Glucosamine	6.6
Phenylalanine	5.85	Lysine	8.13		
Histidine	2.43	Arginine	2.77		
Proline	4.47				

Amino acid and sugar analysis revealed 17 amino acids and six monosaccharides, and the concentrations of protein and carbohydrate were 81.3% and 17.9%, respectively, as summarized in Table 5.2.

The GC-MS analysis of hydrolysates obtained from acid-catalyzed hydrolysis of LGP-2 and its methylated product showed the β-(1→2)-, β-(1→3)-, and β-(1→4)-linkages of L-arabinose, β-(1→2)-, β-(1→3)-, and β-(1→4)-linkages of D-glucose, β-(1→3)-, β-(1→4)-, and β-(1→6)-linkage of D-mannose in the lacquer glycoprotein.

The pyrolysis gas chromatograph/mass spectrometry analysis of lacquer glycoprotein was performed [8], and the total ion chromatogram (TIC) is shown in Figure 5.7. Peaks 1, 2, 3, 4, 5, 6, and 7 were assigned as the structures of toluene, styrene, phenol, 4-methylphenol, 2,3-dihydrobenzofuran, indole, and 3-methyl-1H-indole, respectively, according to the mass spectra.

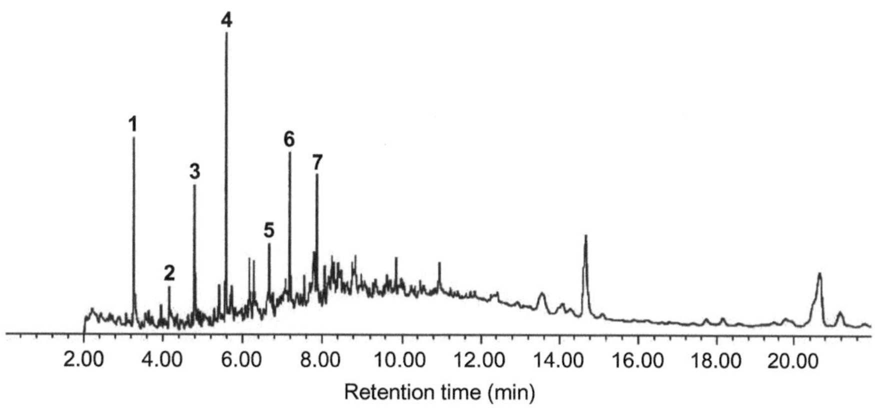

FIGURE 5.7

Total ion chromatogram of lacquer glycoprotein.

Because a glycoprotein consists of saccharide and protein, it can be considered that these pyrolysis products were obtained from the thermal decomposition of protein and saccharide, such as toluene, is from phenylalanine or tyrosine, and indoles from tryptophan. More detailed structural analyses of lacquer glycoprotein and its role and responsibility in original lacquer sap are ongoing research projects in our lab.

REFERENCES

[1] H. Yoshida, Chemistry of lacquer (urushi), Part I, J. Chem. Soc. 43 (1883) 472–486.
[2] S. Hirano, Titration of urushiol using Baryta aqueous solution as indicator, Kogyō Kagaku Zashi (J. Chem. Ind.) 27 (1927) 381.
[3] E. Matsui, Chemistry of lacquer, Urushi Kagaku 63 (1963) 1–178.
[4] J. Kumanotani, M. Achiwa, R. Oshima, in: Cultural Property and Analytical Chemistry—International Symposium on the Conservation and Restoration of Cultural Property, National Research Institute of Cultural Properties, Tokyo, 1979, pp. 51–62.
[5] J. Kumanotani, Urushi—cultural relations, Kagakutokougyo 36 (1983) 151–154.
[6] J. Kumanotani, Urushi (oriental lacquer)—a natural aesthetic durable and future-promising coating, Prog. Org. Coat. 26 (1995) 163–195.
[7] R. Lu, T. Yoshida, T. Miyakoshi, Oriental lacquer: a natural polymer, Polym. Rev. 53 (2013) 153–191.
[8] T. Miyakoshi, K. Nagase, T. Yoshida, Progress of Lacquer Chemistry, IPC, Japan, 1999.

CHAPTER

Laccase

6

6.1 INTRODUCTION

Laccase (*p*-diphenol: oxygen oxidoreductase, EC 1.10.3.2), a copper-containing oxidase obtained from the lacquer tree was first discovered in 1883 by Yoshida [1], and Bertrand partially purified the enzyme about 10 years later in 1893 [2]. More recently laccase has been prepared using different methods by Nakamura [3], Omura [4], Peisach and Levine [5], Osaki and Walaas [6], and Reinhammar [7]. Study of laccase has revealed that only about 55% of laccase molecules are amino acid residues, which accounts for the low nitrogen concentration of 10.4%. The copper concentration of laccase is 0.23%, which corresponds to four copper atoms per protein molecule with a molecular weight of 1.1×10^5 [7]. The structure and a photo of laccase are shown in Figure 6.1.

Laccase is a most important component for the polymerization of lipid components in lacquer sap, and copper ions play a significant role in laccase activity. There are three types of copper ions in laccase, namely type 1, a blue copper center with characteristic spectral features; type 2, a mononuclear copper center with normal spectral features; and type 3, a binuclear copper that is antiferromagnetically coupled through a bridging ligand, hence EPR silent. The laccase catalyst follows a ping-pong mechanism and was reviewed by Thurston in 1994 [8]. The catalyst process requires four coordinated copper ions to transmit the electron. In the oxidation state, laccase first takes a pair of electrons from the substrate by type 1 and type 2 copper ions, then transfers them to type 3 copper ions to restore the oxidation state, and then takes a pair of electrons from the substrate again to form a fully reduced state. The fully reduced state copper cluster site reacts with oxygen to generate the peroxy intermediate [9], then, it loses a water molecule and returns to the fully oxidized state, as shown in Scheme 6.1 [10,11]. The spectroscopic and catalytic properties of *Toxicodendron vernicifluum* laccase depleted of type 2 copper were investigated, and the results showed only small changes in the type 1 copper properties when type 2 copper was removed; meanwhile, the concentration of type 3 copper did not change. Reoxidation of types 1 and 3 copper and the formation of the oxygen intermediate are the same process in the native and type 2-depleted laccases, suggesting that type 2 copper is not necessary for the stabilization of this intermediate [12]. Studies of the effects of proteins and polysaccharides on the activity of *T. vernicifluum* laccase showed that most proteins and polysaccharides, except laccase proteins, are not only incapable of catalyzing the oxidation of urushiol but can inhibit the activity of

Lacquer Chemistry and Applications. http://dx.doi.org/10.1016/B978-0-12-803589-4.00006-7

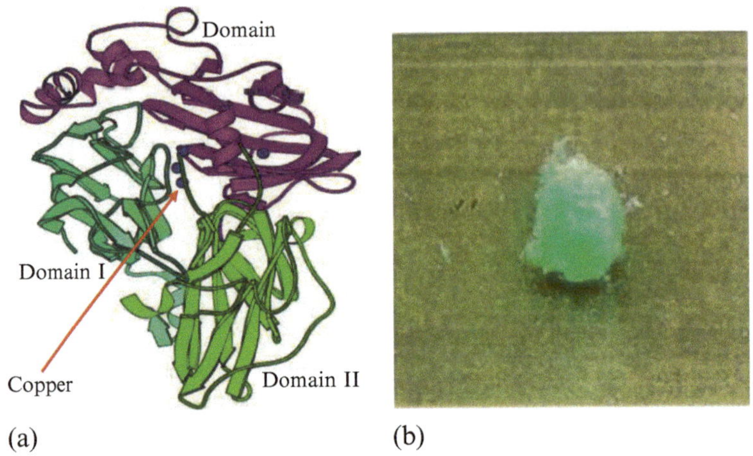

FIGURE 6.1

Structure of laccase (a) and photo (b).

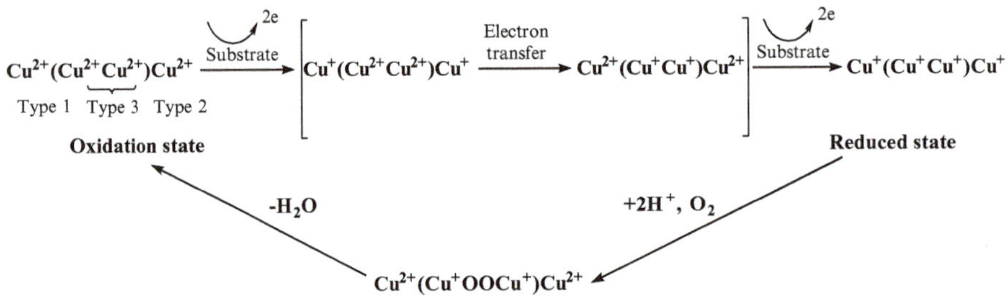

SCHEME 6.1

Oxidation-reduction mechanism of copper ions in laccase.

laccase to varying extents [13], and this result was confirmed by our recent report [14]. Laccase is active in the oxidation of monophenolic compounds such as eugenol and isoeugenol [15]. The oxidation of *O*-phenylenediamine [16], isoeugenol [17], the kinetics of catalyzed oxidation polymerization of catechol [18], and transformation of catechol in the presence of laccase [19] also have been examined.

6.2 SEPARATION AND PURIFICATION OF LACQUER LACCASE

In the natural state, there are hundreds of enzymes in a cell, so it is necessary to isolate and purify an enzyme for study. Many methods for separation and purification of lacquer laccase were developed since it was known that lacquer sap contains lacquer laccase, and they can be divided into four methods according to the characteristics of the method.

(1) Before 1950, the lipid component was removed by an organic solvent such as alcohol or acetone. The acetone insoluble (acetone powder, AP) was dissolved in water, the glycoprotein was removed by precipitation with ammonium sulfate, and a crude laccase blue solution was obtained.

(2) Between 1950 and 1960, Nakamura [3], Omura [4], Peisach and Levine [5], and Osaki and Walaas [6] used Amberlite XE-64, DEAE-Sephadex, Sephadex A-100, and Sephadex G-100 column chromatography and electrophoresis to purify the crude laccase. Although purified laccase was obtained by these methods, they are time consuming, expensive, and the quantity purified is small.

(3) In the early 1970s, Reinhammar [7] successfully used a combination of CM-Sephadex C-50, DEAE-Sephadex A-50, and Sephadex G-25 columns to separate and purify laccase and stellacyanin from Japanese lacquer. Because this method is simple and convenient, and a large amount of highly pure laccase can be obtained, it is still the classic method for separation and purification of lacquer laccase.

(4) After 1980, high-performance liquid chromatography (HPLC) with high-performance column TSK-gel CM-3SW was used in the separation and purification of lacquer laccase.

Although HPLC can obtain very pure laccase, the yield is very small, so the Reinhammar method [7] is often used to separate a large amount of highly pure lacquer laccase. The detailed procedure is as follows:

A: Isolate the AP from lacquer sap. Add four to eight volumes of cold acetone to 1000 g raw lacquer sap, stir, and filter, and then wash the AP with cold acetone several times until no urushiol is washed away. The AP is dried in a cold environment (under 5 °C).
B: Extraction of laccase. Suspend 100 g AP in 2 L 0.01 M cold potassium phosphate buffer (pH 6.0). Stir overnight and filter the suspension through a Buchner funnel. The filtrate is turbid and has a greenish color.
C: Cation-exchange chromatography. Apply the filtrate to a column of CM-Sephadex C-50 equilibrated with 0.01 M buffer and elute with about 1 L 0.01 M buffer. A yellow turbid fraction mainly due to lacquer oligosaccharide and polysaccharides is collected. A blue band is formed at the top of the column.
D: Then wash the column with 1 L 0.05 M buffer until the eluate is no longer absorbed at 250 nm. In this step, monosaccharides, nucleic acids, and low-molecular-weight proteins are eluted, and the blue band due to laccase remains at the top of the column.
E: Increase the buffer concentration to 0.1 M, and the blue band, still at the top of the column, is now divided. One part of it remains at the top of the column while the other part is eluted. Ahead of the blue band, which was found to be peroxidase, is a yellow-light blue and grayish-turbid fraction.
F: Increase the buffer concentration to 0.15 M, and the blue band will slowly go down. After the yellow-light blue peroxidase band is completely eluted, the following blue band will be around the lower middle of the column.

G: Increase the buffer concentration to 0.2 M and collect the blue band.
H: Dialyze the blue lacquer laccase solution collected from step G against 0.01 M pH 6.0 phosphate buffer overnight.
I: Anion-exchange chromatography. Apply the dialysate to a column of DEAE-Sephadex A-50 equilibrated with the same buffer. Laccase is eluted unretarded while a yellow band remains at the top of the column.
J: To concentrate the laccase solution collected in step I, apply it to a column of CM-Sephadex C-50 equilibrated with 0.01 M buffer and directly elute it with 0.25 M pH 6.0 phosphate buffer. Dialyze the concentrated laccase solution against 0.01 M pH 6.0 phosphate buffer overnight to remove most of the salts.
K: Apply the dialysate of step I to a column of Sephadex G-25 equilibrated with deionized water, and elute it with deionized water to completely remove the salts. Freeze-dry to obtain pure blue laccase.

Because laccase is a protein, methods for the detection of protein purity are also applicable to laccase, and ultrahigh-speed centrifugation, gel electrophoresis, isoelectric focusing electrophoresis, and solubility are commonly employed. However, because enzyme has substrate specificity, a catalyst also can be used to detect its purity. Because it may be difficult to determine the purity of an enzyme using only a single method, a combination of several methods is usually used.

6.3 MOLECULAR STRUCTURE OF LACQUER LACCASE

Laccase is a glycoprotein composed of 55% protein and 45% saccharides, and contains about 0.23% copper according to Reinhammar [7]. The molecular weight is about 1.1×10^5 as examined by polyacrylamide gel electrophoresis (PAGE) and column chromatography. The protein part consists of 18 amino acids as summarized in Table 6.1.

We did a rough analysis of the polysaccharide part of lacquer laccase in our laboratory, and galactose, glucose, mannose, arabinose, and fructose were found. The detailed analysis of the polysaccharide part of lacquer laccase is underway.

6.4 LACQUER LACCASE AND FUNGAL LACCASE

Although both lacquer laccase and fungal laccase can oxidize phenol and amino benzene compounds, and belong to the copper-containing oxidases, laccases from different sources have different properties. In the same genus, different species of lacquer tree have different laccase properties.

6.4.1 STRUCTURE AND COMPOSITION

Laccases from different sources have different copper concentrations, copper type, molecular weight, sugar content, and amino acid composition, as summarized in Table 6.2. Differences in laccase structure are mainly manifested in the sugar moiety. Because laccase is a kind of heterogeneous glycoprotein, structural changes of sugar

Table 6.1 Amino acid composition in lacquer laccase (by Reinhammar)

Amino acid	Residue (g/100 g protein)	Residue (mol/1.1 × 10^5 g protein)	Nitrogen (g/100 g protein)
Aspartate	6.28 ± 0.20	60.0 ± 1.9	0.764 ± 0.024
Threonine	4.57 ± 0.10	49.7 ± 1.1	0.633 ± 0.014
Serine	2.93 ± 0.02	37.0 ± 0.3	0.471 ± 0.003
Glutamate	4.30 ± 0.18	36.6 ± 1.5	0.467 ± 0.020
Proline	3.31 ± 0.10	37.5 ± 1.1	0.478 ± 0.014
Glycine	1.87 ± 0.10	36.0 ± 1.9	0.459 ± 0.025
Alanine	2.52 ± 0.11	39.0 ± 1.7	0.496 ± 0.022
Valine	3.87 ± 0.08	43.0 ± 0.9	0.547 ± 0.011
Cysteine	0.64 ± 0.03	6.9 ± 0.3	0.088 ± 0.004
Methionine	1.28 ± 0.03	10.7 ± 0.2	0.137 ± 0.003
Isoleucine	3.20 ± 0.06	31.1 ± 0.6	0.396 ± 0.007
Leucine	3.53 ± 0.06	34.3 ± 0.6	0.437 ± 0.007
Tyrosine	3.62	24.4	0.311
Phenylalanine	4.04 ± 0.20	30.2 ± 1.5	0.385 ± 0.017
Lysine	3.03 ± 0.06	26.0 ± 0.5	0.662 ± 0.013
Histidine	2.10 ± 0.04	16.8 ± 0.3	0.643 ± 0.012
Arginine	1.78 ± 0.08	12.5 ± 0.6	0.638 ± 0.029
Tryptophan	1.06 ± 0.06	6.3 ± 0.4	0.159 ± 0.009

Table 6.2 Molecular weight, sugar, and copper atoms of laccases

Source	MW	Sugar (%)	Copper atoms
T. vernicifluum	11-14.1 × 10^4	45	4
T. succedanea	13 × 10^4	-	6
G. usitata	11 × 10^4	43	4-6
Peach	7.3 × 10^4	-	2
Sycamore	9.7 × 10^4	45	4
Polyporus versicolor I	6.4 × 10^4	10-14	4
Polyporus versicolor II	6.47 × 10^4	10	4
Podaspora anserine I	39 × 10^4	24	16
Podaspora anserine II	7 × 10^4	25	4
Podaspora anserine III	8 × 10^4	23	4
Neurospora crassa	6.48 × 10^4	11	3-4
Coriolus hirstus	6.3 × 10^4	15	4
Agaricus bisporus	10 × 10^4	15	2

moieties easily lead to changes in laccase properties. Galactose, glucose, mannose, arabinose, and fructose were found in lacquer laccase, as described above, and xylose, mannose, glucosamine, and fucose were found in fungal laccases.

It was found that even if the amino acids are different in laccases from different sources, the amino acids and their arrangements are similar in the active laccase center, even if the copper concentration and copper type are not the same. There are one type 1, one type 2, and two type 3 copper ions, for a total of 4 in lacquer laccase. In fungal laccase, the situation is more complicated. *Podaspora anserine I* has four subspecies, and each subspecies has the same number of copper ions and type of laccase, so the total number of copper ions is 16. Laccases isolated from *Prunus persica* and *Agaricus bisporus* only have two copper ions. These copper ions can be removed easily. In general, the activity can be restored to a lacquer laccase by the addition of copper ions, but it cannot be restored to a fungal laccase.

6.4.2 THERMAL STABILITY AND OPTIMUM pH

Laccase from different sources have different pH optima, and lacquer laccases from different lacquer tree species also have different optimum pH. In general, the optimum pH of lacquer laccase from *T. vernicifluum* is about 6.5-7.5, from *T. succedanea* is about 6.5, and from *Gluta usitata* is about 4.5. In contrast, fungal laccases have the optimum pH in the low range, pH 3.5-4.5 for *Coriolus hirstus*, pH 6.2 for *Sycamore*, pH 6.0 for *Rhizoctonia praticola*, and pH 4.0-5.0 for *Polyporus versicolor*.

In general, the half-life of lacquer laccase activity is 12 h at 50 °C, although that of laccase from *A. bisporus* is 3 h. The thermal stability of laccase from microbial fermentation is different with different bacteria. The thermal stability of fungal laccase generally is better than that of lacquer laccase. Of course, an immobilized laccase also has better thermal stability and optimum pH than a free laccase.

Recently, we immobilized lacquer laccase on water-soluble chitosan and chitosan microspheres, and their chelation properties were compared with laccase immobilized on a transitional metal (Fe^{3+}) [20]. The results showed that compared with the free laccase, immobilized laccase displayed a lower specific activity but has a similar substrate affinity with improved stability of various parameters, such as thermal stability, pH, and storage time.

In addition, lacquer laccase-catalyzed polymerization of phenylpropanoid compounds [21], and lignocatechols were demonstrated in our laboratory for the first time [22]. The polymerization mechanism was examined by IR and pyrolysis GC-MS measurements. From the results, we estimated that the polymerization of lignocatechol with laccase proceeded mainly through the quinone radical intermediate and is proposed as shown in Scheme 6.2.

6.4.3 LACCASE-CATALYZED OXIDATION

A lacquer film forms by the oxidative polymerization of lacquer lipid components (urushiol, laccol, and thitsiol), which is catalyzed by laccase, and the construction of cross-links on the long aliphatic unsaturated side chain by autoxidation. The polymerization

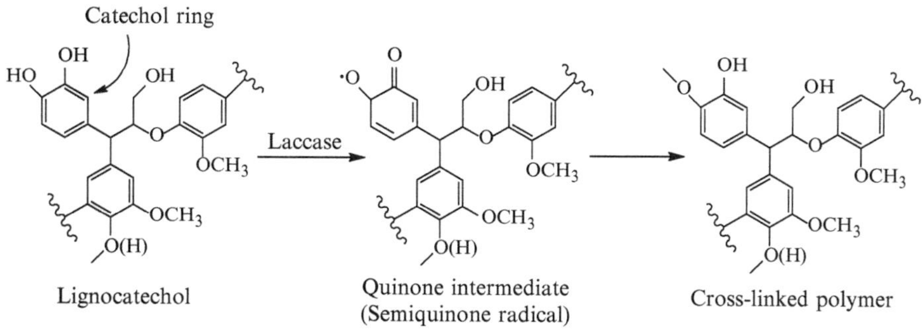

SCHEME 6.2

Proposed polymerization mechanism of lignocatechol catalyzed by laccase.

mechanism of lacquer sap catalyzed by laccase was first reported by Kumanotani [23], and the explanation was supplemented and refined by our later works [24,25].

Laccase is a dehydrogenase that includes Cu^{2+}, written as En-Cu^{2+}. Initially, En-Cu^{2+} is reduced by urushiol at the interface of a water/oil emulsion to produce En-Cu^{+} and urushiol radicals. En-Cu^{+} is oxidized by oxygen to En-Cu^{2+} with the accompanying formation of water, and urushiol radicals are transferred to give semiquinone radicals. The urushiol radical may undergo reaction in two ways: (i) attacking the urushiol nucleus to give biphenyl urushiol, some of which is further oxidized to dibenzofuran compounds; or (ii) attacking the side-chain methylene group inserted between conjugated diolefines and monoolefines to give heptatriene cations that may subsequently produce a nuclear side-chain structure, as shown in Scheme 6.3.

6.4.4 AUTOXIDATION

After the urushiol monomer decreases to less than 30%, autoxidation begins to occur in the side chain [26,27]. The double bond in the side chain is autoxidized by oxygen to produce allyl radicals, and rearrangement of the remaining double bonds takes place to give a stable radical site. Oxygen reacts with the radicals to form peroxy radicals. Cross-linking occurs when the peroxy radicals attack double bonds or allyl radicals in other side chains. The resulting polyperoxide is relatively stable but sometimes is decomposed by heat or light to form oxyradicals, which will cross-link with ether. Generally, the autoxidation reaction in lacquer systems progresses after enzymatic polymerization due to the presence of phenolic hydroxyl groups of urushiol that act as the antioxidants. Both enzymatic reaction and autoxidation are repeated to form a durable network polymer, as shown in Scheme 6.4.

Recently, during analysis of the thitsiol dimer structure produced with a laccase catalyst [28] as described in Chapter 3, we found that no route (ii) reaction occurred; only route (i) occurred to form a nuclear-nuclear structure. That is, no nuclear-side chain (C–O) coupling, but only a nuclear-nuclear (C–C) biphenyl phenol-type structure exists in thitsiol dimer. The reason should be that the side chain in thitsiol is bigger than that in urushiol and there are very few olefin structures in side chains.

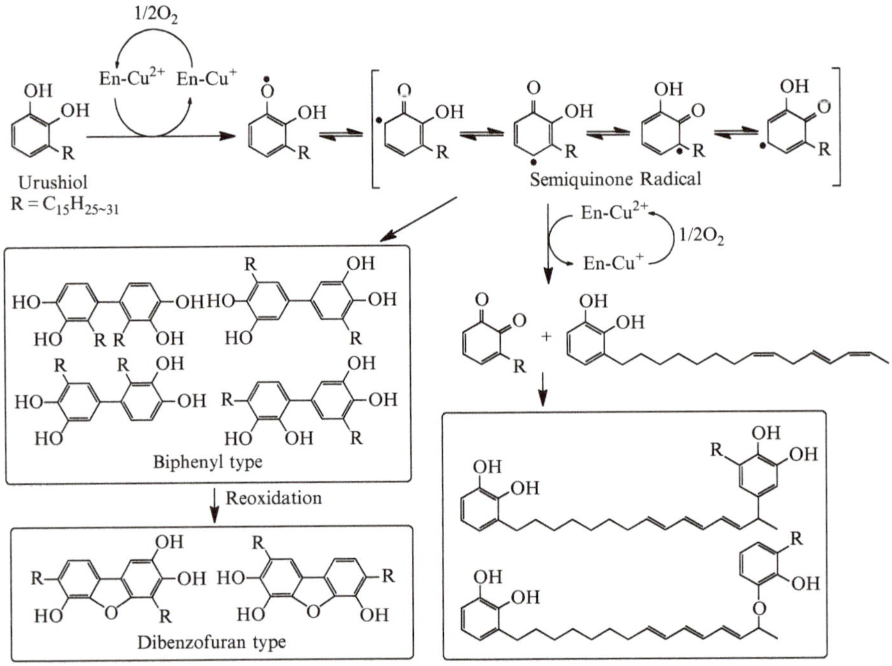

SCHEME 6.3

Oxidation of urushiol with laccase.

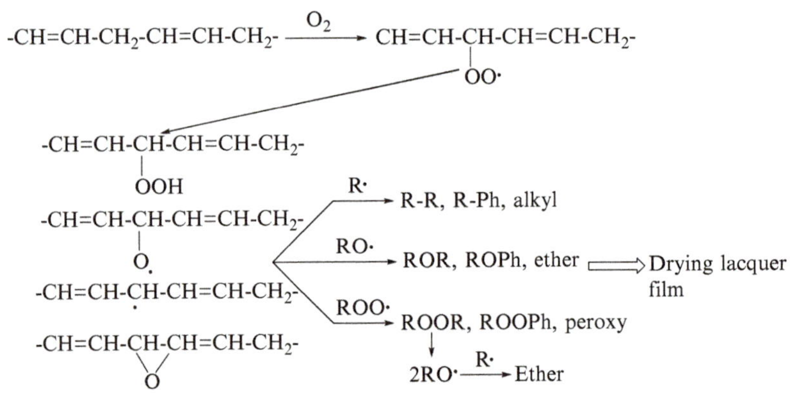

SCHEME 6.4

Autoxidation of an urushiol unsaturated side chain.

REFERENCES

[1] H. Yoshida, Chemistry of lacquer (urushi), Part I, J. Chem. Soc. 43 (1883) 472–486.
[2] G. Bertrand, Chimie industrielle—Sur le latex de l'arbre à laque, Compt. Rendus 118 (1894) 1215–1218.

[3] T. Nakamura, Purification and physico-chemical properties of laccase, Biochim. Biophys. Acta 30 (1958) 44–52.
[4] T. Omura, Studies on laccases of lacquer trees: I. Comparison of laccases obtained from *Rhus vernicifera* and *Rhus succedanea*, J. Biochem. 50 (1961) 264–272.
[5] J. Peisach, W.G. Levine, A comparison of the enzymic activities of pig ceruloplasmin and *Rhus vernicifera* laccase, J. Biol. Chem. 240 (1965) 2284–2289.
[6] S. Osaki, O. Walaas, A new simple and rapid chromatographic method for the purification of laccase and stellacyanin from *Rhus vernicifera*, Arch. Biochem. Biophys. 123 (1968) 638–639.
[7] B. Reinhammar, Purification and properties of laccase and stellacyanin from *Rhus vernicifera*, Biochim. Biophys. Acta 205 (1970) 35–47.
[8] C.F. Thurston, The structure and function of fungal laccases, Microbiology 140 (1994) 19–26.
[9] W. Shin, U.M. Sundaram, J.L. Cole, H.H. Zhang, B. Hedman, K.O. Hodgson, E.I. Solomon, Chemical and spectroscopic definition of the peroxide-level intermediate in the multicopper oxidases: relevance to the catalytic mechanism of dioxygen reduction to water, J. Am. Chem. Soc. 118 (1996) 3202–3215.
[10] U.M. Sundaram, H.H. Zhang, B. Hedman, K.O. Hodgson, E.I. Solomon, Spectroscopic investigation of peroxide binding to the trinuclear copper cluster site in laccase: correlation with the peroxy-level intermediate and relevance to catalysis, J. Am. Chem. Soc. 119 (1997) 12525–12540.
[11] S.K. Lee, S.D. George, W.E. Antholine, B. Hedman, K.O. Hodgson, E.I. Solomon, Nature of the intermediate formed in the reduction of O_2 to H_2O at the trinuclear copper cluster active site in native laccase, J. Am. Chem. Soc. 124 (2002) 6180–6193.
[12] B. Reinhammar, Y. Oda, Spectroscopic and catalytic properties of *Rhus vernicifera* laccase depleted in type 2 copper, J. Inorg. Biochem. 11 (1979) 115–127.
[13] D.F. Zhan, Y.M. Du, B.G. Qian, Effects of proteins and polysaccharides on the activity of *Toxicodendron vernicifera* laccase, Chem. Ind. For. Prod. 11 (1991) 111–116.
[14] Y.Y. Wan, R. Lu, K. Akiyama, K. Okamoto, T. Honda, Y.M. Du, T. Yoshida, T. Miyakoshi, C.J. Knill, J.F. Kennedy, Effects of lacquer polysaccharides, glycoproteins and isoenzymes on the activity of free and immobilized laccase from *Rhus vernicifera*, Int. J. Biol. Macromol. 47 (2010) 76–81.
[15] T. Sakurai, Laccase activates monophenols, eugenol and isoeugenol, J. Pharmacobio-Dyn. 14 (1991) s-114.
[16] D.F. Zhan, Y.M. Du, B.G. Qian, Oxidation product of *O*-phenylenediamine catalysed by *Toxicodendron vernicifera* laccase, Chem. Ind. For. Prod. 11 (1991) 13–16.
[17] T. Shiba, L. Xiao, T. Miyakoshi, C.L. Chen, Oxidation of isoeugenol and coniferyl alcohol catalyzed by laccases isolated from *Rhus vernicifera* Stokes and *Pycnoporus coccineus*, J. Mol. Catal. B: Enzym. 10 (2000) 605–615.
[18] N. Aktaş, A. Tanyolaç, Kinetics of laccase-catalyzed oxidative polymerization of catechol, J. Mol. Catal. B: Enzym. 22 (2003) 61–69.
[19] M.Y. Ahn, C.E. Martinez, D.D. Archibald, A.R. Zimmerman, J.M. Bollag, J. Dec, Transformation of catechol in the presence of a laccase and birnessite, Soil Biol. Biochem. 38 (2006) 1015–1020.
[20] Y.Y. Wan, Y.M. Du, X.W. Shi, J. Li, R. Lu, T. Miyakoshi, B. Chen, Immobilization and characterization of laccase from Chinese *Rhus vernicifera* on modified chitosan, Process Biochem. 41 (2006) 1378–1382.
[21] Y.Y. Wan, R. Lu, K. Akiyamo, T. Miyakoshi, Y.M. Du, Enzymatic synthesis of bioactive compounds by *Rhus* laccase from Chinese *Rhus vernicifera*, Sci. China, Ser. B: Chem. 50 (2007) 179–180.

[22] Y. Takashi, R. Lu, S.Q. Han, K. Hattori, T. Katsuta, K. Takeda, K. Sugimoto, M. Funaoka, Laccase-catalyzed polymerization of lignocatechol and affinity on proteins of resulting polymers, J. Polym. Sci., Part A: Polym. Chem. 47 (2009) 824–832.

[23] J. Kumanotani, Urushi (oriental lacquer)—a natural aesthetic durable and future-promising coating, Prog. Org. Coat. 26 (1995) 163–195.

[24] R. Lu, S. Harigaya, T. Ishimura, K. Nagase, T. Miyakoshi, Development of a fast drying lacquer based on raw lacquer sap, Prog. Org. Coat. 51 (2004) 238–243.

[25] R. Lu, T. Ishimura, K. Tsutida, T. Honda, T. Miyakoshi, Development of a fast drying hybrid lacquer in a low relative-humidity environment based on *Kurome* lacquer sap, J. Appl. Polym. Sci. 98 (2005) 1055–1061.

[26] K. Nagase, Y. Kamiya, T. Kimura, K. Hodumi, T. Miyakoshi, The relationship between the change of progress time in the urushi liquid by the enzymic polymerization and the natural drying property occurring under a low humidity environment, Nippon Kagaku Kaishi 10 (2001) 587–593 (in Japanese).

[27] K. Nagase, R. Lu, T. Miyakoshi, Studies on the fast drying hybrid urushi in low humidity environment, Chem. Lett. 33 (2004) 91–92.

[28] R. Lu, D. Kanamori, T. Miyakoshi, Characterization of thitsiol dimer structures from *Melanorrhoea usitata* with laccase catalyst by NMR spectroscopy, Int. J. Polym. Anal. Charact. 16 (2011) 86–94.

CHAPTER

Lacquer aroma components

7

A smell is caused by a volatile substance stimulating the nasal olfactory cells. Volatile substances are mainly alcohols, acids, esters, phenols, and terpenes. Because natural products often contain these volatile substances, they often have a smell.

Lacquer sap is tapped from lacquer trees and contains a mixture of urushiol, plant gum, laccase, nitrogen substance, and water. The compositions and proportions of lacquer sap differ with age, species, growth status, environment, climate, and season. In addition, the quality of lacquer sap is affected by storage, preservation, and transport. Therefore, gloss, color, smell, and viscosity are usually used to identify the quality of a lacquer sap at the tapping location. To determine quality, smell is one of the most important indicators. Raw lacquer sap has a characteristic sour smell, and the smell is related to the quality in human experience; however, the compound that produces the lacquer aroma component has been unclear.

Majima and Okazaki [1] and Majima and Takayama [2] reported that lacquer sap contains acetic acid, and Majima and Takayama [2] described the smell of Japanese raw lacquer sap. He found that unsaturated C15 carbohydrates were decomposed during the process of Japanese lacquer distillation, and named the aromatic component "urusene." Sato [3] also isolated sesquiterpenes such as α-gurjunene and calarene from *Gluta usitata* lacquer sap.

In order to identify the lacquer aromatic compounds, we analyzed *Toxicodendron vernicifluum* lacquer sap by headspace gas chromatography (GC) and mass spectrometry (MS) [4]. The headspace, meaning "the top space," contains those ingredients that exist at the top of a liquid or solid sample, especially low boiling point components. This method is usually used to analyze the flavor components in food, smell components in chemical products, and toxic volatile compounds in environmental water. In general, solvent extraction, vacuum distillation, and continuous distillation are usually used to extract the aroma compounds from a flower or a fruit. These methods include crushing, squeezing, and heating of aroma components, and would change the structure of aromatic compounds. Therefore, the methods to collect aromatic compounds without changing their structure have been studied, and the headspace method is attracting attention. The headspace method consumes smaller amounts of the sample, does not destroy the sample, and allows easy collection and quick analysis of the volatile components. A column of the porous absorbent polymer Tenax is used for the headspace method, and aroma compounds are first adsorbed and concentrated by the Tenax column, then the column is heated to desorb the aroma compounds and send them to the GC and MS analyzer. An image of the system is shown in Figure 7.1.

Lacquer Chemistry and Applications. http://dx.doi.org/10.1016/B978-0-12-803589-4.00007-9

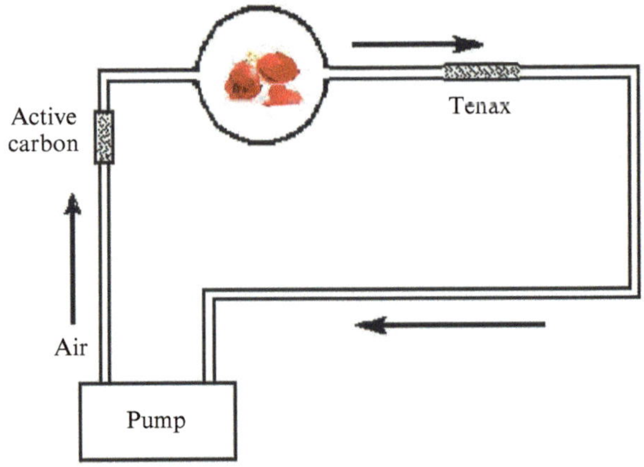

FIGURE 7.1

Column for collecting aroma compounds.

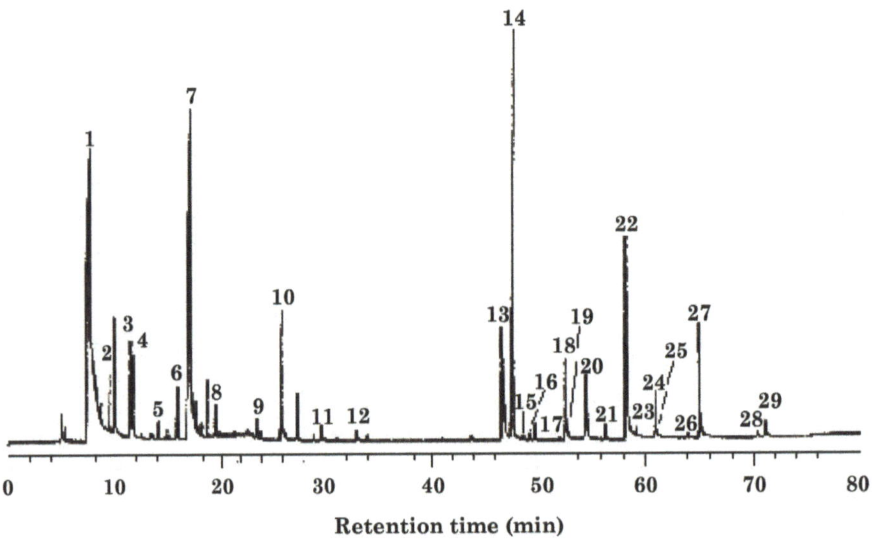

FIGURE 7.2

Mass chromatogram of volatile compounds of *T. vernicifera* lacquer sap.

The GC conditions for analysis of lacquer aroma are as follows: both injection and detector temperatures are 230 °C. After keeping the temperature of the thermostatic bath at 35 °C for 5 min, the temperature is raised to 200 °C at the rate of 2 °C/min. Helium is used as the carrier gas, the flow rate is 30 cm/s, and the column used is a DB-Wax (60 m × 0.25 mm) column.

A mass chromatogram of volatile compounds of *T. verniciffuum* lacquer sap is shown in Figure 7.2.

There are 29 real peaks in the total ion chromatogram (TIC) of volatile compounds of *T. vernicifluum* lacquer sap. The results of analysis of each peak of the mass spectrum are summarized in Table 7.1.

The results showed that alcohols, esters, organic acids, monoterpene hydrocarbons, and sesquiterpene hydrocarbons were contained in the aroma.

Majima and Okazaki [1] and Majima and Takayama [2] named the aroma of lacquer sap "urusene," and suggested that it is an unsaturated hydrocarbon because it can be hydrogenated and absorbs bromine. The molecular formula is $C_{15}H_{28}$ (MW208)

Table 7.1 GC analyzed results of volatile compounds of *T. vernicifluum* lacquer sap

Peak no.	Retention time (min)	Peak area (%)	Structure
1	7.1	49.3	Ethyl formate
2	9.4	1.5	Ethyl acetate
3	11.6	6.5	2-Propanol
4	11.8	4.8	Ethanol
5	14.0	0.3	MHC
6	15.8	0.7	MHC
7	16.8	31.4	1-Propanol
8	19.4	0.5	MHC
9	23.7	0.1	MHC
10	25.9	4.3	*n*-Butanol
11	29.9	0.4	MHC
12	33.0	0.2	Limonene
13	46.8	7.0	Acetic acid
14	47.7	30.4	α-Longipinene
15	48.2	0.5	α-Ylangene
16	48.4	0.5	α-Copanene
17	48.9	0.9	SHC
18	53.0	6.3	Propionic acid
19	53.1	2.9	α-Cedrene
20	54.9	4.0	Isobutyric acid
21	56.3	1.6	α-Bergamotene
22	58.4	3.4	Butyric acid
23	59.1	0.8	α-Himachalene
24	61.2	3.4	Isovaleric acid
25	61.3	3.0	α-Humulene
26	64.1	0.9	β-Selinene
27	65.6	11.7	Valeric acid
28	70.4	0.3	δ-Cadinene
29	71.2	3.0	Caproic acid

MHC, monoterpene hydrocarbon; SHC, sesquiterpene hydrocarbon.

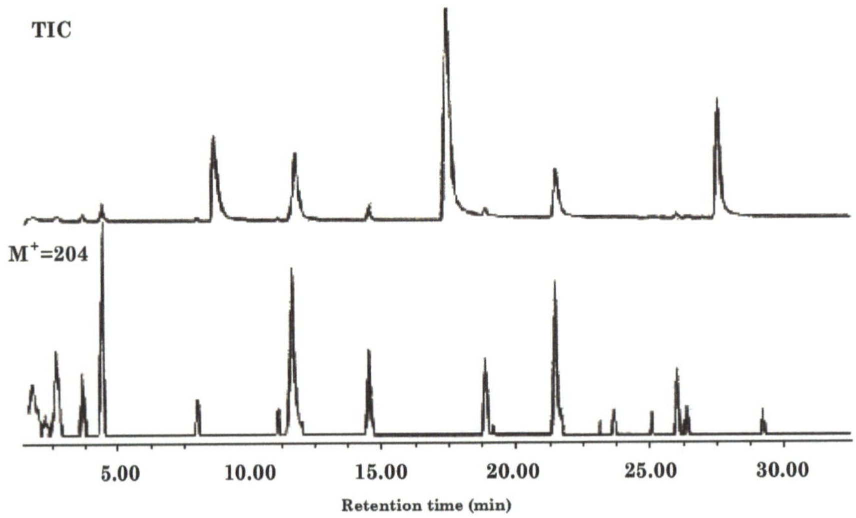

FIGURE 7.3

TIC and mass chromatogram ($M^+=204$) of sesquiterpene.

or $C_{15}H_{26}$ (MW206). However, we detected sesquiterpene (MW204) in the aroma compounds during GC/MS analysis (Figure 7.3), and α-longipiene, α-cedrene, and α-humulene were detected according to the mass spectra.

This difference is due to the different methods of molecular weight determination. Majima used freezing-point depression of benzene to measure the molecular weight of urusene. This method cannot obtain the exact molecular weight and generally has a ±5% error. Taking into account the ±5% error, the MW of urusene is very close to the MW of sesquiterpene, so we speculate that the unsaturated hydrocarbon urusene isolated from lacquer sap by Majima was sesquiterpene hydrocarbon.

REFERENCES

[1] R. Majima, Y. Okazaki, Zur kenntnis des 2.3-dioxy-toluols (isohomobrenzcatechins) und über die nitroderivate seiner methyläther, Ber. Deut. Chem. Ges. 49 (1916) 1482–1496.

[2] R. Majima, G. Takayama, Über den hauptbestandteil des japanlacks, VII.: der urushiol-monomethyläther und der mechanismus der oxydation des urushiols, Ber. Deut. Chem. Ges. 53 (1920) 1907–1916.

[3] M. Sato, Two sesquiterpene hydrocarbons and an azurene derivative isolated from *Melanorrhoea usitata* Wall, Shikizai 43 (1970) 130–134 (in Japanese).

[4] T. Miyakoshi, K. Nagase, T. Yoshida, Progress of Lacquer Chemistry, IPC, Japan, 1999, pp. 131–132.

CHAPTER

Lacquer allergy

8

The main component of lacquer sap is the lipid component, which is called urushiol for *Toxicodendron vernicifluum*, laccol for *T. succedanea*, and thitsiol for *Gluta usitata*. The lipid components are catechol derivatives having long side chains of 15 or 17 carbon atoms with one, two, or three double bonds. Because the main component of lacquer has a catechol ring structure, it has caused dermatitis in lacquer workers [1–3]. Therefore, studies on the external manifestations, allergic mechanism, medical treatment, and prevention are very important.

8.1 MECHANISM OF LACQUER ALLERGY

Lacquer allergy usually refers to an allergic contact dermatitis that is caused by the lipid components contained in lacquer sap. The lacquer lipid component is a hapten from the viewpoint of immunology [4]. It is generally believed that the *O*-quinones derived from the oxidation of urushiol are susceptible to nucleophilic attack at the catechol ring in positions 4, 5, and 6 by proteins to form a complete antigen [5], as shown in Figure 8.1. The mechanism of allergic contact dermatitis has been described in detail by Polak [6] and Benezra [7], as shown in Figure 8.2.

The sensitization of inflammation can be divided into induction and elicitation phases. In the induction phase, the hapten penetrates the skin and becomes a protein–hapten complex, which is a complete antigen that can be incorporated into Langerhans cells (internalization) and present antigen information to T lymphocytes. These T lymphocytes then migrate to the lymph nodes where they differentiate into *effector* cells and *memory* cells and proliferate. At this point, the individual is hypersensitive and has immunologically prepared T lymphocytes that can react to a later attack by the same hapten.

In the elicitation phase, the hapten once again comes into contact with the skin and the previously formed protein–hapten complex. The effector cells that were formed during the first contact and circulate in the blood now come into contact with the complex. These effector cells release chemical mediators called lymphokines whose properties have been studied extensively [8]. Lymphokines are aimed at different targets: macrophages, lymphocytes, and other cell types (cytotoxic for L cells particular). These lymphokines are responsible for the pathologic manifestations of allergic contact dermatitis, erythema, and swelling [9,10].

Lacquer Chemistry and Applications. http://dx.doi.org/10.1016/B978-0-12-803589-4.00008-0

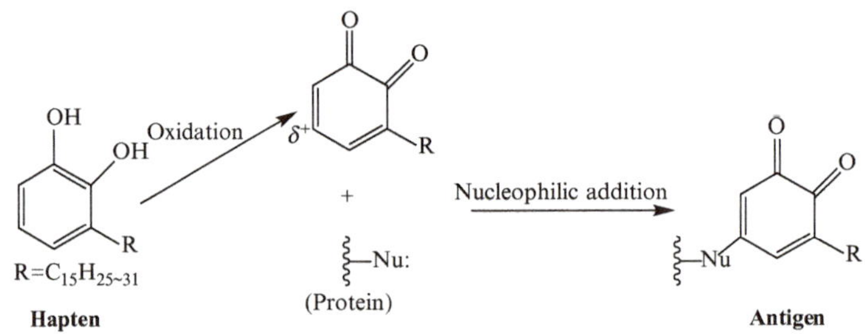

FIGURE 8.1

Reaction between urushiol and protein in skin.

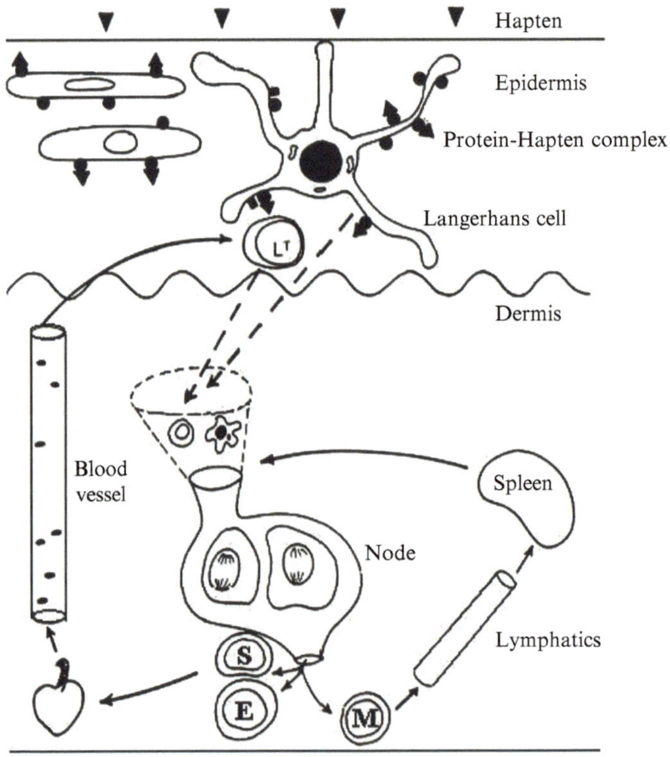

FIGURE 8.2

Mechanism of allergic contact dermatitis. L^T, T lymphocyte; S, suppressor cells; M, memory cells; E, effector cells.

In the case of lacquer, the lipid components of lacquer, urushiol, laccol, and thitsiol are not electrophiles and cannot react with electrophilic proteins to form an antigen [5]. When these lipid molecules penetrate the skin, an enzyme-catalyzed oxidation occurs and converts the lipids into orthoquinones, which are authentic electrophiles, and reacts with keratin or protein in the cellular membrane to form a complete antigen, as shown in Figure 8.3.

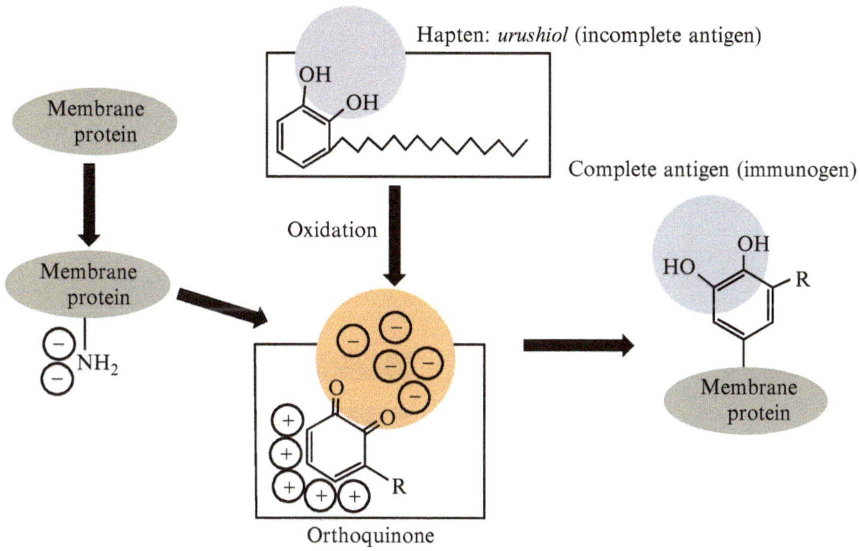

FIGURE 8.3

Mechanism of urushiol allergy.

This urushiol antigen is captured by Langerhans cells, presents antigen information to T lymphocytes, and induces sensitization of T cells. In the initial sensitization process, interleukin-1β (IL-1β) and major histocompatibility complex (MHC) class II are produced in Langerhans cells [5]. The keratinocytes produce urushiol-generated cytokines such as tumor necrosis factor α (TNF-α), IL-1α, and granulocyte/macrophage colony-stimulating factor (GM-CSF), which further activate the Langerhans cells [6].

In order to study urushiol molecular dynamics during the allergic process, we have synthesized many urushiol analogues and tested them in humans, obtaining many valuable results. The details follow.

8.1.1 STUDY USING COMPOUNDS THAT REPLACE THE CATECHOL RING HYDROXYL GROUP

As described above, we consider that after entering the epidermis, urushiol is oxidized to *O*-quinones. In order to validate this process, 3-pentadecylcatechol (PDC) [7], 3-pentadecylcatechol diacetate (PDCDA), and 3-heptadecylveratrole (HDV) [8] were synthesized and subsequently used these three compounds for a patch test *in vivo* and lymphocyte stimulation test *in vitro*, respectively. The chemical structures of PDC, PDCDA, and HDV are shown in Figure 8.4.

Patch testing is a way of identifying whether a substance that comes in contact with the skin causes inflammation of the skin. A plaster patch with the antigen is attached to the back or upper arm skin. After 48 h, the allergic reaction is observed.

Figure 8.5 shows the results of a patch test. The sample concentrations were 0.1% and 0.01%, respectively. The PDC and PDCDA caused almost the same positive reaction at the same concentration, but HDV did not cause a positive reaction at any concentration.

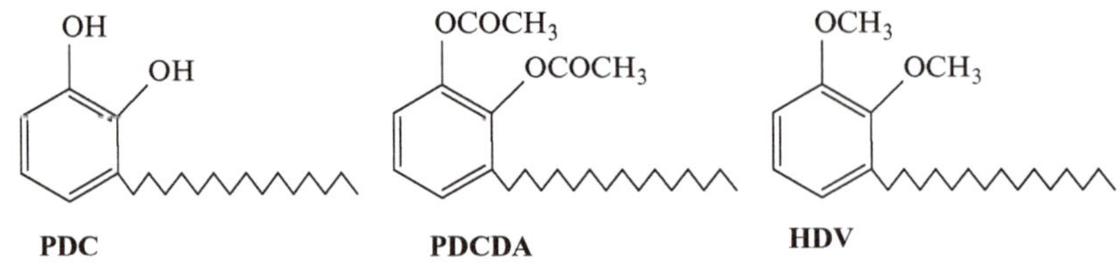

FIGURE 8.4

Chemical structures of PDC, PDCDA, and HDV.

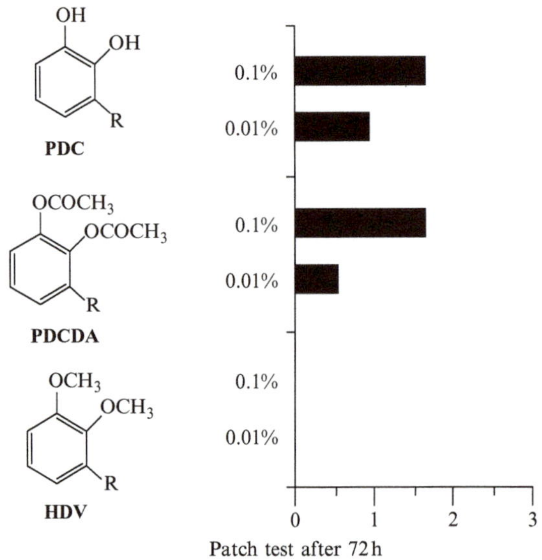

FIGURE 8.5

Results of patch test.

A lymphocyte stimulation test is a method that first isolates and cultures the lymphocytes collected from urushiol allergy sufferers and then adds urushiol in various concentrations to measure the reactivity of lymphocytes. A stimulation index (S.I. %) is used to represent the ratio of control group and experimental group activity and is calculated by the following formula:

$$\text{S.I.}\,\% = \frac{\text{activity of test sample}}{\text{activity of control}} \times 100$$

Because lymphocytes that reacted with urushiol are present in the cultured lymphocytes, if a reaction between lymphocytes and urushiol occurs when a sample of urushiol is added, the sample has antigenicity to urushiol.

Figure 8.6 is the result of lymphocyte stimulation test. The PDC and PDCDA showed almost the same positive reaction at concentrations from 0.002 to 20 μg/ml.

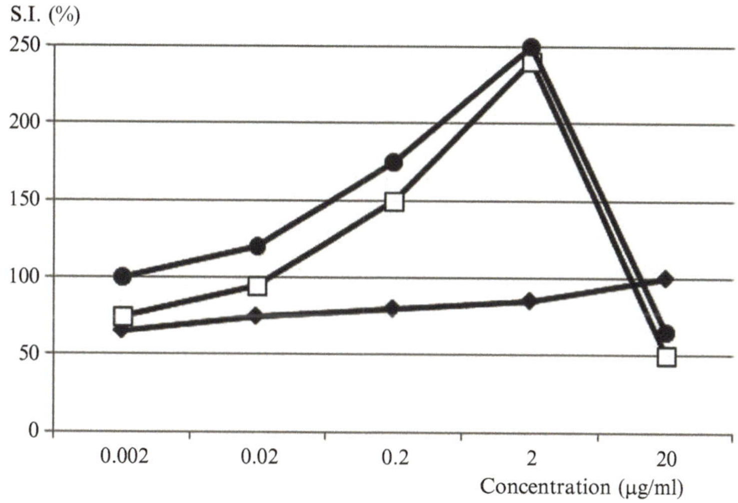

FIGURE 8.6

Results of lymphocyte stimulation test.

However, an obviously positive reaction of HDV was not observed at concentrations lower than 2 μg/ml.

8.1.2 STUDY USING PROTEIN BINDING SITE-SUBSTITUTED COMPOUNDS

The results of *in vitro* test to combine urushiol with protein showed that the protein binding site is the 4-, 5-, or 6-position of the catechol ring. In general, a protein amino group (NH_2–) selectively binds to 5-position and thiol group (SH–) at the 6-position of a catechol ring. In order to reveal the binding mechanism between urushiol and a protein, we synthesized 4-Me-PDC, 5-Me-PDC, 4,5-diMe-PDC, and 5-butyl-PDC (Figure 8.7), and used them in patch tests and lymphocyte stimulation tests.

The patch test results are shown in Figure 8.8. 4-Me-PDC or 5-Me-PDC, that is, compounds in which only one position of the catechol ring was methylated, produced almost the same positive reaction as PDC in the patch test. On the other hand, when the 2-position of the catechol ring was a methylated compound, 4,5-diMe-PDC, or replaced by a relatively big butyl group, 5-butyl-PDC, it produced a very weak positive reaction in the patch test.

A similar result was obtained from the lymphocyte stimulation test (Figure 8.9). These results can be considered due to the fact that if only one position of a catechol ring is substituted, it still has two protein binding sites. For example, for 4-Me-PDC, protein can bind to the 5- and 6-positions; for 5-Me-PDC, protein can bind to the 4- and 6-positions. If two positions of a catechol ring are substituted, it only has one protein binding site (4,5-diMe-PDC). For 5-butyl-PDC, even if only one position of a catechol ring is substituted, the large butyl group has a steric hindrance that prevents

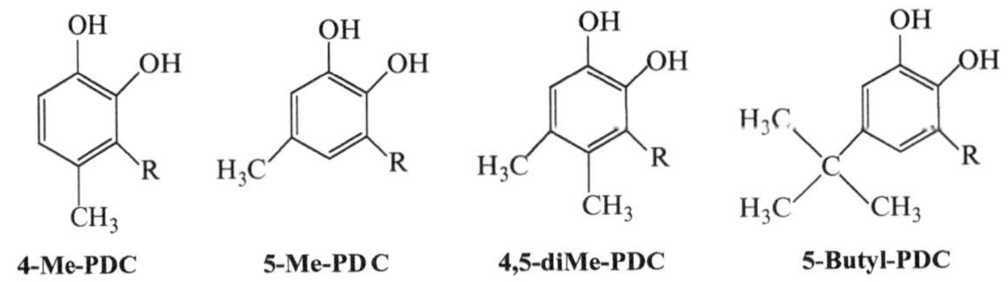

FIGURE 8.7

Chemical structures of urushiol derivatives.

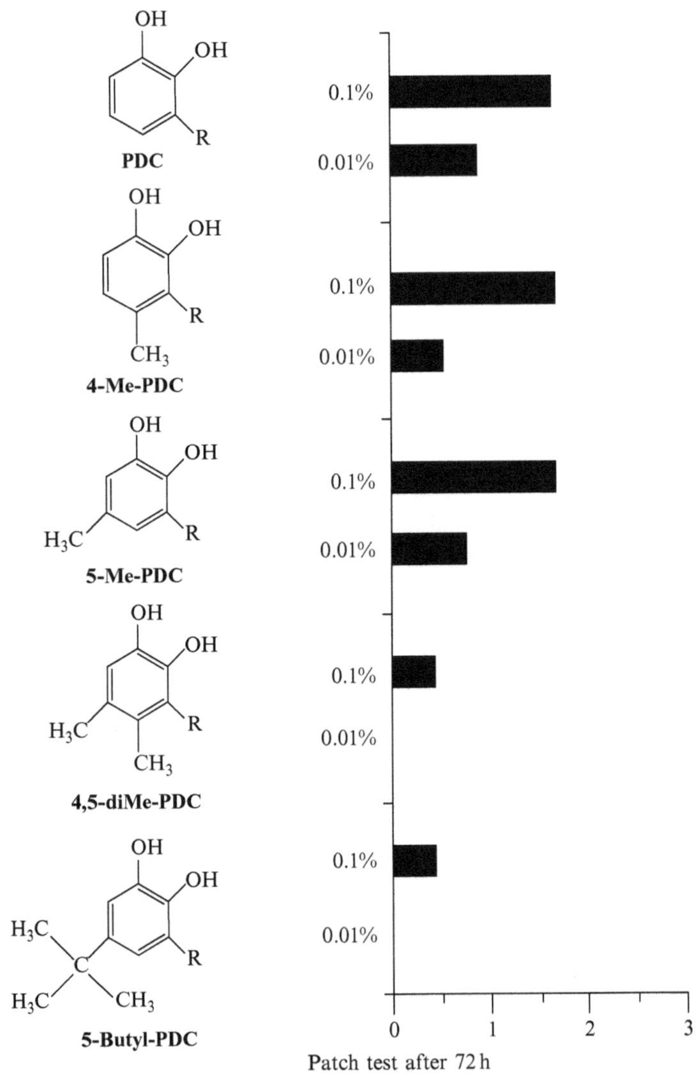

FIGURE 8.8

Results of patch test.

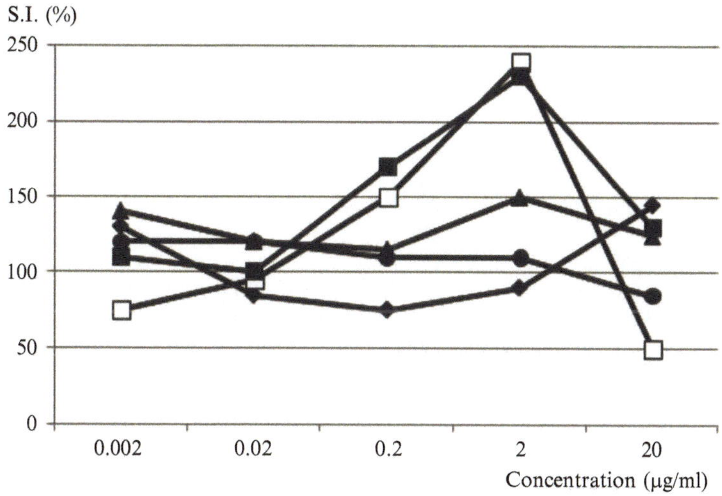

FIGURE 8.9

Results of lymphocyte stimulation test.

another group from binding to the ring and results in a reduced binding site. If it cannot bind to protein, urushiol is not able to become an antigen. Based on the results, it can be considered that after urushiol invaded the skin epidermis, it was oxidized to *O*-quinone and then combined with a protein at one of the 4, 5, or 6 catechol rings to form an antigen and cause allergic contact dermatitis.

8.1.3 STUDY USING HYDROCARBON CHAIN-SUBSTITUTED COMPOUNDS

There is a long (C15) hydrocarbon side chain at the 3-position of urushiol and a C17 side chain at the 4-position of thitsiol. Compared to urushiol, it has been said that thitsiol is less allergenic than urushiol. In this study, we used 3-PDC and 4-PDC as test reagents in the patch test and lymphocyte stimulation test. It can be predicted that 4-PDC has fewer protein binding sites than 3-PDC, as shown in Figure 8.10.

Figure 8.11 shows the test results. 4-PDC produced a weaker positive reaction than 3-PDC in both patches test and lymphocyte stimulation test, as predicted.

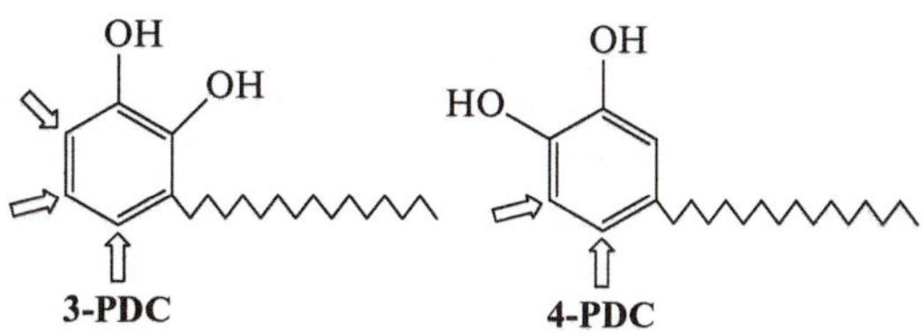

FIGURE 8.10

Protein binding sites of 3-PDC and 4-PDC.

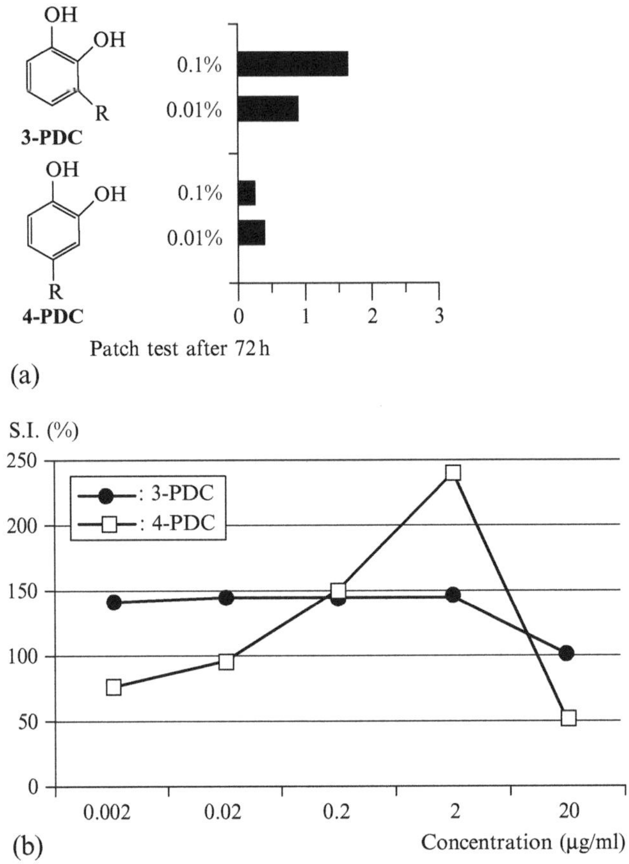

FIGURE 8.11

Results of patch (a) and lymphocyte stimulation (b) tests with 3-PDC and 4-PDC.

In order to confirm these results, further animal experiments using guinea pigs were carried out. It was known that precoating guinea pigs' ears with urushiol causes urushiol allergy. In this test, we used six guinea pigs per group and treated the ears with 3-PDC and 4-PDC. After treatment for 1 week, urushiol at the concentration of 1% in acetone applied to ear-flank method caused an allergic reaction on the flanks of the guinea pigs. In animals that were sensitized to 4-PDC, the results of a challenge with 3-PDC (after pretreatment with 3-PDC) were similar to the reaction to 4-PDC (after pretreatment with 4-PDC). Therefore, lymphocytes from the animals sensitized to 4-PDC are thought to be able to recognize 3-PDC. On the other hand, in animals that were sensitized to 3-PDC, the challenge reaction to 3-PDC was significantly stronger than that to 4-PDC. These results suggested that sensitization to 3-PDC may sensitize three kinds of lymphocytes (e.g., at the 4-, 5-, or 6-position), while sensitization to 4-PDC may sensitize two kinds of lymphocytes (e.g., at 5- or 6-position), as shown in Figure 8.10. However, we do not have enough data to establish this hypothesis yet,

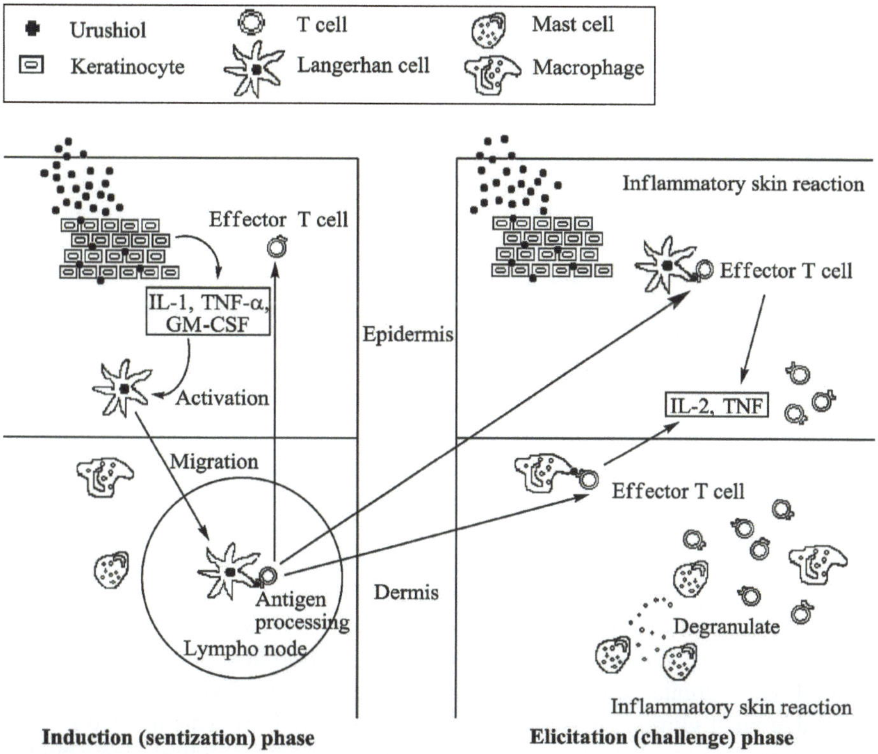

FIGURE 8.12

Induction of sensitization of urushiol allergy.

and further research is required. In fact, several types of urushiol-specific T lymphocytes cell (T-cell) have been recognized.

These results support the assumption that catechol-containing molecules such as urushiol and PDC conjugate with larger proteins via a reactive *O*-quinone intermediate in urushiol-sensitive humans. The urushiol–protein complexes are then recognized as foreign antigens by T cells.

After establishment of sensitization to urushiol, contact dermatitis occurs with the reinvasion of urushiol antigen, as shown in Figure 8.12. On the other hand, sometimes when the human major MHC class II presents antigen information by T lymphocytes, T cells do not react to the antigen as they should, which is called "anergy," and this phenomenon also is considered to be involved in ending contact dermatitis.

8.2 PREVENTION

Urushiol allergy is categorized as a delayed hypersensitivity response on the basis of the time course of the reaction [5]. Many earlier studies reported that CD4+ cells are for delayed-type hypersensitivity reactions effectors in contact dermatitis [7,8], but the results of Gocinski *et al.* [9] and Bour *et al.* [10] showed that CD8+ cells are the major effectors in *in vivo* experiments. In a study of the processing of urushiol hapten by both endogenous and exogenous pathways for presentation to T cells *in vitro*, the processing of urushiol for presentation to CD8+ T cells was inhibited by azide, monensin, and brefeldin A, suggesting that urushiol is processed by the endogenous pathway. On the other hand, presentation of urushiol to CD4+ T cells was inhibited by monensin, but not by brefeldin A, suggesting compatibility with antigen processing by the endosomal pathway [11]. CD8+ cells can differentiate into effector cells in the absence of CD4+ [12]. Lopez *et al.* reported that CD8+ T cells are the effectors of the contact dermatitis induced by urushiol, while CD4+ T cells play an important regulatory role in the onset of the response [13]. A study of cross-reactions between mango contact allergens and urushiol pointed out that the allergenicity of urushiol depends on the number of side-chain double bonds in the third position in catechol; that is to say, a diolefin is more allergenic than a monoolefin in urushiol [14]. Xia *et al.* believed that lipoxygenase plays an important role in the contact dermatitis induced by urushiol [15]. Although many immunology [16–18] and clinical [19–21] studies have attempted to reveal the mechanism of contact dermatitis induced by urushiol, no clear conclusion has been accepted yet.

There are many reports of the treatment of poison ivy and poison oak allergies [22–25]. Because the allergens of poison ivy/oak have the same structure as urushiol, the treatment methods and medicines also can be used to treat dermatitis caused by urushiol. Guin pointed out that barrier creams, protective clothing, and especially disposable vinyl gloves are useful in reducing exposure [26]. This method is effective not only for persons with outdoor occupations such as forest firefighters but also for lacquer workers. Chinese, Japanese, and Korean lacquers contain urushiol, while Vietnamese and China Taiwanese area lacquers contain laccol, and Myanmar, Thai, and Laotian lacquers, thitsiol. In 1590, in the Ming Dynasty of China, Li Shizhen found that China fir soup, purple perilla soup, sealwort soup, and crab soup were useful for the treatment of contact dermatitis induced by lacquer sap [27]. Santucci *et al.* found that when 0.01% aqueous α-methylene-χ-butyrolactone was used to treat a contact dermatitis induced by *Alstroemeria ligtu*, a significant improvement in clinical manifestations was observed [28]. Forest workers in Thailand have discovered an effective medicine made from an extract of boiled teak leaf [29]. Baldwin *et al.* showed that contact sensitivity in response to urushiol was downregulated following treatment with a monoclonal antibody reacting with the allergen urushiol [30]. Yarbrough *et al.* developed a medicine using ethoxylate in combination with sodium lauryl sarcosinate and believed that this combination binds to the available urushiol receptors, rendering them inactive [31]. Although antihistamines, benzocaine derivatives, and corticosteroids are commonly used to treat dermatitis, when applied to the skin, they can cause superimposed forms of contact dermatitis in many people. Thus, if the rash worsens after treatment, the affected person should suspect an allergic reaction to the treatment and see a physician in a timely fashion.

There is an English adage "Leaves of three, let them be." That is to say, avoidance of poison ivy, oak, and sumac is obviously the best method for preventing allergic contact dermatitis. Learning to identify members of the *Toxicodendron* genus is a must. However, for lacquerware workers, direct contact with lacquer sap is inevitable, and can become a serious occupational skin disease. Establishment of spontaneous hyposensitization has been researched by Kligman [2] and Kawai [1]. According to their traditional folk medicine, Koreans believed ingestion of lacquer to be effective for the treatment of gastrointestinal disease or simply be good for health in general, but Park *et al.* [32] showed that ingestion of lacquer can cause severe systemic or cutaneous reaction accompanied by acute itching, even in those not previously known to be allergic to it. For a sensitized person, avoiding direct contact with unhardened resin in furniture and fumes during the process of resin purification and staying away from lacquer trees and sap are recommended. It is necessary for a lacquerware worker to wash exposed areas as soon as possible after exposure to lacquer sap. Most urushiol can be removed with soap and water within 15 min, but after that period the urushiol is chemically bonded to the skin. Urushiol may be transferred to the skin from clothing or tools, so it is important to wash clothing and other materials with soap and water after exposure to urushiol. Because urushiol is oily and not very water soluble, pretreating clothes with alcohol may help remove the urushiol. There is an inexpensive commercial product, Tecnu poison oak-poison ivy cleanser, which is somewhat better than alcohol at removing urushiol.

Toxicodendron dermatitis caused by lacquer has the characteristics of delayed-type hypersensitivity and is known to be the result of sensitization to the structure of a catechol derivative, urushiol. Contact hypersensitivity to urushiol is widespread, resulting in various degrees of reaction. Although there are many studies of contact dermatitis due to urushiol in immunology and cell biology, the allergic mechanism is not yet thoroughly clear. There is currently no special treatment for the contact dermatitis caused by urushiol that provides immediate relief. Avoidance of lacquer trees and lacquer sap is recommended. Workers whose occupations involve lacquer must wear clothing with long sleeves, long pants, and gloves, and wash with soap and water immediately after exposure to urushiol.

REFERENCES

[1] K. Kawai, M. Nakagawa, K. Kawai, K. Konishi, F-M. Liew, H. Yasuno, Y. Shimode, Hyposensitization to urushiol among Japanese lacquer craftsmen. *Contact Dermatitis* 24 (1991) 146–147.

[2] A.M. Kligman, Hyposensitization against *Rhus dermatitis*, Arch. Dermatol. 78 (1958) 47–72.

[3] Y. Ikeda, H. Yasuno, T. Miyakoshi, A. Sato, K. Kawai, Induction of tolerance to pentadecylcatechol in the guinea pig by intravenous injection of 1-O-α-D-glucopyranosyl-3-pentadecylcatechol, Environ. Dermatol. 2 (1995) 27–31.

[4] Y. Ikeda, H. Yasuno, A. Sato, K. Kawai, Oral and epicutaneous desensitization in urushiol contact dermatitis in guinea pigs sensitized by two methods of different sensitizing potency, Contact Dermatitis 39 (1998) 286–292.

[5] L.P. Schornic, P.D. Togni, S. Mariathasan, J. Goellner, J.S. Schoenberger, R.W. Karr, T.A. Ferguson, D.D. Chaplin, Mice deficient in IL-1beta manifest impaired contact hypersensitivity to trinitrochlorobenzone, J. Exp. Med. 183 (1996) 1427–1436.

[6] L. Polak, Immunological aspects of contact sensitivity, Monographs in allergy, 15, Basel, S Karger, 1980.
[7] C. Benezra, G. Ducombs, Y. Sell, J. Foussereau, Immunologic and molecular aspects of plant contact dermatitis: Plant Contact Dermatitis, Decker BC, Inc., Toronto, 1985, pp. 21–28.
[8] H.A. Waldorf, L.J. Walsh, N.M. Schechter, G.F. Murphy, Early cellular events in evolving cutaneous delayed hypersensitivity in humans, Am. J. Pathol. 138 (1991) 477–486.
[9] B.L. Gocinski, R.E. Tigelaar, Roles of CD4+ and CD8+ T cells in murine contact sensitivity revealed by *in vivo* monoclonal antibody depletion, J. Immunol. 144 (1990) 4121–4128.
[10] H. Bour, E. Peyron, M. Gaucherand, J.L. Garrigue, C. Desvignes, D. Kaiserlian, J.P. Revillard, J.F. Nicolas, Major histocompatibility complex class I-restricted CD8+ T cells and class II-restricted CD4+ T cells, respectively, mediate and regulate contact sensitivity to dinitrofluorobenzene, Eur. J. Immunol. 25 (1995) 3006–3010.
[11] S. Kalish, A. Wood, A. Laporte, Processing of urushiol (poison ivy) hapten by both endogenous and exogenous pathways for presentation to T cells *in vitro*, J. Clin. Invest. 93 (1994) 2039.
[12] M. Kalergis, C.B. López, I. Becker, I. Diaz, J. Sein, A. Garbarino, E. Ioannes, Modulation of fatty acid oxidation alters contact hypersensitivity to urushiols: role of aliphatic chain β-oxidation in processing and activation of urushiols, J. Invest. Dermatol. 108 (1997) 57.
[13] B. López, M. Kalergis, I. Becker, A. Garbarino, E. Ioannes, CD8+ T cells are the effectors of the contact dermatitis induced by urushiol in mice and are regulated by CD4+ T cells, Int. Arch. Allergy Immunol. 117 (1998) 194.
[14] K. Oka, F. Saito, T. Yasuhara, A. Sugimoto, A study of cross-reactions between mango contact allergents and urushiol, Contact Dermatitis 51 (2004) 292.
[15] Z. Xia, T. Miyakoshi, T. Yoshida, Lipoxygenase-catalyzed polymerization of phenolic lipids suggests a new mechanism for allergic contact dermatitis induced by urushiol and its analogs, Biochem. Biophys. Res. Commun. 315 (2004) 704.
[16] S. Kalish, L. Johnson, Enrichment and function of urushiol (poison ivy)-specific T lymphocytes in lesions of allergic contact dermatitis to urushiol, J. Immunol. 145 (1990) 3706.
[17] S. Kalish, A. Wood, Induction of hapten-specific tolerance of human CD8+ urushiol (poison ivy)-reactive T lymphocytes, J. Invest. Dermatol. 108 (1997) 253.
[18] T. Wakabayashi, L. Hu, Y. Tagawa, K. Sekikawa, Y. Iwakura, K. Hanada, A. Nakane, IFN-γ and TNF-α are involved in urushiol-induced contact hypersensitivity in mice, Immunol. Cell Biol. 83 (2005) 18.
[19] L. Roberts, An outbreak of contact dermatitis from Japanese lacquer tree, Contact Dermatitis 37 (1997) 237.
[20] H. Oh, R. Haw, H. Lee, Clinical and immunologic features of systemic contact dermatitis from ingestion of *Rhus (Toxicodendron)*, Contact Dermatitis 48 (2003) 251.
[21] W. Myers, M. Newman, B. Katz, B. Gottlieb, Ability to develop *Rhus* allergic contact dermatitis in a patient with psoriasis receiving etanercept, J. Am. Acad. Dermatol. 55 (2006) S127.
[22] L. Epstein, Topical prevention of poison ivy/oak dermatitis, Arch. Dermatol. 125 (1989) 499.
[23] M. Williford, F. Sherertz, Poison ivy dermatitis, Arch. Fam. Med. 3 (1994) 184.
[24] G. Marks, F. Fowler, F. Sherertz, L. Rietschel, Prevention of poison ivy and poison oak allergic contact dermatitis by quaternium-18 bentonite, J. Am. Acad. Dermatol. 33 (1995) 212.

[25] E. Paulsen, P. Christensen, E. Andersen, Dermatitis from common ivy (*Hedera helix* L. subsp. *helix*) in Europe: past, present, and future, Contact Dermatitis 62 (2010) 201.

[26] D. Guin, Treatment of toxicodendron dermatitis (poison ivy and poison oak), Skin Therapy Lett. 6 (2001) 3.

[27] S. Li, Compendium of Materia Medica (Bencao Gangmu), Foreign Language Press, Beijing, 2003.

[28] B. Santucci, M. Picardo, Occupational contact dermatitis to plants, Clin. Dermatol. 10 (1992) 157.

[29] P. Kullavanijaya, S. Ophaswongsi, A study of dermatitis in the lacquerware industry, Contact Dermatitis 36 (1997) 244.

[30] R. Baldwin, J. Clegg, A. Curran, E. Austin, T. Khan, Y. Ma, B. Gunn, F. Hudecz, V. Byers, J. Lepoittevin, M. Price, Regulation of the contact sensitivity response to urushiol with anti-urushiol monoclonal antibody ALG 991, Arch. Dermatol. Res. 291 (1999) 652.

[31] W. Yarbrough, C. Schroeter, Urushiol induced contact dermatitis and method of use, United States Patent, US 6423746 B1, 2002.

[32] S. Park, S. Lee, J. Chun, S. Cha, Clinical features of 31 patients with systemic contact dermatitis due to the ingestion of *Rhus* (lacquer), Br. J. Dermatol. 142 (2000) 937.

CHAPTER

Synthesis of Lacquer

9

The main component of lacquer sap is the lipid component (urushiol, laccol, and thitsiol); it is a catechol derivative having a C15 or C17 unsaturated hydrocarbon chain consisting of a mixture of monoenes, dienes, and trienes mainly at 3- or 4-position of catechol ring. A cross-linking of the lipid component (urushiol, laccol, and thitsiol) takes place with laccase catalysis under the air and humidity to produce an insoluble polymeric film. That is to say, lacquer is an enzyme-catalyzed reaction-type coating material. However, because lacquer is a natural product, the production is very low and price is very high, and synthesis of lacquer derivatives becomes an important research. Of course, the synthesis of lacquer means the synthesis of derivatives of lipid components such as urushiol, laccol, and thitsiol derivatives.

The first purpose of synthesized urushiol is to explain the structure of natural urushiol, especially to confirm the number and position of olefins in unsaturated side chain. The second purpose is to study the mechanism of laccase polymerization and lacquer allergy. The biggest purpose is to attempt to produce lacquer coating material in laboratory. This book will introduce some advanced results in the following section.

9.1 SYNTHESIS OF URUSHIOL DERIVATIVES

In the early study of the structure of lacquer sap, Majiam and Tahara synthesized hydrourushiol from a 2,3-dimethoxyphenylpropionic acid chloride [1], as shown in Scheme 9.1.

About 20 years later, in order to determine the structure of renghol, which exists in Renghas fruit, Backer and Haack successfully synthesized hydrorenghol (hydrourushiol) by a Grignard reaction based on 2,3-dimethoxybenzaldehyde [2], as shown in Scheme 9.2.

With an improvement of the Backer and Haack method, Mason [3] and Dawson et al. [4] synthesized hydrourushiol with a good yield. However, in the demethylation process, strong acid reagents such as dry hydrogen chloride [3], aluminum chloride [4], pyridine hydrochloride [5,6], boron tribromide [7], and iodotrimethylsilane [8] were used. There is no problem deprotecting methyl groups when the urushiol has a saturated side chain, but in the case of urushiol with an unsaturated side chain, polymerization and isomerization of the olefins in the side chain occur and make the demethylation reaction very difficult. Therefore, in the synthesis of urushiol with a

Lacquer Chemistry and Applications. http://dx.doi.org/10.1016/B978-0-12-803589-4.00009-2

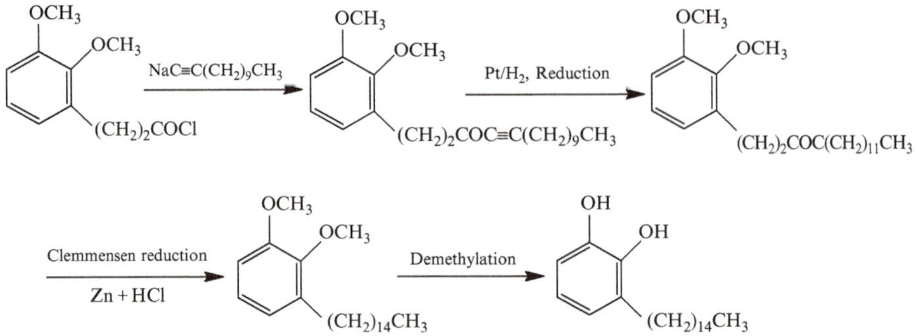

SCHEME 9.1

Synthesis of hydrourushiol (by Majima and Tahara).

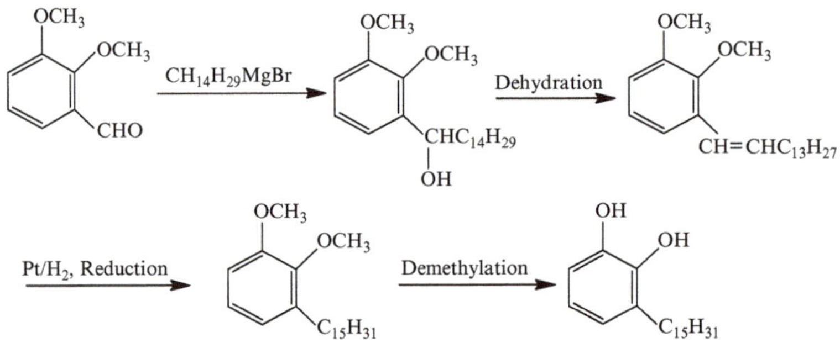

SCHEME 9.2

Synthesis of hydrorenghol (by Backer and Haack).

monoene, diene, and/or triene in the side chain, it is necessary to consider using a protecting group that can maintain the steric structure of olefins. Loev and Dawson [9] selected benzyl ether, which easily deprotects as a hydroxyl-protecting group to synthesize urushiol with a monoene in the side chain. On the other hand, Tyman et al. [10,11] used dimethoxybenzene aldehyde as the starting material to synthesize monoenyl urushiol. The method is to extend the side chain in the aldehyde group to form 3-(7-bromoheptyl)catechol and then react it with 1-octyl lithium acetylene to form monoenyl urushiol, as shown in Scheme 9.3.

The main diene components in urushiol are 3-(8′Z,11′Z-pentadecadienyl)catechol. During the synthesis process, a synthetic pathway must be considered for the position and steric structure of the double bonds in side chain. In addition, it is very important to select the hydroxyl-protecting group in the catechol ring. The Wittig reaction is valid when synthesizing a compound with a complex side chain. In 1990, Bestmann and Hessenauer [12] reported a method for synthesis of 3-(8′Z,11′Z-pentadecadienyl)catechol using ethoxyethyl as the protection group of hydroxyl of the catechol ring. On the other hand, Sargent and Wangchareontrakui [13] also

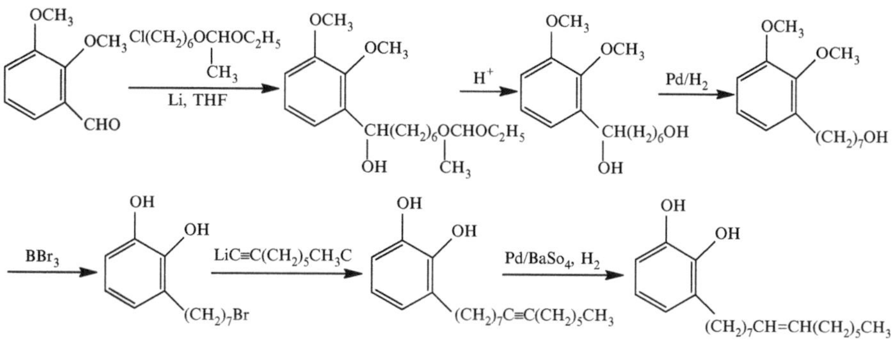

SCHEME 9.3

Synthesis of monoenyl urushiol (by Tyman).

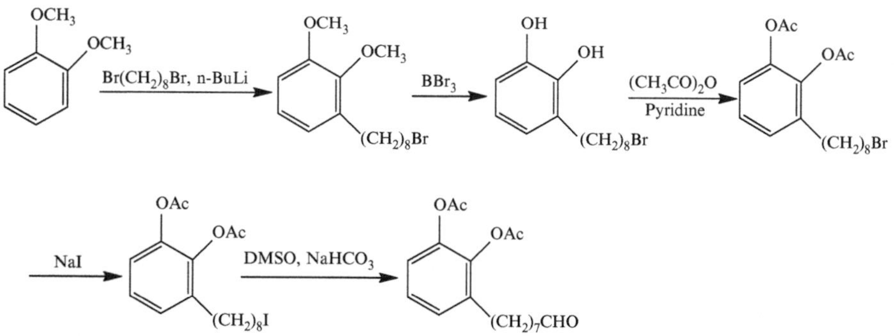

SCHEME 9.4

Synthesis of aromatic part of urushiol.

reported synthesis of 4-(8′Z, 11′Z-heptadecadienyl)catechol, which is the one component of thitsiol, via the Wittig reaction.

Trienyl urushiols are the main components of lacquer sap, and almost all are 3-(8′Z,11′E,13′Z-pentadecatrienyl)catechol, 3-(8′Z,11′E,13′E-pentadecatrienyl)catechol, or 3-(8′Z,11′Z,14-pentadecatrienyl)catechol. These trienyl components play a very significant role in the formation of a lacquer film. Therefore, it is important to establish a method to synthesize trienyl urushiol that can be used as a standard material in the analysis of lacquer film, research on chemical changes during the lacquer drying process, and also in the development of an artificial lacquer. Previously, our laboratory successfully synthesized trienyl urushiol in three steps [14].

First, the aromatic moiety was synthesized using 1,2-dimethoxybenzene as the starting material, as shown in Scheme 9.4.

Second, the unsaturated side chains were synthesized according to their different steric structures [15–17]. Finally, the aromatic moiety and unsaturated side chain were combined by the Wittig reaction to obtain trienyl urushiol as shown in Scheme 9.5.

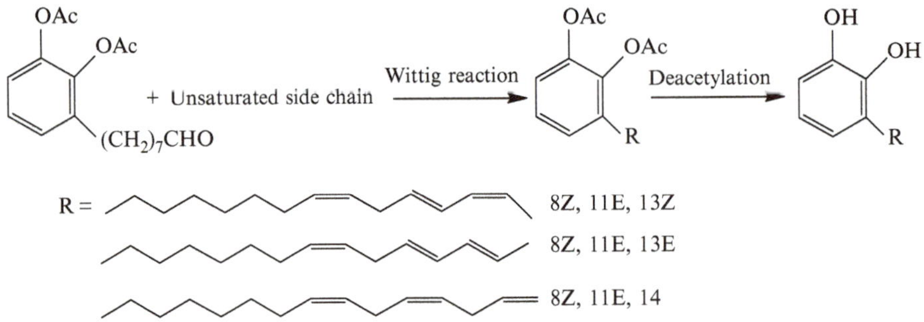

SCHEME 9.5

Synthesis of urushiol by Wittig reaction.

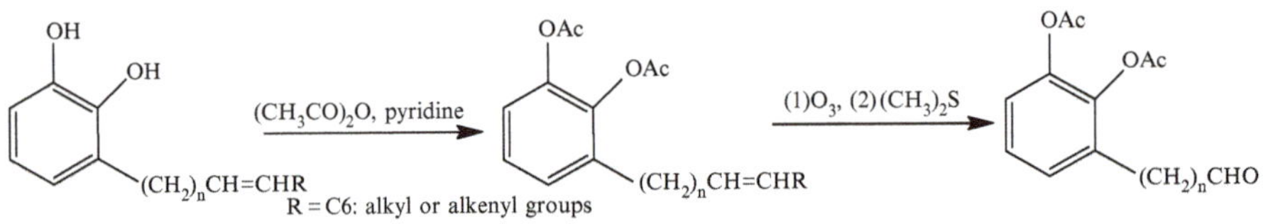

Natural urushiol: compound **7** (n=7)
Natural laccol: compound **10** (n=9)

Compound **8** (n=7)
Compound **11** (n=9)

Compound **9** (n=7)
Compound **12** (n=9)

SCHEME 9.6

Synthesis of aromatic part, compounds **9** and **12**.

In addition, in order to analyze and determine the laccol components from *T. succedanea* lacquer saps, we also systematically synthesized monoenyl, dienyl, and trienyl laccol, and phenol derivatives by the Wittig reaction and the composition and structure of laccol components were identified [18].

Moreover, an efficient one-pot synthesis of substituted catechols from the corresponding phenols has also been reported [19], as shown in Scheme 9.6. However, the product of this method cannot be called an urushiol derivative due to the side chain structures.

As described in Section 3.1 of Chapter 3, urushiol, a major component of the sap tapped from *T. vernicifluum* lacquer tree, is a 3-substituted catechol compound with saturated and unsaturated side chains. The structures of urushiol have been investigated by NMR, HPLC, IR, and GPC measurements and were concluded as shown in Figure 3.5.

Although the urushiol compounds **1**, **3**, **5**, and **6** (Figure 9.1) were synthesized by Halim et al. [8], Bestmann and Hessenauer [12], and Sargent and Wangchareontrakui [13], their synthesis routes are long and the yield is low. In this section, we introduce effective synthesis routes for the monoenyl and dienyl urushiols **1–6** based on adapting the ozonolysis method by the Witting reaction from an ylide derived from a phosphonium salt as the side chain with 3-(8-oxo-1-octyl) catechol diacetate or 3-(10-oxo-1-decyl) catechol diacetate as the aromatic part is followed by the removal of the acetate groups.

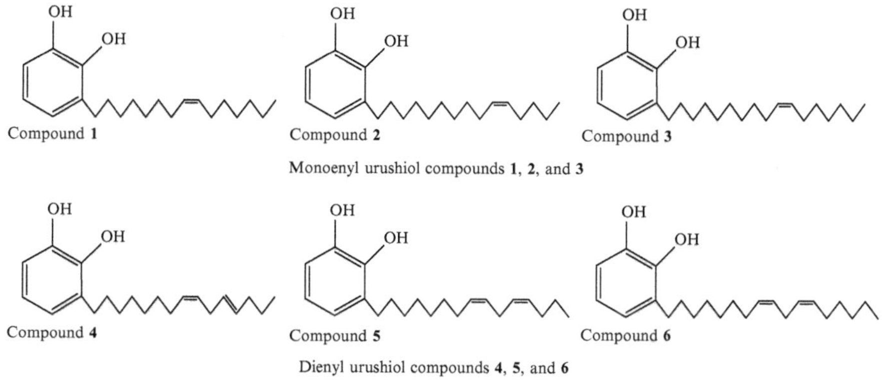

FIGURE 9.1

Monoenyl and dienyl urushiol compounds.

In 1991, we synthesized compound **9**, 3-(8-oxo-1-octyl) catechol diacetate, by a stepwise procedure based on the repeated protection and deprotection of a catechol ring hydroxyl group using veratrol as the starting material [14]. Because the reaction was more complex and yield was not ideal, in this section, we introduce an improved method to synthesize compound **9** by ozonolysis of the double bond of the natural urushiol side chain.

The hydroxyl groups of the urushiol isolated from silica gel column chromatogram were protected by acetic anhydride to give *O*-diacetyl urushiol compound **8**. The double bond of *O*-diacetyl compound **8** was ozonated and the ozonide was then reduced to give 3-(8-oxo-1-octyl) catechol diacetate (compound **9**). Compound **12**, 3-(10-oxo-1-decyl) catechol diacetate, was also synthesized by the same method using natural laccol components that have a double bond at the 10-position of the side chain as the starting material (Scheme 9.6).

Side chain parts, six alkyl- and alkenyl-triphenylphosphonium iodides, were synthesized from various alcohols and carboxylic acid as shown in Scheme 9.7.

Compounds **13** and **14** were synthesized according to the literature [14,15]. The phosphonium salt compound **18** was synthesized by reducing (*E*)-3-heptenoic acid compound **15** with $LiAlH_4$, and iodination of the hydroxyl group of compound **16** and then compound **17** was reacted with triphenylphosphine. Compound **21** was synthesized by iodination of the hydroxyl group of compound **19** to obtain compound **20**, and compound **20** was then reacted with triphenylphosphine. Compound **28** was synthesized from 3-butyne-1-ol 22 as the starting material. After the hydroxyl group of compound **22** was protected by DHP, compound **24** was prepared by the Grignard reaction of compound **23** with n-dipropyl sulfate. Compound **24** was reduced to Z-form with a palladium catalyst to give compound **25**. The THP group of compound **25** was deprotected by *p*-TsOH/MeOH, followed by iodination of the hydroxyl group to give compound **27**, and compound **27** was reacted with triphenylphosphine to obtain compound **28**.

Compounds **1**, **4**, **5**, and **6** were produced by the *cis* selective Wittig reaction of ylides derived from the alkyl- and alkenyl-phosphonium iodide compounds **14**, **18**, **21**, and **28** with aromatic parts of compound **9** and removal of the acetate groups, respectively.

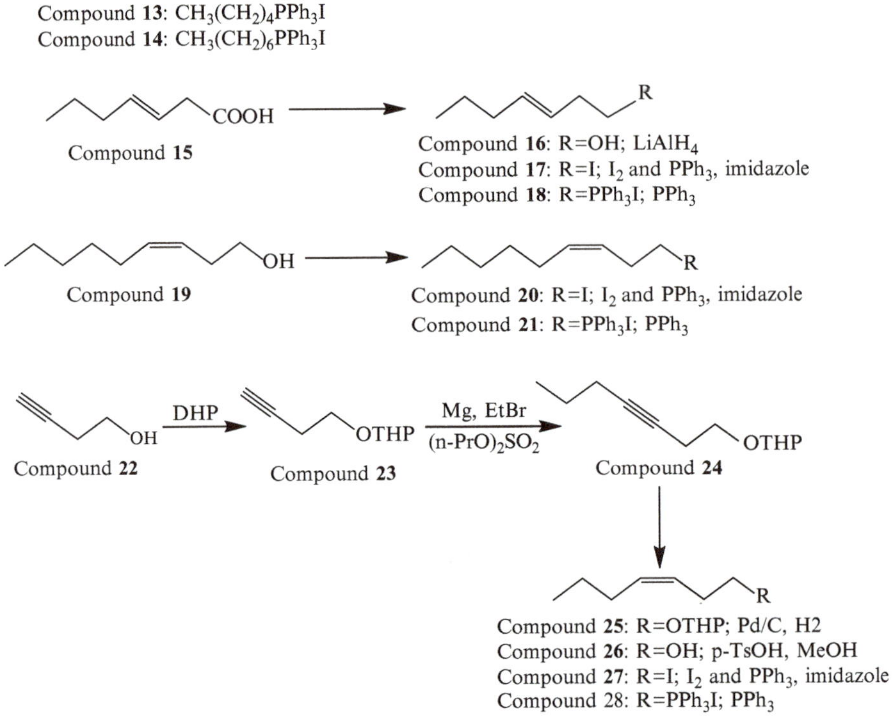

SCHEME 9.7

Synthesis of side chain part from phosphonium salts.

Compound **2** and **3** were synthesized by the *cis* selective Wittig reaction of ylides derived from the alkyl- and alkenyl-phosphonium iodide compounds **13** and **14** with aromatic part of compound **12** and removal of the acetate groups. The conditions of the *cis*-selective Wittig reaction included two methods—a free-salt reaction type [10] and a low-temperature reaction type (at −78°C in THF) [13]. Both Wittig reactions produced a Z double bond more than 80% by GC and NMR analysis. The detailed process and analysis results are as follow.

Compounds **18**, **21**, **23**, **26**, and **28** were synthesized according to our previous report [14]. Compounds **30**–**34** were synthesized according to the Sargent method [13].

3-ALKENYLCATECHOL DIACETATE 8

Compound **8** was synthesized from natural urushiol according to our previously reported method [15], and the yield was 73.3%.

R_f=0.35-0.50 (hexane/ethyl acetate=4:1).

^{1}H NMR: δ (ppm)=2.27(*s*, 3H, −OAc), 2.51(*t*, 2H, J=7.0 Hz, $-CH_2$-Ph), 7.02–7.27 (m, 3H, Ph).

^{13}C NMR: δ (ppm)=168.2, 169.2 (C-15, 17)

3-(8-OXO-1-OCTYL)CATECHOL DIACETATE 9

Ozone was passed into a solution of compound **8** (7.00 g, 19.3–19.7 mmol) and dry methylene chloride (105 ml) at room temperature and then stirred until the thin layer chromatography (TLC) spots of compound **8** disappeared. Dimethyl sulfide (3.80 g, 58.5 mmol) was added to the reaction mixture and stirred for 12 h. The mixture was washed with water and the organic layer was dried on anhydrous Na_2SO_4. After concentration, the residue was purified by column chromatography over silica gel with hexane and ethyl acetate (4:1) to give 3.95 g of 3-(8-oxo-1-octyl) catechol diacetate **9** in 62.6% yield.

3-ALKENYLCATECHOL DIACETATE 11

3-Alkenylcatechol diacetate **11** was synthesized from natural urushiol according to our previous report [15] and the yield was 61.0%.

R_f=0.35-0.45 (hexane/ethyl acetate=4:1).

^{1}H NMR: δ (ppm)=2.26 (s, 3H, –OAc), 2.31 (s, 3H, –OAc), 2.51 (*t*, 2H, J=7.0 Hz, $-CH_2$-Ph), 7.01-7.18 (m, 3H, Ph).

^{13}C NMR: δ (ppm)=168.2, 169.2 (C-17, 19)

3-(8-OXO-1-OCTYL) CATECHOL DIACETATE 12

Compound **12** was synthesized from compound **11** in a manner similar to compound **9** with 64.0% yield.

R_f=0.20 (hexane/ethyl acetate=4:1).

^{1}H NMR: δ (ppm)=1.29 (m, 10H, $-CH_2-\times5$), 1.60 (m, 4H, $-CH_2$–C–Ph, $-CH_2$–C–CHO), 2.27 (s, 3H, –OAc), 2.31(s, 3H, –OAc), 2.41(*t*, J=7.4 Hz, 2H, $-CH_2$-Ph), 7.01–7.26 (m, 3H, Ph), 9.75 (s, 1H, –CHO).

^{13}C NMR: δ (ppm)=20.3, 20.7 [C-18, 20], 22.0–30.0 [C-7-14], 43.6 [C-13], 120.7 [C-3], 126.0 [C-2], 136.4 [C-6], 140.4 [C-5], 168.0, 168.1 [C-17, 19], 202.6 [C-16].

(E)-3-HEPTENE-1-OL 16

Compound **15** (6.99 g, 46.8 mmol) was dissolved in 10.0 ml dry THF and then was added dropwise to a stirred solution of $LiAlH_4$ (2.66 g, 70.2 mmol) in dry THF (10 ml). The mixture was stirred at 70 °C for 5 h. Excess $LiAlH_4$ in the reaction mixture was treated with methanol, and the mixture was poured into a solution of ice water with 1 N HCl. The solution was decanted with ethyl ether, and the organic layer was dried on anhydrous Na_2SO_4. The organic solvent was removed under reduced pressure (20 °C/80 mmHg) to give the crude product. The crude product was purified by column chromatography over silica gel with hexane and ethyl acetate (4:1) to give 3.92 g of (E)-3-heptene-1-ol compound **16** in 73.3% yield.

R_f=0.35 (hexane/ethyl acetate=4:1).

^{1}H NMR: δ (ppm)=0.87-0.92 [*t*, J=7.42, 6.92 Hz, 3H, $-CH_3$], 1.96-2.05 [quarter, J=6.43, 6.92, 8.91 Hz, 2H, $-C{=}C-CH_2-$], 2.23-2.30 [quarter, J=6.93, 2H,

$-CH_2$–C–O–], 3.59–3.65 [*t*, *J*=5.44, 5.94 Hz, 2H, $-CH_2$–O–], 5.33–5.61 [m, 2H, –CH=CH–].

^{13}C NMR: δ (ppm)=13.5 [C 7], 22.5 [C-6], 34.7 [C-5], 35.9 [C-2], 62.0 [C-1], 125.2 [C-3], 133.0 [C-4].

(E)-1-IODO-3-HEPTENE 17

Compound **16** (2.40 g, 21.0 mmol) was added to a solution of triphenylphosphine (8.26 g, 31.5 mmol) with I_2 (6.93 g, 27.3 mmol), imidazole (4.29 g, 63.0 mmol), and benzene (135 ml) and stirred at room temperature for 30 min. The benzene layer was isolated from the viscous yellow product and washed with ethyl ether. After concentration, cold hexane was added to the residue to remove the triphenylphosphine oxide by-product. The hexane solution was filtered and concentrated to give a residue. The residue was purified by column chromatography over silica gel using hexane as the eluent to give 3.17 g of compound **17** in 67.3% yield.

R_f=0.53 (hexane).

^{1}H NMR: δ (ppm)=0.87-0.92 [*t*, *J*=7.42 Hz, 3H, $-CH_3$], 1.32-1.46 [sextet, *J*=6.92, 7.42 Hz, 2H, –C=C–CH_2–], 2.50-2.58 [qua, *J*=6.93 Hz, 2H, $-CH_2$–C–O–], 3.12–3.17 [*t*, *J*=7.42, 2H, $-CH_2$–O–], 5.26–5.61 [m, 2H, –CH=CH–].

^{13}C NMR: δ (ppm)=6.2 [C-1], 13.6 [C-7], 22.4 [C-6], 34.5 [C-5], 36.8 [C-2], 128.4 [C-3], 133.4 [C-4].

(E)-3-HEPTENYLTRIPHENYLPHOSPHONIUM IODIDE 18

Compound **18** was synthesized according to our previous report [14] in 78.8% yield.

^{1}H NMR: δ (ppm)=0.79-0.85 [t, *J*=6.93, 7.42Hz, 3H, $-CH_3$], 1.20-1.33 [sextet, *J*=7.42, 6.93Hz, 2H, –C=C–C–CH_2–], 1.83-1.96 [quartet, *J*=6.43, 6.92, 7.42Hz, 2H, –C=C–CH_2–], 2.36-2.49 [m, 2H, $-CH_2$–C–P], 3.67-3.78 [m, 2H, $-CH_2$-P], 3.67-3.78 [m, 2H, $-CH_2$-P], 5.39-5.60 [m, 2H, –CH=CH-], 7.70-7.99 [m, 15H, Ph3].

^{13}C NMR: δ (ppm)=13.6 [C-7], 22.1 [C-6], 22.9, 23.7 [d, C-1], 25.5, 25.7 [d, C-2], 34.2 [C-5], 128.2 [C-3], 133.8 [C-4], 117.3-135.1 [Ph_3]

(Z)-IODO-3-NONENE 20

Compound **20** was synthesized from compound **19** in the same method as the synthesis of compound **17** in 85.3% yield.

R_f=0.53 (hexane).

^{1}H NMR: δ (ppm)=0.87-0.92 [T, *J*=6.43, 6.92Hz, 3H, $-CH_3$], 1.22-1.39 [m, 6H, $-CH_2$–×3], 1.98-2.05 [quartet, *J*=6.43, 6.93Hz, 2H, –C=C–CH_2–P], 2.59-2.67 [quartet, *J*=6.92, 7.42Hz, 2H, $-CH_2$–C–P], 3.11-3.16 [t, *J*=7.42HZ, 2H, $-CH_2$-P], 5.26-5.59 [m, 2H, –CH=CH–].

^{13}C NMR: δ (ppm)=5.5 [C-1], 14.0 [C-9], 22.5 [C-6], 27.4, 29.2 [C-2, 5], 31.4, 31.5 [C-7,8], 127.7, 132.7 [C-3,4].

(Z)-3-NONENYLTRIPHENYLPHOSPHONIUM IODIDE 21

Compound **21** was synthesized according to our previous report [14] in 88.4% yield.

^{1}H NMR: δ (ppm) = 0.81-0.86 [t, J = 6.43, 7.42Hz, 3H, $-CH_3$], 1.10-1.29 [m, 6H, $-CH_2-\times 3$], 1.74-1.90 [quartet, J = 6.43, 7.41, 2H, $-C{=}C-CH_2-$],2.38-2.51 [m, 2H, $-CH_2-C-P$], 3.65-3.73 [m, 2H, $-CH_2-P$], 5.36-5.61 [m, 2H, $-CH{=}CH-$],7.70-7.87 [m, 15H, Ph_3].

^{13}C NMR: δ (ppm) = 13.9 [C-9], 20.2 [C-2], 22.3 [C-8], 23.0, 23.7 [d, C-1], 27.1 [C-5], 28.3, 31.3 [C-6, 7], 132.8 [C-3], 132.8 [C-4], 117.2-135.2 [Ph_3].

2-(3-BUTYNYLOXY)TETRAHYDRO-2H-PYRAN 23

Compound **23** was synthesized from compound **22** in a similar method according to our previous report [14].

2-(3-BUTYNYLOXY)TETRAHYDRO-2H-PYRAN 24

Ethyl bromide (14.1 g, 129 mmol) was dissolved in 10 ml dry THF and then stirred while slowly adding 3.15 g (129 mmol) magnesium to the solution at room temperature under dry nitrogen. After the magnesium had reacted, a solution of compound **23** (10.0 g, 64.8 mmol) in dry THF (10 ml) was dropped into the mixture at room temperature. The reaction mixture was stirred at 40 °C for 1 h. A solution of di-*n*-propyl sulfate (28.3 g, 155 mmol) in THF (25 ml) and HMPA (20 ml) was added to the reaction mixture at room temperature and continuously stirred for 12 h at 70 °C. Sufficient 3 N HCl was added to the reaction mixture for hydrolysis. After cooling to room temperature, the mixture was extracted with diethyl ether. The organic layer was dried with anhydrous Na_2SO_4 and the organic solvent was removed under vacuum to give the residue. The residue was purified by column chromatography over silica gel with hexane and ethyl acetate (95:5) as eluent to give 6.97 g of 2-(3-heptynylox)tetrahydro-2H-pyran compound **24** in 54.8% yield.

R_f = 0.50 (hexane/ethyl acetate = 95:5).

^{1}H NMR: δ (ppm) = 0.94-0.99 [t, J = 7.17Hz, 3H, $-CH_3$], 1.46-1.76 [m, 8H, $CH_2\times 3$ in pyran, $-C{=}C-C-CH_2-$], 2.09-2.20 [m, 2H, $-C{=}C-CH_2-$], 3.28-3.59 [m, 2H, $-O-CH_2-$ in pyran], 3.72-3.92 [m, 2H, $-O-CH_2-$], 4.60-4.65 [s, 1H, $-O-CH-O-$].

^{13}C NMR: δ (ppm) = 13.6 [C-12], 19.2 [C-3], 20.3 [C-10], 20.9 [C-7], 22.7 [C-11], 25.3 [C-2], 30.4 [C-4], 62.2 [C-6], 65.6 [C-1], 69.3 [C-9], 81.2 [C-8], 98.8 [C-5].

(Z)-2-(3-HEPTENYLOXY) TETRAHYDRO-2H-PYRAN 25

Compound **24** (5.88 g, 30.0 mmol) was added to a solution with a Pd catalyst [Pd (5%)-$CaCO_3$-PbO (3%), 0.67 g] in 60 ml methanol. The reaction flask was connected to a hydrogen tank and was alternately evacuated and filled with hydrogen. The mixture was stirred until the hydrogen ceased to be absorbed. The mixture was filtered to remove the Pd catalyst. After concentration, the residue was purified by column chromatography over silica gel using hexane and ethyl acetate (95:5) as eluent to give 4.61 g of (Z)-2-(3-heptenyloxy) tetrahydro-2H-pyran compound **25** in 77.5% yield.

R_f=0.67 (hexane/ethyl acetate=95:5).

^{1}H NMR: δ (ppm)=0.88-0.93 [t, J=7.42Hz, 3H, $-CH_3$], 1.26-1.48 [quartet, J=6.92, 7.42Hz, 2H, −C−C C CH_2−], 1.50-1.90 [m, 6H, $-CH_2-\times 3$], 2.00-2.10 [quartet, J=6.43, 7.92Hz, 2H, $-C{=}C{-}CH_2-$], 2.32-2.39 [quartet, J=6.43, 6.92Hz, 2H, $-C{=}C{-}CH_2-$], 3.42-3.75 [m, 2H, $-O{-}CH_2-$ in pyran], 3.50-3.90 [m, 2H, $-O{-}CH_2-$], 4.6 [t, 1H, −O–CH–O–], 5.35-5.50 [m, 2H, −CH=CH–].

^{13}C –NMR: δ (ppm)=13.7 [C-12], 19.2 [C-3], 22.7 [C-11], 25.3 [C-2], 27.9 [C-10], 29.3 [C-7], 30.4 [C-4], 62.1 [C-6], 67.0 [C-1],98.6 [C-5], 125.0 [C-9], 131.6 [C-8].

(Z)-3-HEPTENE-1-OL 26

Compound **26** was synthesized from compound **25** by the modified procedure [14] with 91.4% yield.

R_f=0.38 (hexane/ethyl acetate=4:1).

^{1}H NMR: δ (ppm)=0.88-0.94 [t, J=7.42Hz, 3H, $-CH_3$], 1.32-1.45 [M, J=7.42Hz, 2H, $-C{=}C{-}C{-}CH_2-$], (s, 1H, −OH)], 2.01-2.09 [quartet, J=6.43, 6.93, 7.42Hz, 2H, $-C{=}{-}C{-}CH_2-$], 2.30-2.37 [quartet, J=6.43, 7.42Hz, 2H, $-C{=}C{-}CH_2-$], 3.62-3.67 [t, J=6.43Hz, 2H, $-CH_2{-}O-$], 5.53-5.6 [m, 2H, −CH=CH–].

^{13}C NMR: δ (ppm)=13.8 [C-6], 29.1 [C-5], 30.7 [C-2], 62.1 [C-1], 125.2 [C-4], 133.0 [C-3].

(Z)-1-IODO-3-HEPTENE 27

Compound **27** was synthesized from compound **26** in a similar method as the synthesis of compound **17** with 64.6% yield.

R_f=0.70 (hexane).

^{1}H NMR: δ (ppm)=0.86-0.91 [t, J=6.92, 7.42Hz, 3H, $-CH_3$], 1.37-1.46 [m, J=6.92, 7.42Hz, 2H, $-C{=}C{-}C{-}CH_2-$], 1.97-2.05 [quartet, J=6.93, 2H, $C{=}C{-}CH_2$], 2.60-2.67 [quartet, J=6.92, 7.42Hz, 2H, $-C{=}C{-}CH_2-$], 3.10-3.15 [t, J=7.42Hz, 2H, $-CH_2{-}P$], 5.26-5.58 [m, 2H, −CH=CH–].

^{13}C NMR: δ (ppm)=5.7 [C-1], 13.8 [C-7], 22.8 [C-6], 29.1 [C-5], 30.7 [C-2], 125.8 [C-4], 133.0 [C-3].

(Z)-3-HEPTENYLTRIPHENYLPHOSPHONIUM IODIDE 28

Compound **28** was synthesized according to our previous report [14] with 81.3% yield.

^{1}H NMR: δ (ppm)=0.76-0.84 [t, J=7.42Hz, 3H, $-CH_3$], 1.20-1.33 [m, J=7.42, 6.92Hz, 2H, $-C{=}C{-}C{-}CH_2-$], 1.73-1.90 [quartet, J=7.42Hz, 2H, $-C{=}C{-}CH_2-$], 2.39-2.50 [m, 2H, $-C{=}C{-}CH_2-$], 3.63-3.72 [m, 2H, $-CH_2{-}P$], 5.35-5.63 [m, 2H, −CH=CH–], 7.71-7.87 [m, 15H, $P{-}PPh_3$].

^{13}C NMR: δ (ppm)=13.6 [C-7], 19.7, 19.8 [d, C-2], 21.8 [C-6], 22.6-23.3 [d, C-1], 28.7 [C-5], 127.8 [C-3], 132.0 [C-4], 117.2-135.2 [Ph_3].

(Z)-O-DIACETYL-3-(8-PENTADECENYL) CATECHOL 29

Compound **29** was synthesized from compounds **9** and **14** in a method similar to our previous report with 45.7% yield [15].

R_f=0.50 (hexane/ethyl acetate=4:1).

^{1}H NMR: δ (ppm)=0.86-0.90 [t, 3H, $-CH_3$], 1.30 [m, 14H, $-CH_2$-×7], 1.58 [m, 2H, Ph-C–CH_2–], 2.02 [m, 4H, –C=C–CH_2–×2], 2.35, 2.36 [s, 6H, $-COCH_3$×2], 2.62 [t, 2H, Ph-CH_2–], 3.53 [m, 2H, –CH=CH–], 6.90-7.19 [m, 3H, Ph].

^{13}C NMR: δ (ppm)=13.7 [C-22, 24], 22.3, 28.5-31.4 [C-8-12, 17–19], 26.8, 32.2 [C-13,16], 30.1 [C-7], 129.9, 130.1 [C-14, 15], 120.4 [C-3], 125.8 [C-1], 126.9 [C-2], 136.3 [C-6], 140.1 [C-5], 142.0 [C-4], 167.8, 167.9 [C-22, 24].

(Z)-O-DIACETYL-3-(10-PENTADECENYL) CATECHOL 30

Compound **30** was synthesized according to our previous report [13] with 33.0% yield.

R_f=0.40 (hexane/ethyl acetate=4:1).

^{1}H NMR: δ (ppm)=0.88-0.89 [t, J=3.40, 3H, $-CH_3$], 1.27-1.34 [m, 16H, $-CH_2$-×8], 1.52-1.60 [m, 2H, $-CH_2$–], 1.99-2.01 [m, 4H, –C=C–CH_2–×2], 2.31, 2.33 [s, 6H, $-COCH_3$×2], 2.56-2.64 [m, 7.91Hz, 2H, Ph-CH_2–], 5.31-5.39 [m, 2H, –CH=CH–], 6.74-7.00 [m, 3H, Ph].

^{13}C NMR: δ (ppm)=14.0 [C-21], 20.6, 21.0 [C-23, 25], 26.9, 27.2 [C-15,18], 29.3-29.5 [C-8-14, 19, 20], 29.9 [C-7], 120.1 [C-3], 126.5 [C-1], 127.3 [C-2], 129.8, 129.9 [C-16,17], 135.9 [C-6],138.3 [C-5], 141.9 [C-4], 169.6, 169.7 [C-22, 24].

(Z)-O-DIACETYL-3-(10-PENTADECENYL) CATECHOL 31

Compound **31** was synthesized according to our previous report [13] with 23.0% yield.

R_f=0.40 (hexane/ethyl acetate=4:1).

^{1}H NMR: δ (ppm)=0.85-0.90 [t, J=6.43, 6.92Hz, 3H, $-CH_3$], 1.15-1.30 [m, 20H, $-CH_2$–×10], 1.54-1.58 [m, 2H, $-CH_2$–], 1.98-2.02 [m, 4H, –C=C–CH_2–×2], 2.25, 2.30 [s, 6H, $-COCH_3$×2], 2.48-2.53 [t, J=7.92, 7.91Hz, 2H, Ph–CH_2–], 5.33-5.38 [m, 2H, –CH=CH–], 6.91-7.20 [m, 3H, Ph].

^{13}C NMR: δ (ppm)=14.1 [C-1], 20.5, 20.6 [C-25, 27], 22.6-29.7 [C-9-12, 17–22], 27.1 [C-13, 16], 29.9 [C-8], 30.0 [C-7], 120.7 [C-3], 125.8 [C-1], 127.2 [C-2], 129.9, 131.5 [C-14,15], 136.6 [C-6], 140.4 [C-5], 142.2 [C-4], 168.3, 168.5 [C-22, 24].

O-DIACETYL-3- [(8Z,11E)-8,11-PENTADECENYL] CATECHOL 32

Compound **32** was synthesized according to our previous report [13] with 30.0% yield.

R_f=0.50 (hexane/ethyl acetate=4:1).

^{1}H NMR: δ (ppm)=0.85-0.91 [t, J=7.42, 6.92Hz, 3H, $-CH_3$], 1.17-1.41 [m, 10H, $-CH_2$–×5], 1.53-1.58 [t, J=6.93, 7.42Hz, 2H, Ph–C–CH_2–], 1.80-2.04 [m, 4H,

–C=C–CH_2–×2], 2.28, 2.31 [s, 6H, –$COCH_3$×2], 2.48-2.54 [t, *J*=7.42, 7.91Hz, 2H, Ph–CH_2–], 2.70-2.74 [t, *J*=5.44, 4.95Hz, 2H, –C=C–CH_2–C=C–], 5.33-5.43 [m, 4II, –CII=CII–×2], 7.01-7.18 [m, 3II, Ph].

^{13}C NMR: δ (ppm)=13.7 [C-21], 20.3, 20.7 [C-23, 25], 22.6, 29.1, 29.3, 29.4, 29.6, 29.8 [C-8~12, 20],27.1 [C-13], 30.1 [C-7], 30.4 [C-16], 34.6 [C-19], 120.8 [C-3], 126.2 [C-1], 127.2 [C-2], 136.7 [C-6], 140.5 [C-5], 142.4 [C-4], 127.8, 128.5, 130.3, 130.6 [C-14, 15, 17, 18], 168.2, 168.4 [C-22, 24].

O-DIACETYL-3-[(8Z,11Z)-8,11-PENTADECENYL] CATECHOL 33

Compound **33** was synthesized according to our previous report [13] with 30.0% yield.

R_f=0.43 (hexane/ethyl acetate=4:1).

^{1}H NMR: δ (ppm)=0.86-0.91 [t, *J*=6.43, 6.93Hz, 3H, –CH_3], 1.25-1.35 [m, 12H, –CH_2–×6], 1.40-1.56 [m, 4H, –CH_2–×2], 2.01-2.08 [quartet, *J*=6.43, 4H, –C=C–CH_2–×2], 2.28, 2.31 [s, 6H, –$COCH_3$×2], 2.48-2.53 [t, *J*=7.42, 7.91Hz, 2H, Ph–CH_2–], 2.75-2.79 [t, *J*=5.44, 5.93Hz, 2H, –C=C–CH_2–C=C–], 5.30-5.40 [m, 4H, –CH=CH–×2], 7.01-7.18 [m, 3H, Ph].

^{13}C NMR: δ (ppm)=14.1 [C-23], 20.3, 20.7 [C-25, 27], 25.6 [C-16], 27.2 [C-13, 19], 29.8 [C-7, 8], 22.6, 29.2, 29.3, 29.3, 29.4, 29.6, 30.1 [C-9~12, 20–22], 127.9, 128.0, 130.1, 130.2 [C-14, 15, 17, 18], 120.8 [C-3], 126.2 [C-1], 127.2 [C-2], 136.7 [C-6], 140.5 [C-5], 142.4 [C-4], 168.4, 169.0 [C-24, 26].

O-DIACETYL-3-[(8Z,11Z)-8,11-PENTADECENYL] CATECHOL 34

Compound **34** was synthesized according to our previous report [13] with 55.8% yield.

R_f=0.48 (hexane/ethyl acetate=4:1).

^{1}H NMR: δ (ppm)=0.85-1.00 [t, *J*=7.42Hz, 3H, –CH_3], 1.22-1.45 [m, 10H, –CH_2–×5], 1.56 [m, 2H, –CH_2–], 2.00-2.07 [m, 4H, –C=C–CH_2–×2], 2.27-2.31 [s, 6H, –$COCH_3$×2], 2.48-2.56 [t, *J*=7.91 Hz, 2H, Ph–CH_2–], 2.75-2.79 [t, *J*=5.44Hz, 2H, –C=C–CH_2–C=C–], 5.32-5.43 [m, 4H, –CH=CH–×2], 6.98-7.18 [m, 3H, Ph].

^{13}C NMR: δ (ppm)=13.8 [C-21], 20.3, 20.7 [C-23, 25], 25.6 [C-16], 27.2 [C-13], 29.2 [C-19], 29.3 [C-8], 30.0 [C-7], 22.8, 27.1, 29.4, 29.6, 29.8 [C-9-12, 20], 127.9, 128.1, 129.9, 130.1 [C-14, 15, 17, 18], 120.8 [C-3], 126.1 [C-1], 127.2 [C-2], 136.7 [C-6], 140.5 [C-5], 142.4 [C-4], 168.2, 168.4 [C-22, 24].

(Z)-3-(8-PENTADECENYL) CATECHOL 1

2N Sodium methoxide (1.60mL, 3.20mmol) was added to a solution of compound **29** in methanol (41.0ml) at 25°C and stirred for 1h. Then, cation exchange resin was added to the solution until it became neutral. The solution was filtered to remove the resin and concentrated in vacuo to give the residue. The residue was purified by column chromatography over silica gel with hexane and ethyl acetate (4:1) as eluent to give 0.36g of compound **1** with 61.5% yield.

(Z)-3-(10-PENTADECENYL) CATECHOL 2

Compound **2** was synthesized by a similar method to compound **1**, and the yield was 43.5%.

^{1}H NMR: δ (ppm)=0.86-0.91 [t, J=6.7Hz, 3H, $-CH_3$], 1.31-1.39 [m, 16H, $-CH_2-\times 8$], 1.54-1.62 [m, 2H, Ph–C–CH_2], 1.98-2.02 [m, 4H, $-C=C-CH_2-\times 2$], 2.55-2.61 [t, J=7.9Hz, 2H, Ph–CH_2–], 5.29-5.41 [m, 2H, –CH=CH–], 5.60 [s, 2H, $-OH\times 2$], 6.71 [s, 3H, Ph].

^{13}C NMR: δ (ppm)=13.9 [C-21], 22.3-31.9 [C-7~14, 19–20], 27.2 [C-15], 31.9 [C-18], 112.9 [C-3], 120.1 [C-1], 122.1 [C-2], 129.4 [C-6], 129.8 [C-16, 17], 141.8 [C-5], 142.9 [C-4].

(Z)-3-(10-HEPTADECENYL) CATECHOL 3

Compound **3** was synthesized by a similar method to compound **1**, and the yield was 69.5%.

3-[(8Z,11E)-8,11-PENTADECENYL] CATECHOL 4

Compound **4** was synthesized by a similar method to compound **1**, and the yield was 55.0%.

R_f=0.44 (hexane/ethyl acetate=4:1).

^{1}H NMR: δ (ppm)=0.86-0.90 [t, J=7.33Hz, 3H, $-CH_3$], 1.24-1.43 [m, 10H, $-CH_2-\times 5$], 1.58-1.66 [m, 2H, Ph–C–CH_2–], 1.92-2.01 [m, 4H, $-C=C-CH_2-\times 2$], 2.58-2.61 [t, J=7.78Hz, 2H, Ph–CH_2–], 2.70-2.74 [t, J=5.34Hz, 2H, $-C=C-CH_2-C=C-$], 5.16 [s, 2H, –OH], 5.32-5.49 [m, 4H, $-CH=CH-\times 2$], 6.71 [s, 3H, Ph].

^{13}C NMR: δ (ppm)=13.7 [C-21], 22.9, 29.2, 29.4, 29.5, 29.6, 29.8 [C-8~12, 20], 29.2 [C-13], 29.8 [C-7], 30.4 [C-16], 34.6 [C-19], 112.9 [C-3], 120.1 [C-1], 122.1 [C-2], 129.3 [C-6], 141.8 [C-5], 142.9 [C-4], 127.8, 128.5, 130.4, 130.6 [C-14, 15, 17, 18].

3-[(8Z,11Z)-8,11-HEPTADECENYL] CATECHOL 5

Compound **5** was synthesized by a similar method to compound **1**, and the yield was 60.0%. Compound **5** was identified by comparison with the previously reported standard product [13].

3-[(8Z,11E)-8,11-PENTADECENYL] CATECHOL 6

Compound **6** was synthesized by a similar method to compound **1** and the yield was 50.0%.

R_f=0.35 (hexane/ethyl acetate=4:1).

^{1}H NMR: δ (ppm)=0.88-0.94 [t, J=7.42Hz, 3H, $-CH_3$], 1.33-1.42 [m, 10H, $-CH_2-\times 5$], 1.55-1.66 [m. 2H, Ph–C–CH_2–], 2.00-2.07 [quartet, J=6.60Hz, 4H, $-C=C-CH_2-\times 2$], 2.57-2.63 [t, J=7.92Hz, 2H, Ph–CH_2–], 2.76-2.80 [t, J=5.44Hz, 2H,

–C=C–CH_2–C=C–], 4.99, 5.10 [s, 2H, –OH×2], 5.30-5.43 [m, 4H, –CH=CH–×2], 6.71 [s, 3H, Ph].

^{13}C NMR: δ (ppm) = 13.8 [C-21], 25.6 [C-16], 27.2 [C-13], 29.2 [C-19], 29.3 [C-8], 29.7 [C-7], 22.8, 29.4-29.6 [C-9 ~ 12. 20], 112.9 [C-3], 120.1 [C-1], 122.1 [C-2], 129.3 [C-6], 141.8 [C-5], 143.0 [C-4], 127.9, 128.1, 129.9, 130.1 [C-14, 15, 17, 18].

In addition, we have studied the synthesis of alkenylcatechol for a long time and obtained many good results. The following will introduce mainly the synthesis of 4-alkenylcatechol, an urushiol derivative, that was employed in our laboratory.

9.2 SYNTHESIS OF 4-ALKENYLCATECHOL

Urushiol is a catechol derivative having a C15 unsaturated hydrocarbon chain consisting of a mixture of monoenes, dienes, and trienes mainly at the 3- or 4-position of the catechol. The cross-linking of urushiol is catalyzed by laccase in humid air to produce an insoluble polymeric film as described in Chapter 1. Due to the prominence of the strong protective action in natural products and in order to confirm its chemical structure and explain the mechanism of laccase polymerization and lacquer allergy, the main component of lacquer, urushiol, was actively synthesized in our laboratories. The majority of the synthetic routes to the cardol and cardanol derivatives rely on the deprotection reaction of the methoxyl and acetyl groups. Because the yield of these synthesis methods is very low, an improvement in the yield or reaction pathway is needed.

The previous synthesis of 4-alkenylcatechol was performed as follows: a natural unsaturated fatty acid was reduced and alcoholized and then catalyzed by a Friedel-Craft catalyst to introduce the unsaturated fatty acid into the catechol ring. However, the carbon double bond of the fatty acid attacked the hydroxyl group of the catechol ring to cause an additional reaction, and the yield was predominantly an ether-type compound. Moreover, the side chain will diverge at the position of the double bond with an alkyl group. Therefore, the subsequent synthesis used the wrasse trawl for the starting material, but the operation is difficult and the cost is high because it is necessary to use BBr_3 for the deprotection of the methoxyl group reaction. For this reason, we recently developed a reaction that uses the easy deprotection of the hydroxyl group in the catechol ring using an agent such as cardanol by an easy two-step synthesis procedure with a high yield. The synthetic route is shown in Scheme 9.8.

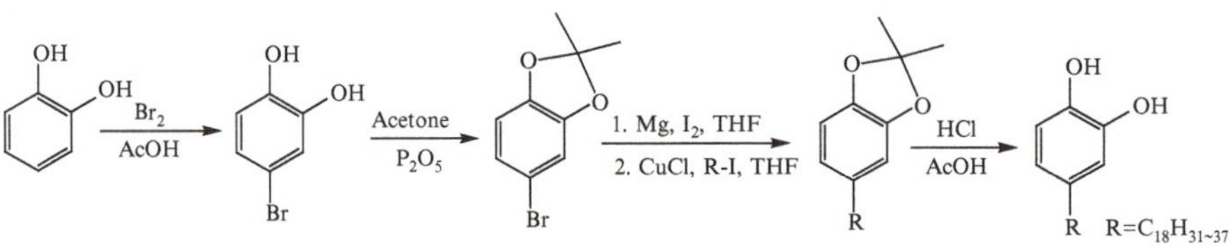

SCHEME 9.8

Synthesis of C18: 4-alkenylcatechols.

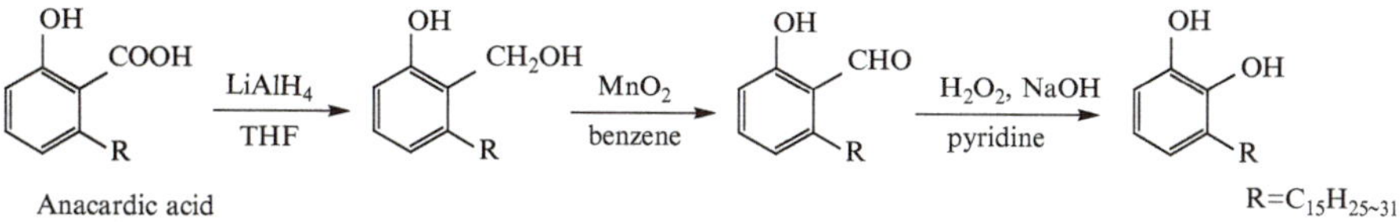

SCHEME 9.9

Synthesis of C15; 3-alkenylcatechols.

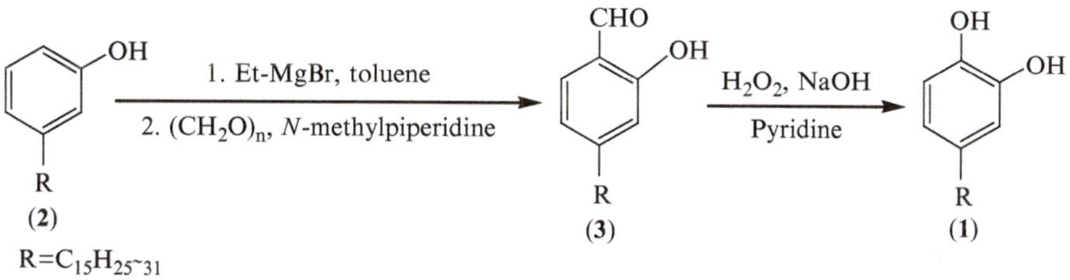

SCHEME 9.10

Synthesis of C_{15}; 4-alkenylcatechols.

However, in this reaction, the subreaction takes priority during the crossing coupling, thus making the yield very low.

The other method, which is the synthesis of the alkenylcatechol from a cashew nut shell liquid (CNSL), was also carried out in our lab [14] as shown in Scheme 9.9.

As described above, the synthesis of 3-alkenylcatechol from CNSL has been reported. However, although the synthesis for the 4-alkenylcatechol was carried out, to the best of our knowledge, only the isomer was produced. We subsequently achieved the synthesis of 4-alkenylcatechol (1) from renewable resources such as cardanol (2), and the detailed synthesis method is described in this section, and the synthetic route is shown in Scheme 9.10.

4-Alkenylcatechol was synthesized from cardanol by an easy two-step synthesis procedure. 4-Alkenylcatechol was prepared by formylation of cardanol with paraformaldehyde and ethylmagnesium bromide in THF, followed by Dakin oxidation with hydrogen peroxide and aqueous NaOH with a 70% total yield (Scheme 9.10). From this synthesis of 4-alkenylcatechol (1), 3-alkenylcatechol was obtained as a by-product with about 4% yield. It can be said that the synthesis of 4-alkenylcatechol from the CNSL in this research is a new reaction. Moreover, because it is a method that uses a comparatively cheap reagent and produces a high yield in a short reaction time, it can be said that it is also a very useful synthesis method from an industrial viewpoint. The detailed process is as follows.

9.2.1 SYNTHESIS OF 4-ALKENYL-2-HYDROXYBENZALDEHYDE

Toluene (40 ml) was added to a 3-neck flask containing nitrogen. Ethyl magnesium bromide (3.70 ml, 11.00 mmol) was added to the flask using a syringe, then cardanol

(3.00 g, 9.97 mmol) was slowly dropped into the flask using a pressure dropping funnel at room temperature, and the reaction occurred. After all the cardanol was added, N-methylpiperidine (1.50 g, 15.13 mmol) and formaldehyde (1.20 g, 40.00 mmol) were added and heated in an oil bath of 100 °C for 0.5 h.

After the reaction had finished, the reaction system was cooled to room temperature. Then 5 ml of water and 5 ml of hydrochloric acid were added and stirred for 10 min to react with the aqueous solutions completely. The crude products were extracted with ether and then dried using anhydrous sulfuric acid sodium. The products were isolated and purified by silica gel column chromatography eluted with a mixture of hexane and ethyl acetate at 9:1 (R_f=0.70) to give 4-alkenyl-2-hydroxybenzaldehyde (2.83 g, 8.61 mmol 86.0%, Scheme 9.11).

Methods for synthesis of aldehydes from phenols have been reported. The typical methods include the Vilsmeier-Haack reaction (DMF, $POCl_3$) and the Duff reaction (hexamethylenetetramine in a strong acid). However, the severe reaction conditions, low yield, easily obtained diformylation products, easy attack of the double bond of the cardanol side chain, and bad selection have limited their use in this research. Therefore, the formylation reaction was carried out in this study (Scheme 9.11). It is a useful reaction for the synthesis of the *o*-formylation of cardanol to obtain an aldehyde in a short time and high yield. The reaction does not occur in the phenols with an electronic acceptor group like NO_2 and COMe, but occurs in the phenols, which have an alkyl substitute like cardanol.

The positional selection reaction of the *o*-formylation of cardanol occurs first in a solvent with low permittivity, such as toluene and xylene, at a high reaction temperature, and the phenol hydroxyl with a high reactivity was controlled; then the chelate of a magnesium reagent was formed using the cations of the Grignard reagent ($MgBr^+$) such as ethyl magnesium bromide, which selectively produces the reaction in the ortho position of cardanol. The formaldehyde that became a carbonium cation electron attacks the magnesium salt formed, and the aroma aldehydes that have the formyl group in the ortho position via the oxidation and reduction process exist as a tertiary amine. The tertiary amine plays a very important role in the forming of *o*-formaldehyde. If the tertiary amine did not exist, the main products would be the couplings of two phenols. The mechanism is shown in Scheme 9.12. This repeated reaction in the faction causes pre-polymerization, similar to the mechanism with the formaldehyde of CNSL.

The target compound was obtained via *o*-formylation by the reaction mechanism shown in Scheme 9.13 with the existence of the tertiary amine. Although the

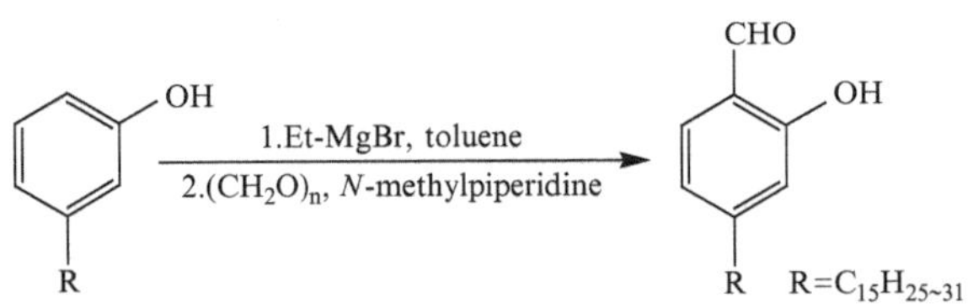

SCHEME 9.11

Synthesis of 4-alkenyl-2-hydroxybenzaldehyde.

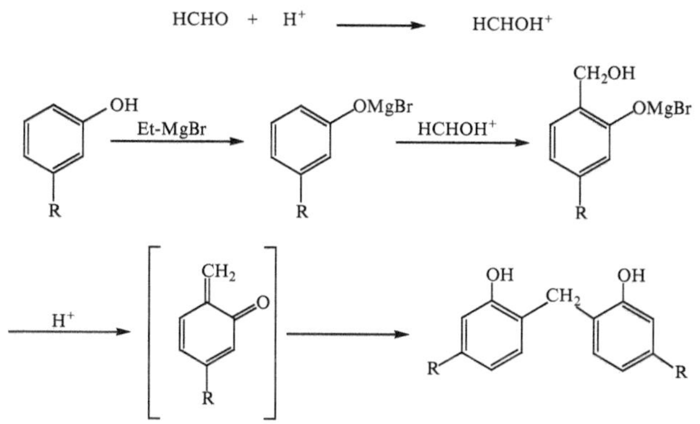

SCHEME 9.12

Reaction mechanism of cardanol with formaldehyde.

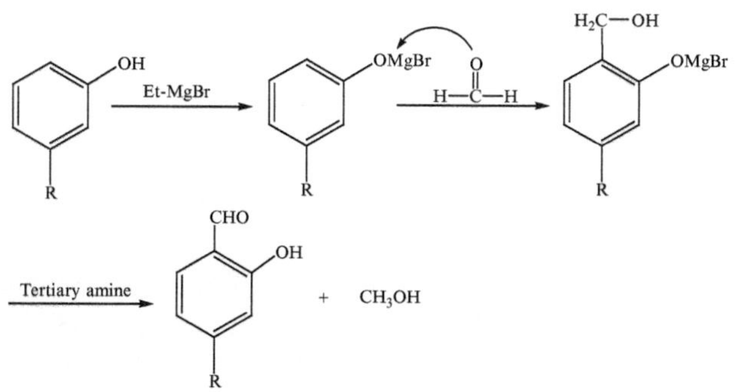

SCHEME 9.13

Mechanism of *o*-formylation of cardanol with tertiary amine.

formaldehyde changed to the carbonyl cation to form the hydroxyl benzyl alcohol at the ortho and/or para positions of cardanol, the main product at a high temperature (100 °C) was the ortho position. The tertiary amine can decrease the acidity of magnesium to decrease the formation of the *p*-quinonmethyl medium, and 4-alkenyl-2-hydroxybenzaldehyde was obtained.

Figure 9.2 is the ^{1}H NMR spectrum of 4-alkenyl-2-hydroxybenzaldehyde. The signal at 9.83 ppm is due to a formyl group, at 11.04 ppm to H at the ortho position, at 2.62 ppm for the methylene group attached to the 4-position, at 1.57 ppm for a methylene group next to the benzylic position, at 1.95–2.08 ppm for the methylene group next to the double bond, and at 2.75-2.85 ppm for the methylene group between the two double bonds. In addition, the peak at 5.30-5.46 ppm is the *cis* structure of a double bond, at 4.96-5.00 ppm (2H; four pile line of the doublet), and at 5.78-5.86 ppm (1H; multiplet) for the double bond structure in the terminal of cardanol side chain, and the signals at 6.80, 6.83, and 7.45 ppm are due to the protons at the 3-, 5-, and 6-positions of the benzene ring.

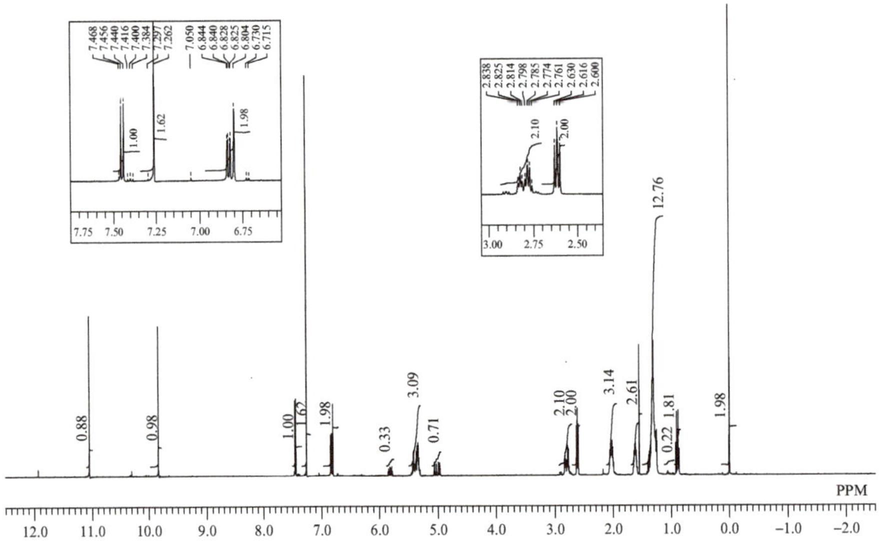

FIGURE 9.2

^1H NMR spectrum of 4-alkenyl-2-hydroxybenzaldehyde.

Figure 9.3 is the mass chromatogram of 4-alkenyl-2-hydroxybenzaldehyde. The side chain showed (1) C15:3 (m/z 326), (2) C15:2 (m/z 328), (3) C15:1 (m/z 330), and (4) C15:0 (m/z 332) determined from the mass spectra for the molecular ion peak of A: m/z=326, B: m/z=328, C: m/z=330, and D: m/z=332, respectively in Figure 9.4. So, the side chain confirmed the existence of C15:3 (43.64%), C15:2 (33.96%), C15:1 (17.09%), and C15:0 (5.30%). From these results, it can be concluded that 4-alkenyl-2-hydroxybenzaldehyde was synthesized without affecting the double bond of the cardanol side chain.

9.2.2 SYNTHESIS OF 4-ALKENYLCATECHOL FROM 4-ALKENYL-2-HYDROXYBENZALDEHYDE

4-Alkenyl-2-hydroxybenzaldehyde (1.65 g, 5.02 mmol), NaOH (0.19 g), and H_2O/THF (9:1, 16 ml) were added to a 100-ml 3-neck flask and stirred. Then 30% H_2O_2 (0.2 ml) was slowly dropped into the flask using a pressure dropping funnel at room temperature and continuously stirred for 4 h.

After the reaction was finished, 10 ml of 10% hydrochloric acid was added to the reaction system and continuously stirred at room temperature for 30 min. The crude products were extracted with ether and dried using anhydrous sulfuric acid sodium. The products were isolated by silica gel column chromatography and eluted with the mixture of hexane and ethyl acetate with the ratio of 7:3 (R_f=0.50) to give 4-alkenylcatechol (1.29 g, 4.07 mmol 81.0%) as shown in Scheme 9.14.

In general, the Dakin and/or Baeyer-Villiger reaction are used for oxidation of an aroma aldehyde and ketone. In this method, 4-alkenylcatechol was synthesized

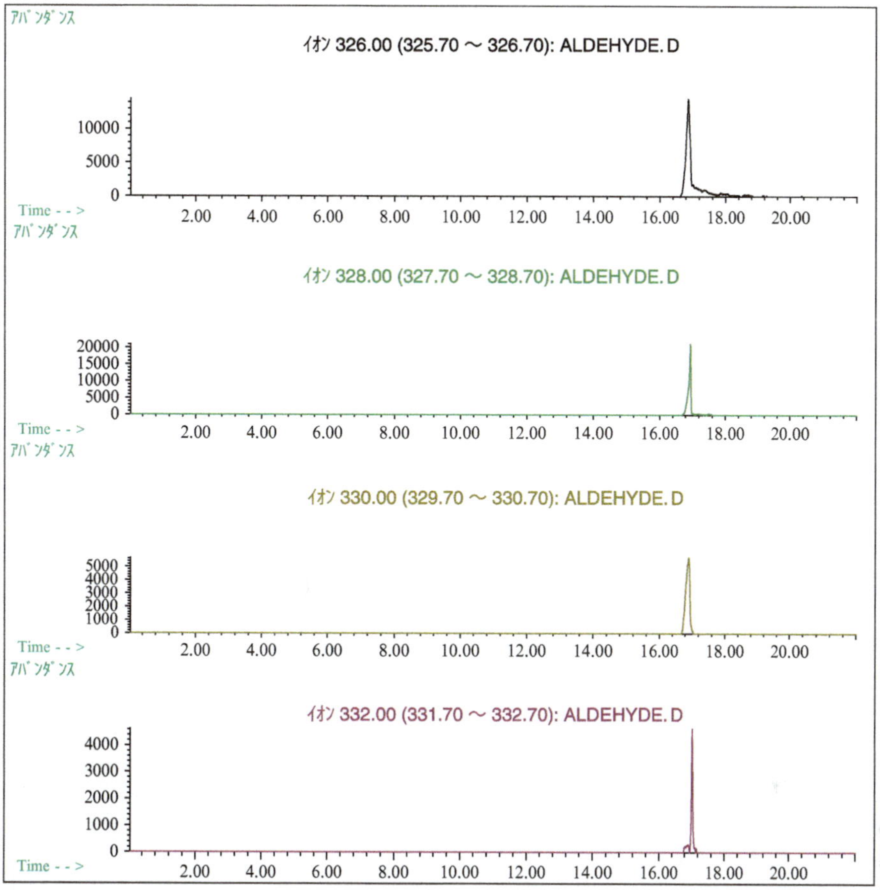

FIGURE 9.3

Mass chromatograms of 4-alkenyl-2-hydroxybenzaldehyde.

by oxidization of the formyl group using the Dakin reaction. 4-Alkenyl-2-hydroxybenzaldehyde was reacted with hydrogen peroxide under basic conditions to form an aryl as the formate and then 4-alkenylcatechol was obtained by the aqueous solution reaction. The suggested mechanism is shown in Scheme 9.15. No reaction occurred with substitution at the metaposition of the 4-alkenyl-2-hydroxybenzaldehyde.

Under alkaline conditions, the hydroxyl and/or amino group at the ortho and/or para position of the aroma aldehyde (A) was reacted with hydrogen peroxide to form product (a), and then the oxygen atom of the aromatic ring moved to form the phenol ester (b). The ester (b) generates the phenol conductor (B) by an alkali resolution reaction.

Figure 9.5 shows the ^{1}H NMR spectrum of the synthesized 4-alkenylcatechol. It has been confirmed that the peaks assigned to the formyl group at 9.83 ppm and 10.30 ppm and the peaks assigned to the united hydrogen of the ortho position at 11.04 and 11.93 ppm in Figure 9.2 have disappeared in Figure 9.5. Also, it was

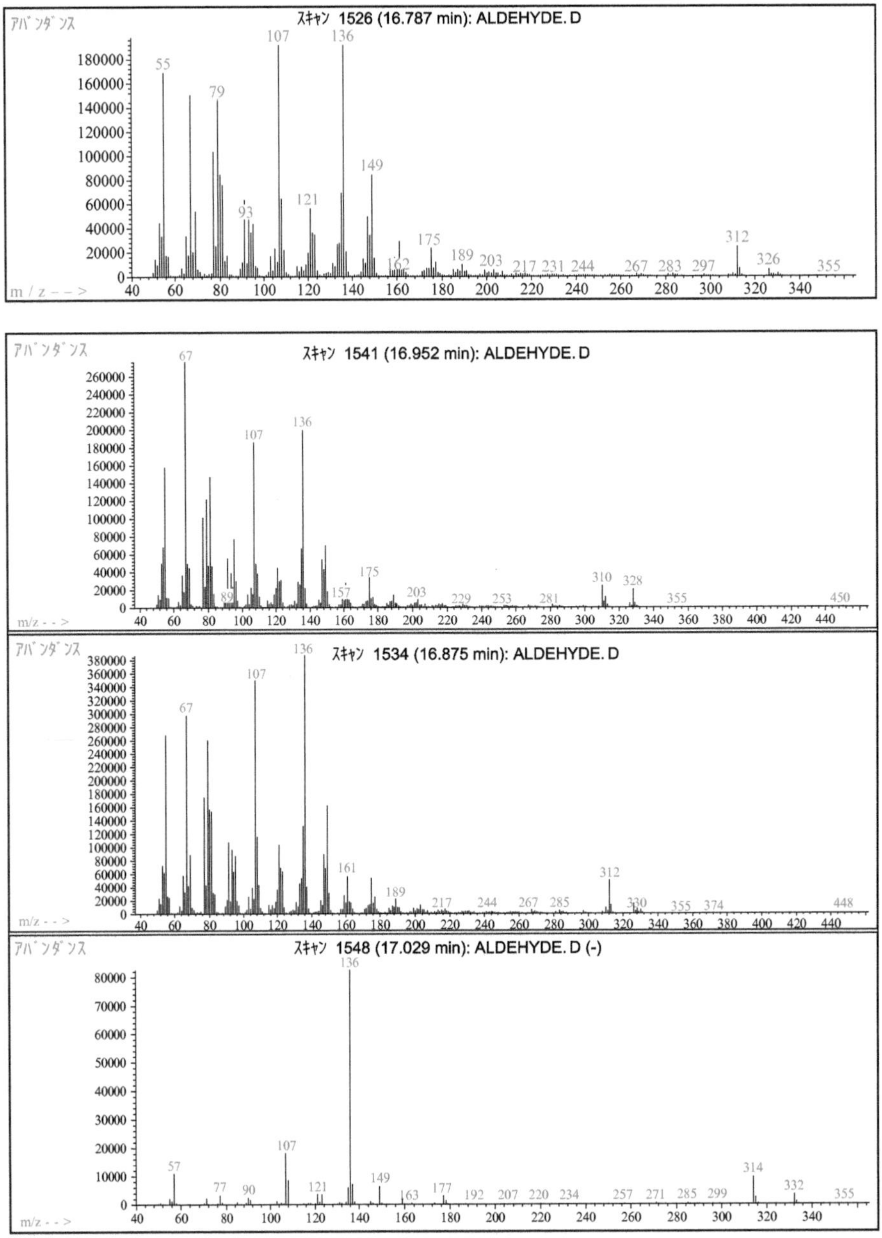

FIGURE 9.4

Mass spectra of 4-alkenyl-2-hydroxybenzaldehyde.

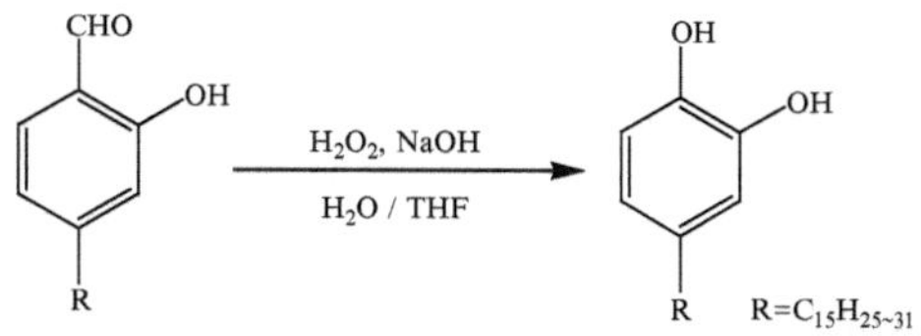

SCHEME 9.14

Synthesis of 4-alkenylcatechol.

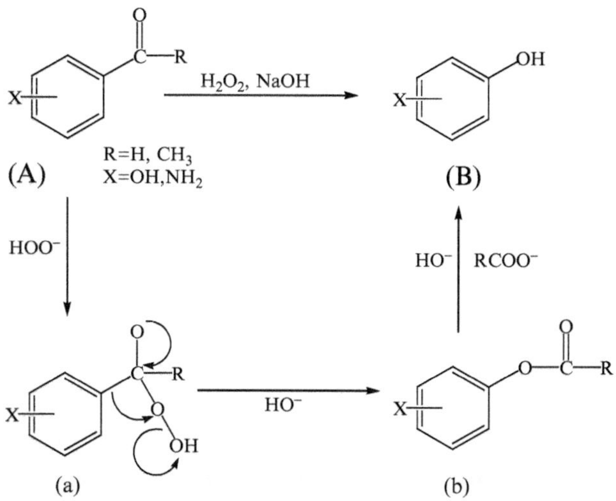

SCHEME 9.15

Mechanism of Dakin oxidation reaction.

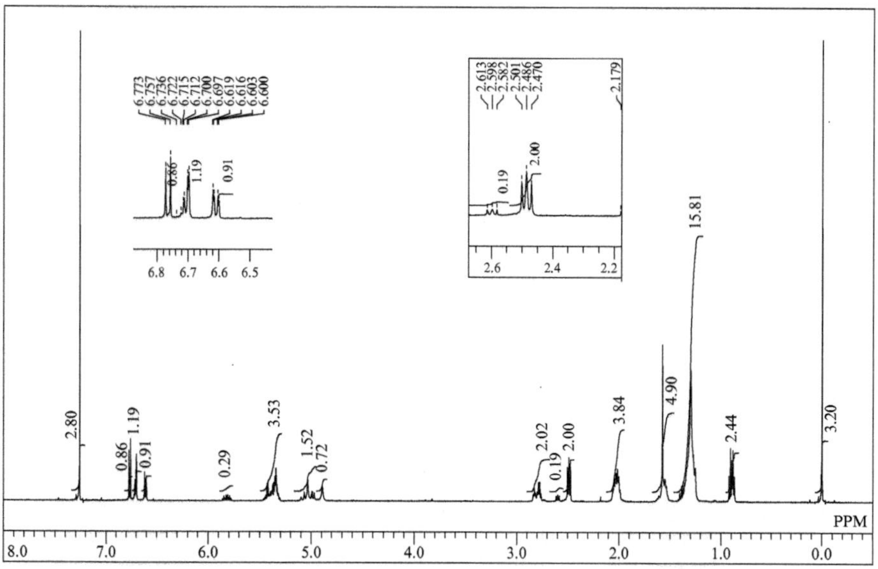

FIGURE 9.5

^1H NMR spectrum of C15: 4-alkenylcatechol.

considered that the peaks at 4.93 and 5.07 ppm are due to the catechol hydroxyl. The threefold line at 2.49 ppm (J-value: 8.0) is the methylene group of the side chain attached at the 4-position of the benzene ring, and 1.62 ppm is the peak of the next methylene group at the side chain. The peaks at 1.98-2.07 ppm were assigned to the methylene group beside the double bond, 2.75-2.85 ppm due to the methylene group between the two double bonds, and the peaks at 5.30-5.86 ppm were due to the *cis* structure of the double bond in the side chain. Moreover, peaks at 6.61 (J-value: 8.0), 6.70, and 6.77 ppm (J-value: 8.0) were due to the proton of the benzene ring.

The slight threefold line at 2.60 ppm (J-value: 8.0) was assigned as the peak of the methylene group attached at the 3-position of benzene ring of the 3-alkenylcatechol by-products.

Figure 9.6 is the mass chromatogram of 4-alkenylcatechol. The side chain showed (1): C15:3 (m/z 314), (2): C15:2 (m/z 316), (3): C15:1 (m/z 318), and (4): C15:0(m/z 320) determined from the mass spectra of the molecular ion peaks of A: m/z 314, B: m/z

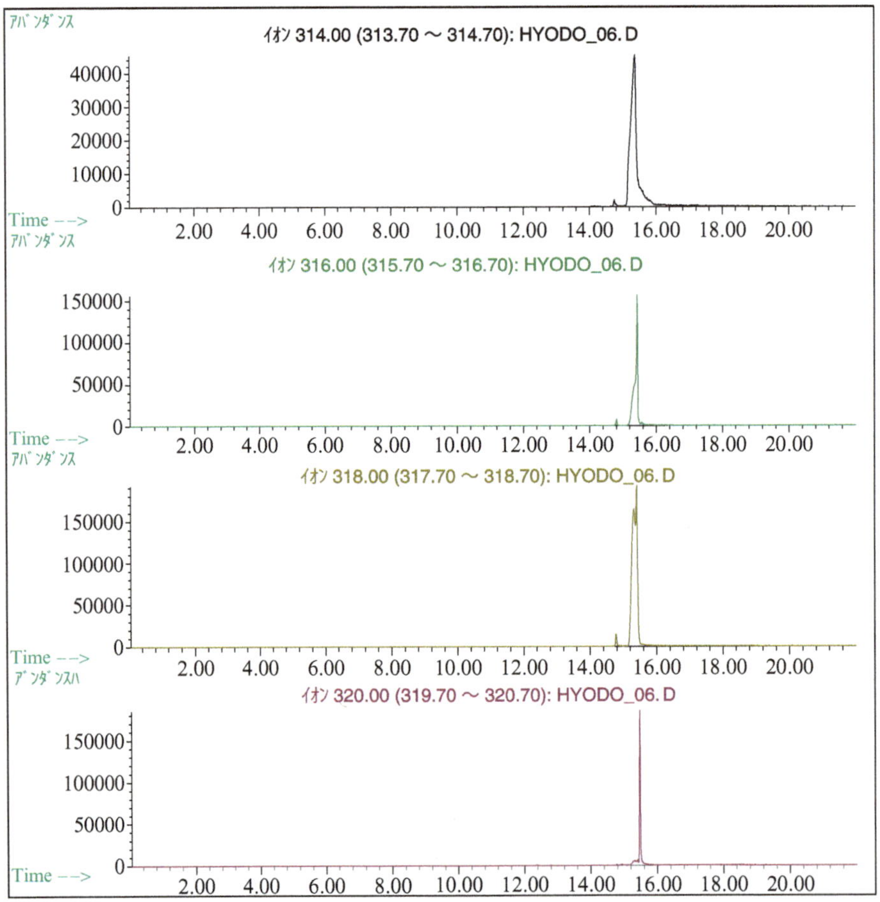

FIGURE 9.6

Mass chromatograms of C15: 4-alkenylcatechol.

316, C: m/z 318, and D: m/z 320 in Figure 9.7, respectively. Therefore, the existence of C15:3 (15.45%), C15:2 (20.66%), C15:1 (54.06%), and C15:0 (9.83%) in the side chain was confirmed. Based on these results, it can be concluded that 4-alkenylcatechol was synthesized without affecting the double bond of the cardanol side chain.

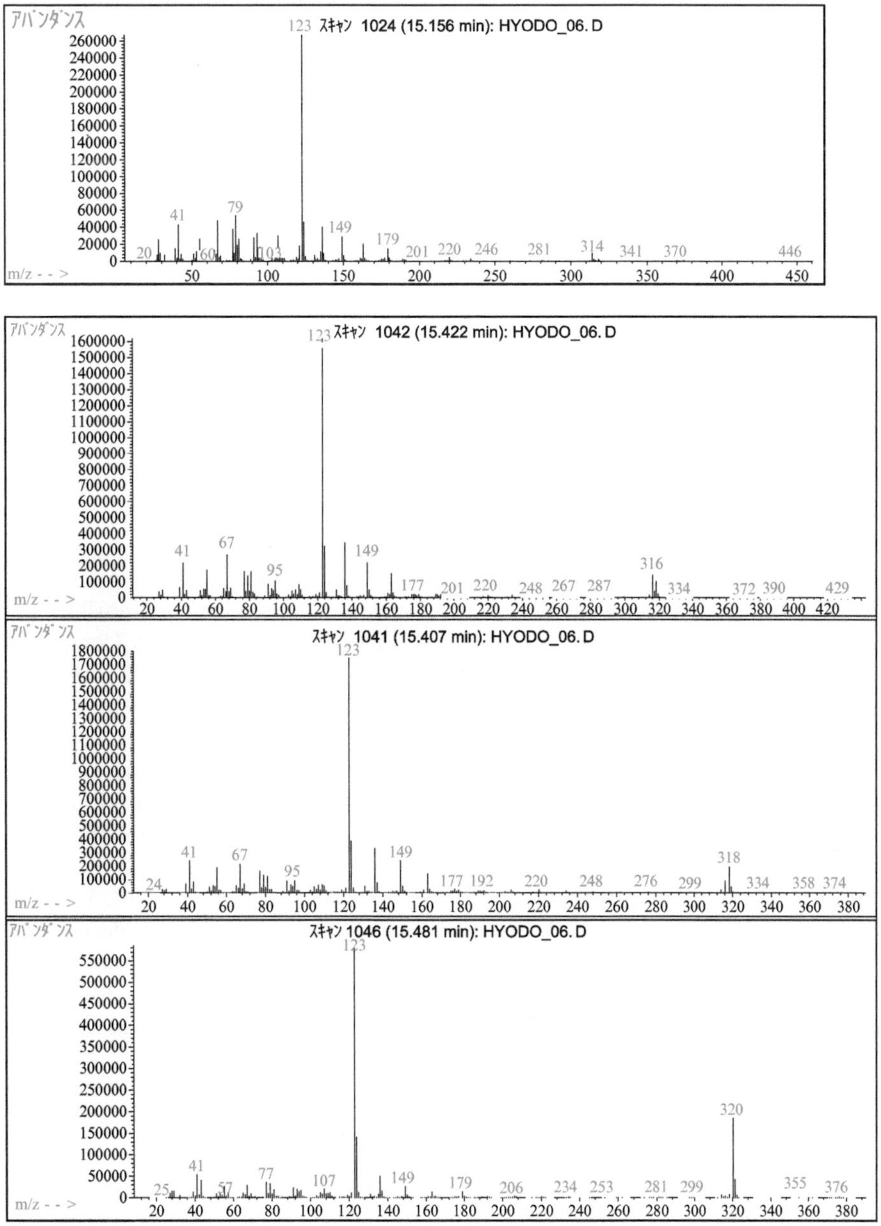

FIGURE 9.7

Mass spectra of C15: 4-alkenylcatechol.

In this synthesis of 4-alkenylcatechol, 3-alkenylcatechol was obtained as a by-product in about 4% yield. To our knowledge, this synthesis of 4-alkenylcatechol from the CNSL is a new reaction. Because it is a method that uses comparatively cheap reagents and obtains a high yield in a short reaction, it is a useful method from the green chemistry and industrial viewpoint.

9.3 SYNTHESIS OF LACCOL

Laccol is the main component of lacquer sap tapped from *T. succedanea* lacquer trees. It is a 3-substituted catechol with saturated and unsaturated side chains. The composition of the olefin side chain of laccol has been investigated using GPC, HPLC, IR, and NMR, and the chemical structure has been revealed. Based on the results, we have synthesized novel laccol compounds as shown in Figure 9.8. Structural identification of the laccol monomer components was also carried out for comparison with standard laccol.

Laccol components **1** and **2** shown in Figure 9.8 were synthesized by hydrogenation of the double bonds of the side chains in natural laccol and urushiol because the major components of natural laccol and urushiol are C17 and C15 unsaturated side chains.

Laccols **3-12**, having unsaturated side chains, were synthesized by a Wittig reaction from a ylide derived corresponding to alkyl and alkenyl-phosphonium iodide **22-29** (Figure 9.9) with aromatic parts **15**, **16**, and **17**, followed by deprotection of the hydroxyl groups of catechol ring as shown in Scheme 9.16.

Aromatic parts **15** and **16** were obtained by acetylation of the hydroxy groups and ozonolysis of the double bonds of the side chains in natural laccol and urushiol because nearly all have double bonds at positions 8 and 10 of the side chains. Compound **17** as an aromatic component was synthesized from veratrol by the same method as synthesis of urushiol derivatives and identified from its ^{1}H and ^{13}C NMR spectra.

Various alkyl- and alkenyl-phosphonium iodides **22-29** as side chains were synthesized from alcohols, carboxylic acid, and alkyl halide similar to the synthesis of urushiol derivatives. Compound **12**, 3-[(10*Z*,13*E*,15*E*)-heptadecatrienyl]catechol, the major component in *T. succedanea* lacquer sap, is a novel laccol component. Several laccol components in Figure 9.8 were previously synthesized by Halim et al. [8] and Sargent and Wangchareontrakui [13], but the yield was low. We have developed an effective and shorter synthesis route to synthesize laccol derivatives. Similarly, phenol derivatives **13** and **14** were also synthesized by a Wittig reaction of the corresponding phosphonium salts **23** and **29** with *O*-acetyl-3-(10-oxo-decyl)phenol **21**, followed by the deprotection of the hydroxy groups of the phenol as shown in *O*-acetyl-3-(10-oxo-decyl)phenol **21**, as shown in Scheme 9.17.

3-(1-DECYL-10-OL)ANISOLE 18

A solution of 3-bromoanisole (2.92 g, 15.6 mmol) in dry THF (10 mL) was added to magnesium (0.37 g, 15.6 mmol) in dry THF (1.5 mL) with stirring at room temperature under dry nitrogen. This reactant was solution **A**. Triethylamine (0.64 g,

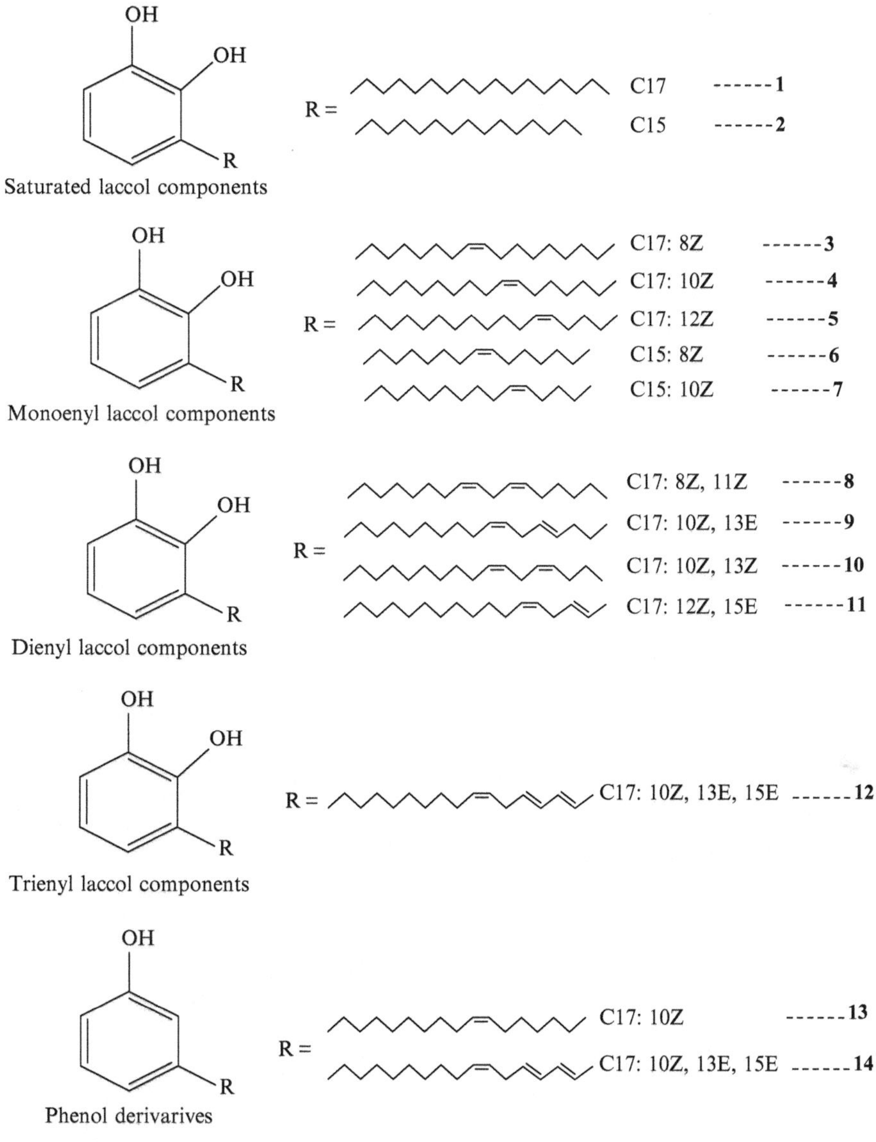

FIGURE 9.8

Structures of synthesized laccol and phenol derivatives.

6.34 mmol) and trimethylsilyl chloride (0.69 g, 6.34 mmol) were added to a solution of 1-iodo-decan-1-ol (1.50 g, 5.28 mmol) in dry THF (8.5 mL) in another flask under dry nitrogen at room temperature. The resultant solution and CuCl (0.42 g, 4.22 mmol) were added by a syringe to solution **A** and stirred for 5 h at reflux. The reaction mixture was treated with saturated ammonium chloride and extracted with

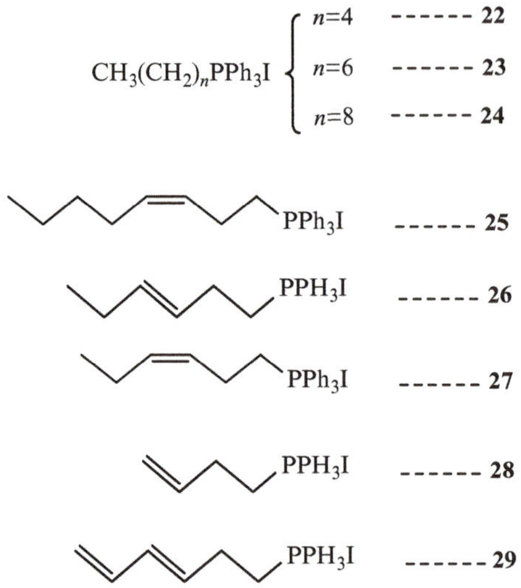

FIGURE 9.9

Structure of synthesized phosphonium salts as side chain.

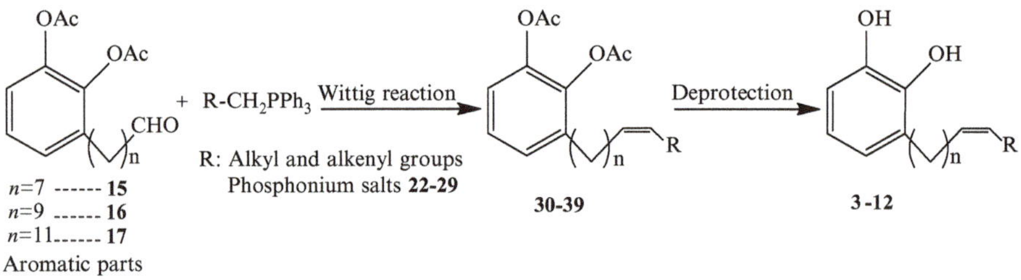

SCHEME 9.16

Synthesis of laccol components.

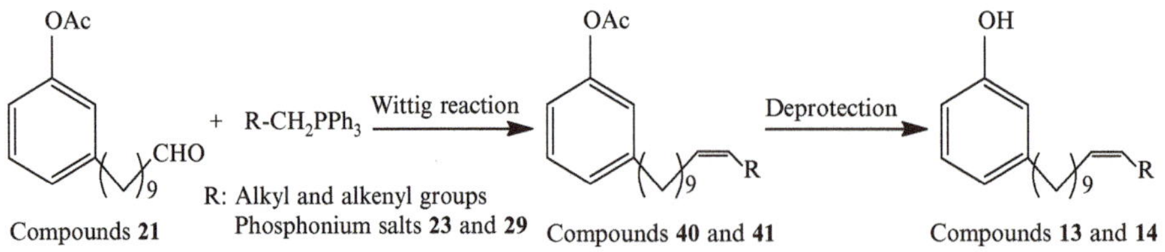

SCHEME 9.17

Synthesis of phenol derivatives.

ether. The organic layer was dried with Na_2SO_4 and the organic solvent was removed under reduced pressure to give a residue. The residue was purified over silica gel (hexane:ethyl acetate = 4:1) to give compound **18** (0.61 g, 43.7%).

R_f = 0.42 (hexane:ethyl acetate = 7:3).

^{1}H NMR: δ (ppm) = 1.29 [m, 15H, $-CH_2-\times 7$ and –OH], 1.51-1.63 [m, 2H, $-CH_2$–C–Ph], 2.54- 2.60 [t, J = 7.9Hz, 2H, $-CH_2$–Ph], 3.61-3.66 [t, J = 6.5Hz, 2H, $-CH_2$–O–], 3.80 [s, 3H, –O–CH_3], 6.71-7.21 [m, 4H, Ph].

^{13}C NMR: δ (ppm) = 25.7-32.8 [C-8 ~ 15], 36.0 [C-7], 55.1 [C-17], 63.1 [C-16], 110.7 [C-6], 114.2 [C-4], 120.8 [C-2], 129.1 [C-5], 144.6 [C-3], 159.5 [C-1].

3-(10-IODO-1-DECYL)PHENOL 19

Twenty seven milliliters of 52–58% hydroiodic acid and 51.0 ml acetic acid were added to compound **18** (0.78 g, 2.95 mmol) and stirred for 2 h at 110 °C. The solution was neutralized with sodium hydride and extracted with ethyl acetate. The organic layer was dried with Na_2SO_4 and the organic solvent was removed under reduced pressure to give a residue. The residue was purified over silica gel (hexane:ethyl acetate = 9:1) to give compound **19** (0.77 g, 72.4%).

R_f = 0.27 (hexane:ethyl acetate = 9:1).

^{1}H NMR: δ (ppm) = 1.28 [m, 12H, $-CH_2-\times 6$], 1.55-1.61 [m, 2H, $-CH_2$–C–Ph], 1.78-1.86 [m, 2H, $-CH_2$–C–I], 2.53-2.57 [t, J = 7.8Hz, 2H, $-CH_2$–Ph], 3.17-3.20 [t, J = 7.0Hz, 2H, $-CH_2$–I], 4.63 [s, 1H, –OH], 6.63-7.26 [m, 4H, Ph].

^{13}C NMR: δ (ppm) = 7.4 [C-16], 28.5-31.2 [C-8 ~ 14], 33.5 [C-15], 35.8 [C-7], 112.4 [C-3], 121.0 [C-2], 129.4 [C-1], 144.9 [C-6], 155.4 [C-4].

O-ACETYL-3-(10-IODO-1-DECYL)PHENOL 20

Compound **19** was added to a mixture of acetic anhydride (1.01 g, 9.9 mmol) and pyridine (0.05 g, 0.7 mmol) and stirred for 5 h at 100 °C. The reaction mixture was extracted with ether and washed with 1 N HCl. The organic layer was dried with Na_2SO_4 and the organic solvent was removed under reduced pressure to give a residue isolated over silica gel (hexane:ethyl acetate = 9:1) as compound **20** (0.54 g, 40.6%).

R_f = 0.55 (hexane:ethyl acetate = 9:1).

^{1}H NMR: δ (ppm) = 1.23-1.34 [m, 12H, $-CH_2-\times 6$], 1.57-1.63 [m, 2H, $-CH_2$–C–Ph], 1.76-1.86 [quint, J = 7.4Hz, 2H, $-CH_2$–C–I], 2.28 [s, 3H, $-COCH_3$], 2.57-2.63 [t, J = 7.4Hz, 2H, $-CH_2$–Ph], 3.15-3.20 [t, J = 7.4Hz, 2H, $-CH_2$–I], 6.88-7.28 [m, 4H, Ph].

^{13}C NMR: δ (ppm) = 7.3 [C-16], 21.1 [C-18], 28.5-31.1 [C-8 ~ 14], 33.5 [C-15], 35.7 [C-7], 118.6 [C-6], 115.3 [C-5], 121.3 [C-2], 125.7 [C-4], 129.0 [C-5], 144.6 [C-3], 150.6 [C-1], 169.5 [C-17].

O-ACETYL-3-(10-OXO-1-DECYL)PHENOL 21

Compound **21** was produced from 3-bromoanisole by Grignard reaction, iodination, deprotection, and protection of the hydroxy group of phenol, and then oxidation with

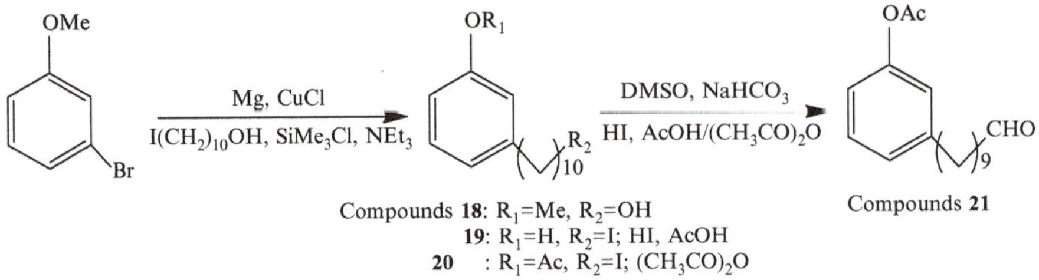

SCHEME 9.18

Synthesis of *O*-acetyl-3-(10-oxo-decyl)phenol **21**.

DMSO. The *cis*-selective Wittig reaction requires a low temperature of −78 °C for THF solvent, as shown in Scheme 9.18.

The olefin from the Wittig reaction was confirmed to be the *Z*-form at more than 80% by GC and NMR spectral analysis.

R_f=0.20 (benzene:ethyl acetate=9:1).

^{1}H NMR: δ (ppm)=1.23 [m, 10H, $-CH_2-\times 5$], 1.58-1.61 [m, 4H, $-CH_2$–C–Ph, $-CH_2$–C–CO–], 2.28 [s, 3H, $-COCH_3$], 2.37-2.43 [dt, *J*=2.0, 7.4Hz, 2H, $-CH_2$–CO–], 2.56-2.62 [t, *J*=7.8Hz, 2H, $-CH_2$–Ph], 6.87-7.26 [m, 4H, Ph], 9.75 [s, 1H, CHO].

^{13}C NMR: δ (ppm)=21.0 [C-18], 21.9-31.0 [C-8 ~ 14], 35.6 [C-7], 43.8 [C-15], 118.6 [C-6], 121.3 [C-2], 125.8 [C-4], 129.0 [C-5], 144.5 [C-3], 150.6 [C-1], 169.4 [C-17], 202.8 [C-16].

Phosphonium salts **22–24** were synthesized from the corresponding alkyl halides [20,21] and confirmed by ^{1}H and ^{13}C NMR spectra. Compounds **25–29** were synthesized by the method of synthesis of urushiol. Among them, compound **28** is a novel phosphonium salt synthesized from 3*E*-pentenoic acid. Similarly, acetyl laccols and acetylphenol derivatives **30–41** were also synthesized by the same method, i.e., Wittig reaction of ylides from phosphonium salts **25–29** with the corresponding aromatic parts **15–18** at a temperature of −78 °C to obtain *cis*-olefins.

(Z)-O-DIACETYL-3-(8-HEPTADECENYL)CATECHOL 30

Compound **30** was synthesized from compounds **24** and **15** like the synthesis of urushiol derivatives and was identified by comparison with the urushiol standard [17].

(Z)-O-DIACETYL-3-(10-HEPTADECENYL)CATECHOL 31

Compound **31** was synthesized from compounds **23** and **16**.

^{1}H NMR: δ (ppm)=0.86-0.91 [t, *J*=6.4Hz, 3H, $-C-CH_3$], 1.13-1.28 [m, 20H, $-CH_2-\times 10$], 1.53-1.58 [m, 2H, $-CH_2$–C–Ph], 1.98-2.02 [m, 4H, $-CH_2-C=C-\times 2$], 2.27, 2.31 [s, 6H, $-CO-CH_3\times 2$], 2.48-2.54 [t, *J*=7.4Hz, 2H, $-CH_2$–Ph], 5.33-5.37 [m, 2H, –CH=CH–], 7.01-7.21 [m, 3H, Ph].

^{13}C NMR: δ (ppm)=14.1 [C-23], 20.3,20.7 [C-25, 27], 27.2 [C-15], 22.7-30.1 [C-7 ~ 14, 19 ~ 22], 31.8 [C-18], 120.8 [C-3], 126.2 [C-1], 127.2 [C-2], 129.8 [C-16], 129.9 [C-17], 136.7 [C-6], 140.5 [C-5], 142.4 [C-4], 168.2, 168.3 [C-24, 26].

(Z)-O-DIACETYL-3-(12-HEPTADECENYL)CATECHOL 32

Compound **32** was synthesized from compounds **22** and **17**.

^{1}H NMR: δ (ppm)=0.89-0.92 [t, J=5.0Hz, 3H, –C–CH_3], 1.26-1.35 [m, 20H, –CH_2–×10], 1.51-1.65 [m, 2H, –CH_2–C–Ph], 1.99-2.08 [m, 4H, –CH_2–C=C–×2], 2.30, 2.32 [s, 6H, –CO–CH_3×2], 2.44-2.50 [t, J=7.4Hz, 2H, –CH_2–Ph], 5.29-5.48 [m, 2H, –CH=CH–], 6.60-7.02 [m, 3H, Ph].

^{13}C NMR: δ (ppm)=13.9 [C-23], 20.5,20.9 [C-25, 27], 26.9 [C-17], 27.1 [C-20], 29.1-31.9 [C-7~16, 21~22], 120.1 [C-3], 126.5 [C-1], 127.3 [C-2], 129.8 [C-18], 129.9 [C-19], 135.9 [C-6], 138.3 [C-5], 141.9 [C-4], 169.1, 169.7 [C-24, 26].

(Z)-O-DIACETYL-3-(8-PENTADECENYL)CATECHOL 33

Compound **33** was obtained from compounds **23** and **15**. Compound **33** was identified by comparison with our previously reported synthesis of the urushiol standard [17].

(Z)-O-DIACETYL-3-(10-PENTADECENYL)CATECHOL 34

Compound **34** was synthesized from compounds **22** and **16**.

^{1}H NMR: δ (ppm)=0.88-0.89 [t, J=3.4Hz, 3H, –C–CH_3], 1.27-1.34 [m, 16H, –CH_2–×8], 1.52-1.60 [m, 2H, –CH_2–C–Ph], 1.99-2.01 [m, 4H, –CH_2–C=C–×2], 2.31, 2.33 [s, 6H, –CO–CH_3×2], 2.56-2.64 [m, 2H, –CH_2–Ph], 5.31-5.39 [m, 2H, –CH=CH–], 6.74-7.00 [m, 3H, Ph].

^{13}C NMR: δ (ppm)=14.0 [C-21], 20.6, 21.0 [C-23, 25], 26.9, 27.2 [C-15, 18], 29.3-29.5 [C-8~14, 19, 20], 29.9 [C-7], 120.1 [C-3], 126.5 [C-1], 127.3 [C-2], 129.8 [C-16], 129.9 [C-17], 135.9 [C-6], 138.8 [C-5], 141.9 [C-4], 169.6, 169.7 [C-22, 24].

O-DIACETYL-3-[(8Z,11Z)-8,11-HEPTADECADIENYL]CATECHOL 35

Compound **35** was synthesized from compounds **25** and **15**. Compound **35** was identified by comparison with our previously reported synthesis of standard [17].

O-DIACETYL-3-[(10Z,13E)-10,13-HEPTADECADIENYL]CATECHOL 36

Compound **36** was synthesized from compounds **26** and **16**.

^{1}H NMR: δ (ppm)=0.86-0.91 [t, J=7.0Hz, 3H, –CH_3], 1.27-1.47 [m, 14H, –CH_2–×7], 1.57-1.66 [m, 2H, –CH_2–C–Ph], 1.93-2.08 [m, 4H, –C=C–CH_2–×2], 2.28, 2.31 [s, 6H, –COCH_3×2], 2.51-2.63 [t, J=7.4Hz, 2H, Ph–CH_2–], 2.71-2.75 [t, J=5.0Hz, 2H, –C=C–CH_2–C=C–, 5.32-5.48 [m, 4H, –CH=CH–×2], 7.01-7.21 [m, 3H, Ph].

O-DIACETYL-3-[(10Z,13Z)-10,13-HEPTADECADIENYL]CATECHOL 37

Compound **37** was synthesized from compounds **27** and **16**.

^{1}H NMR: δ (ppm)=0.85-0.91 [t, J=7.4Hz, 3H, –CH_3], 1.28-1.48 [m, 14H, –CH_2–×7], 1.56-1.64 [m, 2H, –CH_2–C–Ph], 2.01-2.08 [m, 4H, –C=C–CH_2–×2],

2.27, 2.30 [s, 6H, $-COCH_3\times2$], 2.56-2.63 [t, J=7.4Hz, 2H, Ph–CH_2–], 2.77-2.80 [t, J=5.4Hz, 2H, –C=C–CH_2–C=C–], 5.36-5.43 [m, 4H, –CH=CH–×2], 7.01-7.22 [m, 3H, Ph].

O-DIACETYL-3-[(12Z,15E)-12,15-HEPTADECADIENYL]CATECHOL 38

Compound **38** was synthesized from compounds **28** and **17**.

^{1}H NMR: δ (ppm)=1.20-1.35 [m, 16H, –CH_2–×8], 1.58-1.64 [m, 2H, –CH_2–C–Ph], 1.64-1.66 [d, J=6.0Hz, 3H, –C=C–CH_3], 2.01-2.08 [m, 2H, –C=C–CH_2–], 2.27, 2.31 [s, 6H, $-COCH_3\times2$], 2.58-2.63 [t, J=7.4Hz, 2H, Ph–CH_2–], 2.70-2.73 [t, J=5.4Hz, 2H, –C=C–CH_2–C=C–], 5.34-5.68 [m, 4H, –CH=CH–×2], 7.01-7.23 [m, 3H, Ph].

O-DIACETYL-3-[(10Z,13E,15E)-10,13,15-HEPTADECATRIENYL] CATECHOL 39

Compound **39** was synthesized from compounds **29** and **16**.

^{1}H NMR: δ (ppm)=1.22-1.27 [m, 10H, –CH_2–×5], 1.53-1.58 [m, 4H, –CH_2–C–Ph, –C=C–C–CH_2–], 1.71-1.74 [d, J=6.0Hz, 3H, –C=C–CH_3], 1.99-2.07 [m, 2H, –C=C–CH_2–], 2.27, 2.31 [s, 6H, $-COCH_3\times2$], 2.48-2.54 [t, J=7.0Hz, 2H, Ph–CH_2–], 2.77-2.82 [t, J=6.5Hz, 2H, –C=C–CH_2–C=C–], 5.35-5.67 [m, 4H, –CH=CH–×2], 5.99-6.01 [m, 2H, –CH=CH–], 7.01-7.21 [m, 3H, Ph].

^{13}C NMR: δ (ppm)=18.0 [C-23], 20.3, 20.7 [C-25, 27], 27.1 [C-15], 29.3-30.1 [C-7~14], 30.3 [C-18], 120.8 [C-3], 126.2 [C-1], 126.9 [C-22], 127.1 [C-20], 127.2 [C-2], 129.3 [C-17], 129.9 [C-21], 130.9 [C-16], 131.5 [C-19], 136.7 [C-6], 140.5 [C-5], 142.4 [C-4], 168.2, 168.4 [C-24, 26].

(Z)-O-ACETYL-3-(10-HEPTADECENYL)PHENOL 40

Compound **40** was synthesized from compounds **23** and **18**.

^{1}H NMR: δ (ppm)=0.86-0.91 [t, J=6.4Hz, 3H, –C–CH_3], 1.28 [m, 16H, –CH_2–×8], 1.57-1.63 [m, 2H, –CH_2–C–Ph], 1.98-2.06 [m, 4H, –CH_2–C=C–×2], 2.28 [s, 3H, –CO–CH_3], 2.57-2.62 [t, J=7.4Hz, 2H, –CH_2–Ph], 5.33-5.36 [m, 2H, –CH=CH–], 7.05-7.33 [m, 4H, Ph].

^{13}C NMR: δ (ppm)=14.1 [C-23], 21.1 [C-25], 22.7-31.8 [C-8~14, 19~22], 27.2 [C-15], 31.8 [C-18], 35.7 [C-7], 120.4 [C-3], 125.8 [C-1], 118.7 [C-6], 122.1 [C-2], 125.9 [C-4], 129.1 [C-5], 129.9 [C-16], 129.9 [C-17], 144.7 [C-3], 150.6 [C-1], 169.5 [C-24].

O-ACETYL-3-[(10Z,13E,15E)-10,13, 15-HEPTADECATRIENYL]PHENOL 41

Compound **41** was synthesized from compounds **29** and **18**.

^{1}H NMR: δ (ppm)=1.27-1.28 [m, 10H, –CH_2–×5], 1.55-1.63 [m, 4H, –CH_2–C–Ph, –C=C–C–CH_2–], 1.71-1.73 [d, J=6.0Hz, 3H, –C=C–CH_3], 2.01-2.04

[m, 2H, –C=C–CH_2–], 2.28 [s, 3H, –$COCH_3$], 2.57-2.63 [t, J=7.4Hz, 2H, Ph–CH_2–], 2.77-2.82 [t, J=6.5Hz, 2H, –C=C–CH_2–C=C–], 5.35-5.60 [m, 4H, –CH=CH–×2], 5.99-6.08 [m, 2H, –CH=CH–], 6.88-7.27 [m, 4H, Ph].

^{13}C NMR: δ (ppm)=18.0 [C-23], 21.1 [C-25], 27.1 [C-15], 29.3-31.2 [C-8~14], 30.3 [C-18], 35.7 [C-7], 118.7 [C-6], 121.4 [C-2], 125.9 [C-4], 126.9 [C-22], 127.1 [C-20], 129.1 [C-5], 129.9 [C-17], 130.4 [C-21], 130.9 [C-16], 131.5 [C-19], 144.7 [C-3], 150.6 [C-1], 169.5 [C-24].

Laccol and phenol derivatives **3–14** were synthesized from compounds **30–41** by deprotection of the hydroxyl group of catechol and phenol with sodium methoxide as described in the synthesis of urushiol section.

(Z)-3-(8-HEPTADECENYL)CATECHOL 3

^{1}H NMR: δ (ppm)=0.86-0.90 [t, J=6.8Hz, 3H, –C–CH_3], 1.26-1.31 [m, 20H, –CH_2–×10], 1.57-1.62 [m, 2H, –CH_2–C–Ph], 1.57-1.62 [m, 4H, –CH_2–C=C–×2], 2.50-2.59 [t, J=7.8Hz, 2H, –CH_2–Ph], 5.15 [s, 2H, –OH×2], 5.30-5.39 [m, 2H, –CH=CH–], 6.71 [s, 3H, Ph].

^{13}C NMR: δ (ppm)=14.1 [C-23], 22.7-29.2 [C-7~12, 17~22], 27.2 [C-13], 31.9 [C-16], 112.8 [C-3], 120.1 [C-1], 122.1 [C-2], 129.3 [C-6], 129.8 [C-14], 130.0 [C-15], 141.8 [C-5], 143.0 [C-4].

(Z)-3-(10-HEPTADECENYL)CATECHOL 4

Compound **4** was identified by comparison with the previously reported standard product [13].

(Z)-3-(12-HEPTADECENYL)CATECHOL 5

^{1}H NMR: δ (ppm)=0.87-0.95 [t, J=3.4Hz, 3H, –C–CH_3], 1.26-1.34 [m, 20H, –CH_2–×10], 1.58-1.61 [m, 2H, –CH_2–C–Ph], 1.99-2.02 [m, 4H, –CH_2–C=C–×2], 2.57-2.61 [t, J=7.8Hz, 2H, –CH_2–Ph], 5.31-5.39 [m, 2H, –CH=CH–], 5.58-5.72 [m, 2H, –OH×2], 6.71 [s, 3H, Ph].

^{13}C NMR: δ (ppm)=13.9 [C-23], 22.2-27.8 [C-7~16, 21~22], 26.9 [C-13], 32.6 [C-20], 112.9 [C-3], 120.1 [C-1], 122.1 [C-2], 129.4 [C-6], 130.3 [C-18], 130.4 [C-19], 141.8 [C-5], 143.0 [C-4].

Compounds **6**, **7**, and **8** are (Z)-3-(8-pentadecenyl)catechol, (Z)-3-(10-pentadecenyl) catechol, and 3-[(8Z,11Z)-8,11-heptadecadienyl]catechol, respectively (Figure 9.15), and were identified by comparison with previous reports of synthesis of urushiol standard products in the literature [8,13,17], respectively.

3-[(10Z,13E)-10,13-HEPTADECADIENYL]CATECHOL 9

^{1}H NMR: δ (ppm)=0.85-0.91 [t, J=7.0Hz, 3H, –C–CH_3], 1.28-1.45 [m, 14H, –CH_2–×7], 1.58-1.66 [m, 2H, –CH_2–C–Ph], 1.92-2.06 [m, 4H, –CH_2–C=C–×2], 2.57-2.62 [t, J=7.4Hz, 2H, –CH_2–Ph], 2.71-2.74 [t, 2H, J=5.0Hz, =C–CH_2–C=], 5.16 [s, 2H, –OH×2], 5.31-5.47 [m, 4H, –CH=CH–×2], 6.71 [s, 3H, Ph].

^{13}C NMR: δ (ppm) = 13.7 [C-23], 22.6 [C-18], 27.1 [C-15], 29.3 [C-21], 29.4-30.6 [C-7 ~ 14, 22], 112.8 [C-3], 120.1 [C-1], 122.1 [C-2], 127.7, 128.5, 130.4, 130.6 [C-16,17,19,20], 129.3 [C 6], 141.8 [C 5], 143.0 [C-4].

3-[(10Z,13Z)-10,13-HEPTADECADIENYL]CATECHOL 10

^1H NMR: δ (ppm) = 0.85-0.91 [t, *J* = 7.4Hz, 3H, $-C-CH_3$], 1.28-1.45 [m, 14H, $-CH_2-\times7$], 1.55-1.62 [m, 2H, $-CH_2-C-Ph$], 2.01-2.07 [m, 4H, $-CH_2-C=C-\times2$], 2.56-2.62 [t, *J* = 7.4Hz, 2H, $-CH_2-Ph$], 2.76-2.80 [t, *J* = 5.4Hz, 2H, $=C-CH_2-C=$], 5.28-5.32 [s, 2H, $-OH\times2$], 5.35-5.43 [m, 4H, $-CH=CH-\times2$], 6.71 [s, 3H, Ph].

^{13}C NMR: δ (ppm) = 13.8 [C-23], 25.6 [C-18], 27.2 [C-15], 29.3 [C-21], 29.4-29.7 [C-7 ~ 14, 22], 112.9 [C-3], 120.1 [C-1], 122.1 [C-2], 127.9, 128.2, 129.9,130.2 [C-16,17,19,20], 129.3 [C-6], 141.8 [C-5], 143.0 [C-4].

3-[(12Z,15E)-12,15-HEPTADECADIENYL]CATECHOL 11

^1H NMR: δ (ppm) = 1.20-1.37 [m, 16H, $-CH_2-\times8$], 1.58-1.63 [m, 2H, $-CH_2-C-Ph$], 1.64-1.66 [d, *J* = 6.0Hz, 3H, $=C-H_3$], 2.01-2.05 [m, 2H, $-CH_2-C=C-$], 2.58-2.62 [t, *J* = 7.4Hz, 2H, $-CH_2-Ph$], 2.70-2.73 [t, *J* = 5.4Hz, 2H, $=C-CH_2-C=$], 5.00,5.09 [s, 2H, $-OH\times2$], 5.34-5.69 [m, 4H, $-CH=CH-\times2$], 6.71 [s, 3H, Ph].

^{13}C NMR: δ (ppm) = 17.0 [C-23], 26.0 [C-20], 29.8 [C-17], 21.0-30.4 [C-7 ~ 16], 112.9 [C-3], 120.1 [C-1], 122.1 [C-2], 127.6,127.8,130.5,130.6 [C-16,17,19,20], 129.3 [C-6], 139.6 [C-5], 143.0 [C-4].

3-[(10Z,13E,15E)-10,13,15-HEPTADECATRIENYL]CATECHOL 12

^1H NMR: δ (ppm) = 1.22-1.42 [m, 10H, $-CH_2-\times5$], 1.58-1.63 [m, 4H, $-CH_2-CPh$, $=C-C-CH_2-$], 1.71-1.73 [d, *J* = 6.0Hz, 3H, $=C-CH_3$], 2.01-2.19 [m, 2H, $-CH_2-C=C-$], 2.56-2.62 [t, *J* = 7.4Hz, 2H, $-CH_2-Ph$], 2.77-2.82 [t, *J* = 5.4Hz, 2H, $=C-CH_2-C=$], 5.12 [s, 2H, $-OH\times2$], 5.34-5.65 [m, 4H, $-CH=CH-\times2$], 5.96-6.04 [m, 2H, $-CH=C-H-$], 6.71 [s, 3H, Ph].

^{13}CNMR: δ (ppm) = 18.0 [C-23], 27.1 [C-20], 29.3-29.8 [C-7 ~ 14], 30.3 [C-18], 112.9 [C-3], 120.1 [C-1], 122.1 [C-2], 126.9 [C-22], 127.2 [C-20], 129.3 [C-17], 129.9 [C-21], 130.4 [C-6], 131.0 [C-16], 131.5 [C-19], 141.8 [C-5], 143.0 [C-4].

(Z)-3-(10-HEPTADECENYL)PHENOL 13

^1H NMR: δ (ppm) = 0.86-0.90 [t, *J* = 5.9Hz, 3H, $-CH_3$], 1.28-1.33 [m, 20H, $-CH_2-\times10$], 1.56-1.66 [m, 2H, $-CH_2-C-Ph$], 1.98-2.02 [m, 4H, $-CH_2-C=C-\times2$], 2.52-2.57 [t, *J* = 8.0Hz, 2H, $-CH_2-Ph$], 4.68 [s, 1H, $-OH$], 5.33-5.37 [m, 2H, $-CH=CH-$], 6.65-7.25 [m, 4H, Ph].

^{13}C NMR: δ (ppm) = 14.1 [C-23], 22.7-29.0 [C-7 ~ 14, 19 ~ 22], 27.2 [C-15], 29.3 [C-18], 112.4 [C-3], 115.3 [C-5], 121.0 [C-2], 129.3 [C-1], 129.8 [C-16], 129.9 [C-17], 145.0 [C-6], 155.4 [C-4].

3-[(10Z,13E,15E)-10,13,15-HEPTADECATRIENYL]PHENOL 14

^{1}H NMR: δ (ppm) = 1.28 [m, 10H, $-CH_2-\times5$], 1.56-1.64 [m, 4H, $-CH_2$–C–Ph, =C–C–CH_2–], 1.71-1.73 [d, J=6.0Hz, 3H, =C–CH_3], 1.99-2.04 [m, 2H, $-CH_2$–C=C–], 2.51-2.57 [t, J=8.0Hz, 2H, $-CH_2$–Ph], 2.77-2.82 [t, J=6.5Hz, 2H, =C–CH_2–C=], 4.68 [s, 1H, –OH], 5.34-5.65 [m, 4H, –CH=CH–×2], 5.96-6.04 [m, 2H, –CH=C–H–], 6.62-7.13 [m, 4H, Ph].

^{13}C NMR: δ (ppm) = 18.0 [C-23], 27.1 [C-15], 29.6 [C-18], 21.0-30.3 [C-7 ~ 14], 112.4 [C-3], 115.3 [C-5], 121.0 [C-2], 126.9, 127.2, 129.9, 130.4, 130.9, 131.5 [C-16,17,19 ~ 22], 129.3 [C-1], 144.9 [C-6], 155.4 [C-4].

9.4 SYNTHESIS OF THITSIOL

The specific lipid components of *G. usitata* lacquer are ω-phenylalkylcatechols and ω-phenylalkylphenols, as shown in Scheme 9.19. These compounds are not present in *T. vernicifluum* and *T. succedanea* lacquer saps. In order to reveal the polymerization mechanism of thitsiol, we synthesized 3-(10-phenyldecyl)catechol and polymerized it with acetone powder containing laccase into the synthesized lacquer film. This film was analyzed by Py-GC/MS, and the resulting data were compared with that of a natural *G. usitata* lacquer film to discuss the molecular structure and polymerization mechanism of the lacquer sap from *G. usitata* lacquer trees.

9.4.1 SYNTHESIS OF 3-(10-PHENYLDECYL)CATECHOL

3-(10-Phenyldecyl)catechol was synthesized by the reaction of the dimethyl ether of catechol and 1-phenyl-10-iododecane with *n*-BuLi, followed by deprotection of the hydroxyl groups of catechol as shown in Scheme 9.20.

A 47% HBr aq. (23.0g, 138mmol) solution was added to a solution of 1,10-decanediol (8.00g, 45.9mmol) in toluene (65ml) and slowly stirred at 110°C for 3h. The reactant was neutralized with NaOH aq. and extracted with toluene. The organic layer was dried

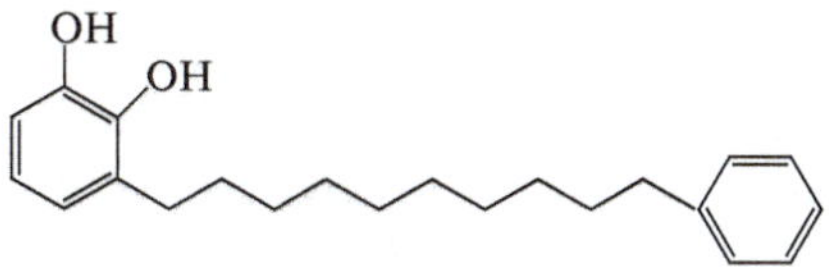

3-(10-phenyldecyl)catechol (**1**)

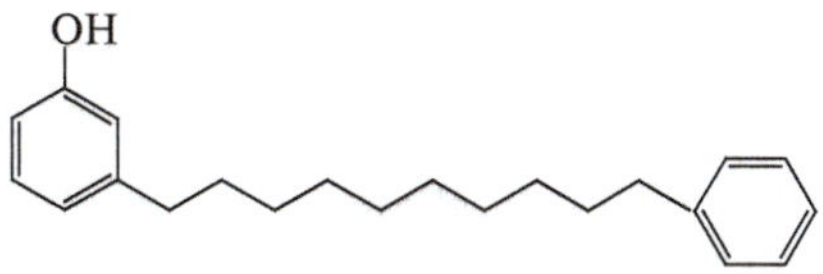

3-(10-phenyldecyl)phenol (**3**)

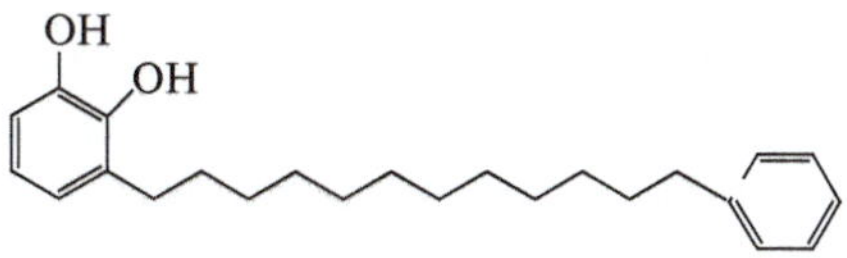

3-(12-phenyldecyl)catechol (**2**)

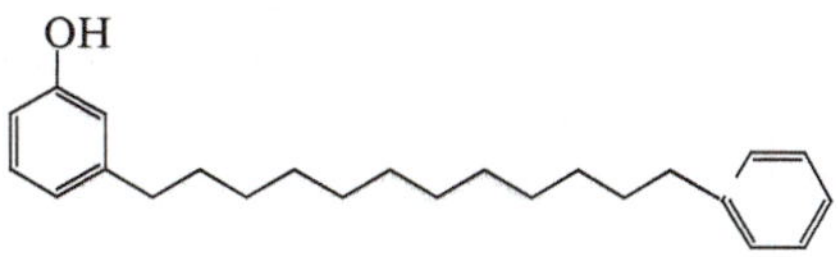

3-(12-phenyldecyl)phenol (**4**)

SCHEME 9.19

Structure of ω-phenylalkylcatechols and ω-phenylalkylphenols.

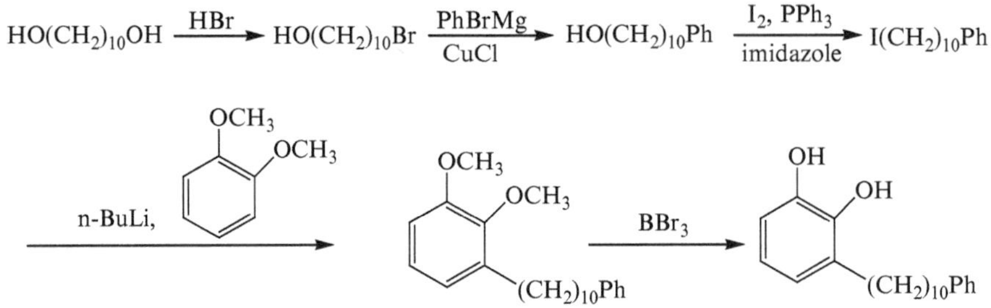

SCHEME 9.20

Synthesis of 3-(10-phenyldecyl)catechol.

with Na_2SO_4 and concentrated in vacuo. The product was isolated by silica gel column chromatography (hexane:ethyl acetate=4:1) to give 10-bromo-1-decanol (3.95 g, 62.6%). R_f=0.70 (hexane:ethyl acetate=2:3).

Bromobenzene (32.8 g, 209 mmol) was added to a solution of magnesium (3.15 g, 129 mmol) and I_2 (small amount) in THF (90.0 ml) and then stirred at room temperature under dry nitrogen. After the magnesium had reacted, 10-bromo-1-decanol (8.26 g, 34.9 mmol) and CuCl (0.225 g, 2.16 mmol) were added to the mixture at room temperature. The reaction mixture was stirred under reflux for 2 h. Then, it was treated with ice water and the organic components were extracted with ethyl acetate. The organic layer was dried with Na_2SO_4, and the organic solvent was removed in vacuo to give the residue. The product was isolated by silica gel column chromatography (hexane:ethyl acetate=9:1) to give 10-phenyl-1-decanol (6.70 g, 82.1%). R_f=0.30 (hexane:ethyl acetate=9:1).

10-Phenyl-1-decanol (2.66 g, 11.4 mmol) was added to a solution of triphenylphosphine (4.48 g, 16.8 mmol), I_2 (3.78 g, 14.9 mmol), imidazole (2.32 g, 34.1 mmol), and benzene (72 ml) and stirred at room temperature for 30 min. The benzene layer was isolated from the viscous yellow product, which was washed with ethyl ether. The ether wash was added to the benzene layer. The organic solvent was then concentrated in vacuo. Cold hexane was added to the residue to remove the triphenylphosphine oxide by-product. The hexane solution was filtered and concentrated in vacuo again to give a residue. The residue was isolated by silica gel column chromatography (hexane:ethyl acetate=4:1) to give 10-phenyl-1-iodeodecane (2.76 g, 70.5%). R_f=0.80 (hexane:ethyl acetate=4:1).

A hexane solution of *n*-BuLi (1.60 M) (8.64 ml, 13.8 mmol) was added to a dry THF (10 ml) solution of veratrol (1.91 g, 13.8 mmol) at room temperature using a syringe. After 15 min, 10-phenyl-1-iodedodecane (81.59 g, 4.61 mmol) was added to the mixture and stirred under reflux for 2 h. The reaction mixture was neutralized with 2 N HCl and extracted with ethyl ether. The organic layer was dried with Na_2SO_4. The organic solvent was removed under reduced pressure to give the crude product. The crude product was isolated by silica gel column chromatography (hexane:ethyl acetate=4:1) to give 3-(10-phenyldecyl)veratrole (1.24 g, 75.7%). R_f=0.70 (hexane:ethyl acetate=4:1).

A dichloromethane (4 ml) solution of BBr_3 (1.62 g, 2.74 mmol) was added to the dichloromethane (6 ml) solution of 3-(10-phenyldecyl)veratrole (0.971 g, 2.74 mmol) and stirred at room temperature for 2 h. The reaction mixture was neutralized with $NaHCO_3$ and extracted with dichloromethane. The organic layer was dried with Na_2SO_4. The organic solvent was removed in vacuo to give the residue. The residue was isolated by silica gel column chromatography (hexane:ethyl acetate = 4:1) to give 3-(10-phenyldecyl)catechol (0.668 g, 77.0%). R_f = 0.24 (hexane:ethyl acetate = 4:1).

9.4.2 SYNTHESIS OF ω-PHENYLALKYLCATECHOL AND ω-PHENYLALKYLPHENOL

As shown in Scheme 9.17, the ω-phenylalkylcatechols **1** and **2** were synthesized by reacting 3-lithioveratrole derived from veratrole with 1-phenyl-10-iodododecane and 1-phenyl-12-iodododecane, followed by deprotection of the methoxyl group. 1-Phenyl-10-iodododecane and 1-phenyl-12-iodododecane were synthesized from 1,10-decanediol and 1,12-dodecanediol by the halogenation using HBr and the cross-coupling reaction with PhMgBr, followed by the iodination of the hydroxyl group, respectively.

The ω-phenylalkylphenols **3** and **4** were synthesized by the cross-coupling reaction of the Grignard reagents from 3-bromoanisole and bromobenzene with 1,10-dibromodecane and 1,12-dibromododecane, followed by deprotection of the methoxyl group as shown in Schemes 9.21 and 9.22. 1,10-Dibromodecane and 1,12-dibromododecane were synthesized from 1,10-decanediol and 1,12-dodecanediol by a well-known method, halogenation of a hydroxyl group using CBr4.

In the reaction of the cross-coupling compound, **18**, **19** and dibromoalkane **16** or **17**, a diphenylalkane and a di(3-methoxyphenyl)alkane, as the by-products were produced. But considering the total yield, this reaction was acceptable because this synthetic route was shorter than that shown in Scheme 9.20 and had a higher total yield.

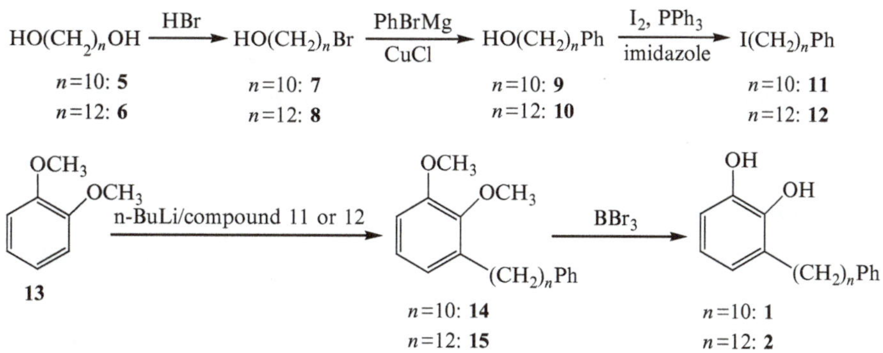

SCHEME 9.21

Synthesis of 3-(ω-phenylalkyl)catechol.

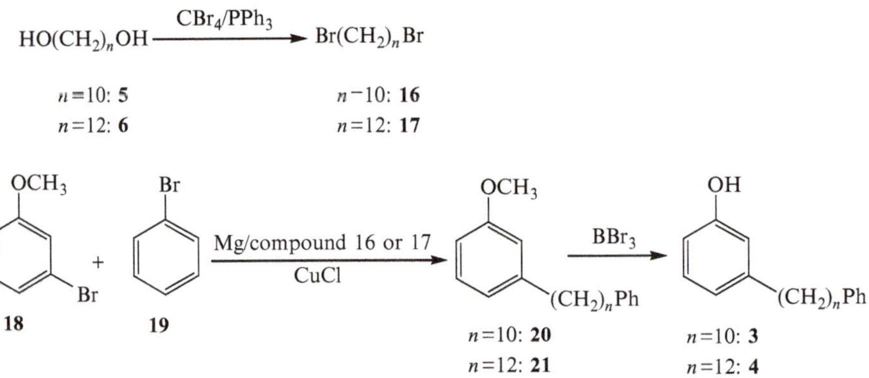

SCHEME 9.22

Synthesis of 3-(ω-phenylalkyl)phenol.

All ^{1}H NMR and ^{13}C NMR spectra were taken in chloroform-d with tetramethylsilane as the internal standard using a JNM-GSX 400 spectrometer (JEOL Ltd.).

3-(10-PHENYLDECYL)CATECHOL 1

A dichloromethane (4ml) solution of BBr_3 (1.62g, 2.74mmol) was added using a funnel to a dichloromethane (6ml) solution of compound 14 (0.971g, 2.74mmol), which was stirred at room temperature for 2h. The reaction mixture was neutralized with $NaHCO_3$ and extracted with dichloromethane. The organic layer was dried with Na_2SO_4. The organic solvent was removed in vacuo to give the residue. The residue was chromatographed over silica gel (hexane–ethyl acetate=4:1) to give 3-(10-phenyldecyl)catechol 1 (0.668g, 77.0%).

R_f=0.24 (hexane–ethyl acetate=4:1).

^{1}H NMR: δ (ppm)=1.29 [m, 12H, $-CH_2-\times 6$], 1.60 [m, 4H, $-CH_2$–C–Ph×2], 2.59 [t, 7.4, 7.92Hz, 4H, $-CH_2$–Ph×2], 5.14 [s, 2H, –OH×2], 6.71 [s, 3H, Ph in catechol], 7.19 [m, 5H, –Ph].

^{13}C NMR: δ (ppm)=29.5~29.7 [C-9~14], 31.4 [C-8,15], 35.9 [C-7,16], 112.7 [C-3], 119.8 [C-2], 121.8 [C-1], 125.4 [C-20], 126.1 [C-6], 128.1 [C-19,21], 128.3 [C-18,22], 142.1 [C-4], 142.8 [C-5], 143.2 [C-17].

3-(10-PHENYLDODECYL)CATECHOL 2

3-(10-Phenyldodecyl)catechol **2** (yield 52.1%) was synthesized by a method similar to that described above for compound **1**.

R_f=0.25 (hexane–ethyl acetate=4:1).

^{1}H NMR: δ (ppm)=1.26 [m, 16H, $-CH_2-\times 8$], 1.61 [m, 4H, $-CH_2$–C–Ph×2], 2.59 [t, 7.4, 7.9Hz, 4H, $-CH_2$–Ph×2], 5.16 [s, 2H, –OH×2], 6.70 [m, 3H, Ph in veratrole], 7.19 [m, 5H, –Ph].

^{13}C NMR: δ (ppm)=29.3, 29.5, 29.6, 29.8 [C-9~16], 31.5 [C-8, 17], 36.0 [C-7,18], 112.9 [C-3], 120.1 [C-2], 122.1 [C-1], 125.5 [C-22], 128.2 [C-6], 128.4 [C-21, 23], 129.4 [C-20, 24], 141.8 [C-4], 142.9 [C-5], 143.0 [C-19].

3-(10-PHENYLDECYL)PHENOL 3

3-(10-Phenyldecyl)phenol **3** (yield 90.1%) was synthesized by a method similar to that described for compound **1**.

R_f=0.50 (hexane–ethyl acetate=4:1).

^{1}H NMR: δ (ppm)=1.26 [m, 12H, $-CH_2-\times 6$], 1.58 [m, 4H, $-CH_2$–C–Ph$\times$2], 2.57 [m, 4H, $-CH_2$–Ph$\times$2], 4.64 [s, 1H, –OH], 6.50 [m, 4H, Ph in anisole], 7.19 [m, 5H, –Ph].

^{13}C NMR: δ (ppm)=29.0~30.0 [C-9~14], 31.2, 31.3 [C-8, 15], 35.8, 35.9 [C-7,16], 112.2 [C-3], 115.3 [C-5], 121.0 [C-1], 125.6 [C-20], 128.4 [C-19, 21], 129.1 [C-2], 129.2 [C-18, 22], 142.9 [C-6], 145.6 [C-17], 155.3 [C-4].

3-(12-PHENYLDODECYL)PHENOL 4

3-(12-Phenyldodecyl)phenol **4** (yield 63.6%) was synthesized by a method similar to that described for compound **1**.

R_f=0.50 (hexane–ethyl acetate=4:1).

^{1}H NMR: δ (ppm)=1.25 [m, 16H, $-CH_2-\times 8$], 1.59 [m, 4H, $-CH_2$–C–Ph$\times$2], 2.57 [m, 4H, $-CH_2$–Ph$\times$2], 4.59 [s, 1H, –OH], 6.70 [m, 4H, Ph in anisole], 7.26 [m, 5H, –Ph].

^{13}C NMR: δ (ppm)=29.1~29.6 [C-9~16], 31.3, 31.5 [C-8,17], 35.8, 36.0 [C-7,18], 112.4 [C-3], 115.3 [C-5], 121.0 [C-1], 125.6 [C-22], 128.4 [C-21, 23], 129.2 [C-20, 24], 129.3 [C-2], 142.9 [C-6], 145.6 [C-19], 155.3 [C-4].

10-BROMO-1-DECANOL 7

Aqueous 47% HBr (23.0g, 138mmol) was added to a toluene (65ml) solution of 1,10-decanediol (5) (8.00g, 45.9mmol) and the mixture was slowly stirred for 3h at 110°C. The reactant was neutralized with NaOH aq and extracted with toluene. The organic layer was dried (Na_2SO_4) and concentrated in vacuo. The product was isolated using silica gel (hexane–ethyl acetate=4:1) to give 10-bromo-1-decanol 7 (3.95g, 62.6%).

R_f=0.70 (hexane–ethyl acetate=2:3).

^{1}H NMR: δ (ppm)=1.30 [m, 12H, $-CH_2-\times 6$], 1.53 [m, 3H, $-CH_2$–C–O– and –OH], 1.85 [quin, J=7.3Hz, 2H,$-CH_2$–C–Br], 3.40 [t, J=6.8Hz, 2H, $-CH_2$–Br], 3.61 [t, J=6.3Hz, 2H, $-CH_2$–OH].

^{13}C NMR: δ (ppm)=25.5~29.3 [C-3~8], 32.2 [C-2], 32.6 [C-9], 33.7 [C-1], 62.4 [C-10].

12-BROMO-1-DODECANOL 8

12-Bromo-1-dodecanol **8** (yield 73.8%) was synthesized by a method similar to that described above.

R_f=0.25 (hexane:–ethyl acetate=4:1).

^{1}H NMR: δ (ppm)=1.27 [m, 17H, $-CH_2-\times 8$ and –OH], 1.55 [quin, J=5.0Hz, 2H, $-CH_2$–C–OH], 1.85 [quin, J=7.3, 6.9Hz, 2H, $-CH_2$–C–Br], 3.41 [t, J=6.4, 6.9Hz, 2H, $-CH_2$–Br], 3.61 [t, J=5.0Hz, 2H, $-CH_2$–OH].

[13]H NMR: δ (ppm)=25.6~29.6 [C-3~10], 32.2 [C-2], 32.6 [C-2], 32.7 [C-11], 33.8 [C-1], 62.6 [C-12].

10-PHENYL-1-DECANOL 9

Bromobenzene (32.8 g, 209 mmol) was added using a funnel to a solution of magnesium (3.15 g, 129 mmol) and I_2 (small amount) in THF (90.0 ml) with stirring at room temperature under dry nitrogen. After the magnesium had reacted, compound **7** (8.26 g, 34.9 mmol) and CuCl (0.225 g, 2.16 mmol) were added to the mixture at room temperature. The reaction mixture was stirred at reflux for 2 h. Then it was cooled in ice water, and the organic components were extracted with ethyl acetate. The organic layer was dried with Na_2SO_4 and the organic solvent was removed in vacuo to give the residue. The product was isolated over silica gel (hexane:ethyl acetate=9:1) to give 10-phenyl-1-decanol **9** (6.70 g, 82.1%).

R_f=0.30 (hexane:ethyl acetate=9:1).

^{1}H NMR: δ (ppm)=1.29 [m, 12H, $-CH_2-\times 6$], 1.56 [m, 5H, $-CH_2$–C–O– and $-CH_2$–C–Ph and –OH], 2.60 [t, J=7.4, 7.9Hz, 2H, $-CH_2$–Ph], 3.64 [t, J=6.4, 6.9Hz, 2H, $-CH_2$–O–], 7.26 [m, 5H, –Ph].

^{13}C NMR: δ (ppm)=25.6~29.4 [C-3~8], 31.3 [C-9], 32.5 [C-2], 35.8 [C-10], 62.3 [C-1], 125.3 [C-14], 128.0 [C-13, 15], 128.1 [C-16], 128.4 [C-12], 142.6 [C-11].

12-PHENYL-1-DODECANOL 10

12-Phenyl-1-dodecanol **10** (yield 73.3%) was synthesized by a method similar to that described above.

R_f=0.40 (hexane–ethyl acetate=4:1).

^{1}H NMR: δ (ppm)=1.19 [m, 16H, $-CH_2-\times 8$], 1.55 [m, 5H, $-CH_2$–C–O– and $-CH_2$–C–Ph and –OH], 2.53 [t, J=7.6Hz, 2H, $-CH_2$–Ph], 3.63 [m, 2H, $-CH_2$–O–], 7.35 [m, 5H, –Ph].

^{13}C NMR: δ (ppm)=25.7~29.6 [C-3~10], 31.5 [C-11], 32.8 [C-2], 36.0 [C-12], 63.1 [C-1], 125.5 [C-16], 128.0 [C-15, 17], 128.2 [C-18], 128.4 [C-14], 143.0 [C-13].

10-PHENYL-1-IODODECANE 11

Compound **9** (2.66 g, 11.4 mmol) was added to a solution of triphenylphosphine (4.48 g, 16.8 mmol), I_2 (3.78 g, 14.9 mmol), imidazole (2.32 g, 34.1 mmol), and benzene (72 ml), which was stirred at room temperature for 30 min. The benzene layer was isolated from the viscous yellow product, which was washed with ethyl ether. The ethyl ether wash was added to the benzene layer. The organic solvent was concentrated in vacuo. Cold hexane was added to the residue to remove the triphenylphosphine oxide by-product. The hexane solution was filtered and concentrated in vacuo again to give a residue. The residue was chromatographed over silica gel (hexane–ethyl acetate = 4:1) to give 10-phenyl-1-iododecane 11 (2.76 g, 70.5%).

R_f = 0.80 (hexane–ethyl acetate = 4:1).

^{1}H NMR: δ (ppm) = 1.35 [m, 12H, $-CH_2-\times 8$], 1.58 [quin, J = 7.0 Hz, 2H, $-CH_2$–C–Ph], 1.82 [m, 2H, $-CH_2$–C–I], 2.60 [t, J = 7.83 Hz, 2H, $-CH_2$–Ph], 3.12 [t, J = 6.8, 7.0 Hz, 2H, $-CH_2$–I], 7.20 [m, 5H, –Ph].

^{13}C NMR: δ (ppm) = 7.17 [C-1], 31.5 [C-11], 28.5 ~ 29.4 [C-3 ~ 8], 31.4 [C-9], 33.5 [C-2], 35.9 [C-10], 125.5 [C-14], 128.1 [C-13,15], 128.3 [C-14,16], 142.6 [C-11].

12-PHENYL-1-IODODODECANE 12

12-Phenyl-1-iodododecane **12** (yield 62.6%) was synthesized by a method similar to that described above.

R_f = 0.88 (hexane–ethyl acetate = 4:1).

^{1}H NMR: δ (ppm) = 1.24 [m, 16H, $-CH_2-\times 8$], 1.59 [m, 2H, $-CH_2$–C–Ph], 1.78 [m, 2H, $-CH_2$–C–I], 2.53 [t, J = 7.6, 7.8 Hz, 2H, $-CH_2$–Ph], 3.12 [t, J = 7.0, 6.8 Hz, 2H, $-CH_2$–I], 7.01 [m, 5H, –Ph].

^{13}C NMR: δ (ppm) = 7.08 [C-1], 28.5, 29.3, 29.34, 29.5, 30.4 [C-3 ~ 10], 31.5 [C-11], 33.5 [C-2], 35.9 [C-12], 125.4 [C-16], 128.1 [C-15, 17], 128.3 [C-14,18], 142.7 [C-13].

3-(10-PHENYLDECYL)VERATROLE 14

A hexane solution of n-BuLi (1.60 M) (8.64 ml, 13.8 mmol) was added using a syringe to a dry THF (10 ml) solution of veratrole (1.91 g, 13.8 mmol) at room temperature. After 15 min, 10-phenyl-1-iododecane 11 (81.59 g, 4.61 mmol) was added to the mixture, which was stirred at reflux for 2 h. The reaction mixture was neutralized with 2 N HCl and extracted with ethyl ether. The organic layer was dried with Na_2SO_4. The organic solvent was removed under reduced pressure to give the crude product. The crude product was chromatographed over silica gel (hexane–ethyl acetate = 4:1) to give 3-(10-phenyldecyl)veratrole **14** (1.24 g, 75.7%).

R_f = 0.70 (hexane–ethyl acetate = 4:1).

^{1}H NMR: δ (ppm) = 1.28 [m, 12H, $-CH_2-\times 6$], 1.60 [m, 4H, $-CH_2$–C–Ph × 2], 2.59 [m, 4H, $-CH_2$–Ph × 2], 3.81, 3.85 [s, 6H, –OMe × 2], 6.77 [m, 3H, Ph in veratrole], 7.19 [m, 5H, –Ph].

^{13}C NMR: δ (ppm) = 29.2, 29.4, 29.5 [C-9 ~ 14], 30.7, 31.4 [C-8, 15], 35.9 [C-7,16], 55.4, 60.4 [C-23, 24], 109.8 [C-3], 114.0 [C-2], 121.8 [C-1], 125.4 [C-20], 128.1 [C-19], 128.2 [C-21], 128.4 [C-18], 128.5 [C-22], 136.5 [C-6], 138.9 [C-4], 147.0 [C-17], 152.6 [C-5].

3-(12-PHENYLDODECYL)VERATROLE 15

3-(12-Phenyldodecyl)veratrole **15** (yield 88.7%) was synthesized by a method similar to that described above.

R_f = 0.73 (hexane–ethyl acetate = 4:1).

^{1}H NMR: δ (ppm) = 1.25 [m, 16H, $-CH_2-\times 8$], 1.61 [m, 4H, $-CH_2$–C–Ph × 2], 2.60 [m, 4H, $-CH_2$–Ph × 2], 3.81, 3.85 [s, 6H, –OMe × 2], 6.97 [m, 3H, Ph in veratrole], 7.26 [m, 5H, –Ph].

^{13}C NMR: δ (ppm) = 29.3 ~ 29.8 [C-9 ~ 16], 30.8, 31.5 [C-8,17], 35.9 [C-7,18], 55.6, 60.5 [C-25, 26], 109.9 [C-3], 114.0 [C-2], 121.9 [C-1], 125.5 [C-22], 128.1 [C-21,23], 128.3 [C-20,24], 136.7 [C-6], 128.5 [C-22], 139.1 [C-4], 147.1 [C-19], 152.7 [C-5].

3-(10-PHENYLDECYL)ANISOLE 20

A solution of bromobenzene (3.37 g, 21.3 mmol) and 3-bromoanisole (3.99 g, 21.3 mmol) in THF (20 ml) was slowly added using a funnel to a suspension of magnesium (1.04 g, 42.8 mmol) in THF (10 ml) at room temperature. After the reaction, a solution of 1,10-dibromodecane (2.15 g, 7.2 mmol) in THF (10 ml) was added to the reaction mixture using a funnel. Then CuCl (0.580 g, 5.86 mmol) was added and the reaction mixture was stirred at reflux for 2 h. The reaction mixture was washed with ice water and then extracted with ethyl ether. The organic layer was dried with Na_2SO_4. The organic solvent was removed in vacuo to give the residue. The reside was chromatographed over silica gel (hexane–ethyl acetate = 9:1) to give 3-(10-phenyldecyl)anisole 20 (1.15 g, 49.8%).

R_f = 0.80 (hexane–ethyl acetate = 4:1).

^{1}H NMR: δ (ppm) = 1.26 [m, 12H, $-CH_2-\times 6$], 1.60 [m, 4H, $-CH_2-C-Ph\times 2$], 2.54 [m, 4H, $-CH_2-Ph\times 2$], 3.81 [s, 3H, –OMe], 6.77 [m, 3H, Ph in anisole], 7.19 [m, 5H, –Ph].

^{13}C NMR: δ (ppm) = 29.2 ~ 29.5 [C-9 ~ 14], 30.7, 31.4 [C-8, 15], 35.9 [C-7,16], 55.0 [C-23], 109.8 [C-3], 114.8 [C-5], 121.5 [C-1], 121.8 [C-2], 125.4 [C-20], 128.4 [C-19, 21], 128.5 [C-18, 22], 136.5, 138.9 [C-6], 147.0 [C-17], 152.6 [C-4].

3-(12-PHENYLDODECYL)ANISOLE 21

3-(12-Phenyldodecyl)anisole **21** (yield 10.1%) was synthesized by a method similar to that described above.

3-(12-Phenyldodecyl)anisole **21** (yield 10.1%) was synthesized by a method similar to that described above.

R_f = 0.63 (hexane–ethyl acetate = 4:1).

^{1}H NMR: δ (ppm) = 1.29 [m, 16H, $-CH_2-\times 8$], 1.60 [m, 4H, $-CH_2-C-Ph\times 2$], 2.57 [m, 4H, $-CH_2-Ph\times 2$], 3.81 [s, 6H, –OMe], 6.78 [m, 4H, Ph in anisole], 7.26 [m, 5H, –Ph].

^{13}C NMR: δ (ppm) = 29.3 ~ 29.6 [C-9 ~ 16], 31.4, 31.5 [C-8, 17], 35.9 [C-7,18], 55.0 [C-25], 110.7 [C-3], 114.1 [C-5], 120.1 [C-1], 125.5 [C-22], 128.4 [C-21, 23], 129.1 [C-20, 24], 129.7 [C-2], 142.9 [C-6], 147.0 [C-19], 159.5 [C-4].

9.5 ENZYME POLYMERIZATION AND CHARACTERIZATION OF SYNTHESIZED LACQUER FILM

4-Alkenylcatechol films were synthesized by adding acetone powder (AP) obtained from a natural lacquer sap and two kinds fungal laccase, NS44002 and NS44003, respectively. The lacquer drying process can be divided into three steps: dust-free drying (DF), touch-free drying (TF), and hardened dry (HD) [21]. The urushiol and its analog film were tested under the drying conditions of 20 °C with an 80% relative humidity using an automatic drying time recorder that measured the drying process of lacquer to DF, TF, and HD. The time until the marks of the needle appear on the glass board is equivalent to DF, the time until the marks of the needle appear on the lacquer film surface is equivalent to TF, and the time until the marks of the needle completely do not appear on the lacquer film surface is equivalent to HD. The drying time results are summarized in Table 9.1 (ND means not dry).

The synthesized 4-alkenylcatechol demonstrated the activity of the laccase (acetone powder) enzyme. However, the synthesized 4-alkenylcatechol films had a slow drying speed and a low hardness compared with the natural urushiol films. It can be considered that although the enzyme oxidation reaction is the main one, polymerization also occurred in the double bond of the side chain. However, the unsaturated percentage of C15 4-alkenylcatechol was 1.97% while the unsaturated percentage of urushiol was 2.55%. Therefore, urushiol has a faster drying time than 4-alkenylcatechol due to the high unsaturated compound content.

On the other hand, the cardanol and cardol films were not dry even after 30 days; this was due to the difference in the polymerization mechanism. The polymerization occurs in an alkenylcatechol such as urushiol due to the oxidation of the catechol hydroxyl by the laccase. However, cardanol does not show revitalization in laccase and its polymerization is due to the oxidation of the unsaturated side chain using the metallic salts as a catalyst. Therefore, it can be said that this synthesis method succeeded in the synthesis of the compound that showed revitalization in the laccase by a two-step reaction using a material that did not show revitalization in the laccase.

The synthesized 4-alkenylcatechol and the natural urushiol were made into films by adding acetone powder isolated from lacquer sap tapped from *T. verniciffluum* trees. The film of the synthesized 4-alkenylcatechol (Figures 9.10A) had a transparent

Table 9.1 Drying properties of natural urushiol and 4-alkenylcatechol (4-AC)

Substrate	Enzyme	Time/hardness										
		1h	2h	3h	6h	1d	2d	3d	7d	10d	20d	30d
Urushiol	AP	TF	HD	H	2H	2H	2H	2H	3H	3H	4H	4H
4-AC	AP	ND	ND	ND	DF	TF	HD	HD	6B	6B	2B	2H
4-AC	NS4002	1h	2h	3h	6h	1d	2d	3d	7d	10d	20d	30d
4-AC	NS4003	ND	ND	ND	DF	TF	HD	6B	6B	3B	HB	H

(a) (b)

FIGURE 9.10

Color of synthesized 4-alkenylcatechol and natural urushiol films.

yellow color, while the natural urushiol films had a dark brown color (Figures 9.10B). These different colors are considered to be due to the different polymerization mechanisms of the synthesized 4-alkenylcatechol and natural urushiol films.

In order to reveal the different polymerization mechanisms, the drying process of the natural and synthetic lacquer films was analyzed by NMR, IR, and UV measurements.

Figures 9.11 and 9.12 are ^{1}H NMR of the native urushiol and synthesis 4-alkenylcatechol films. The peaks at 6.6–6.8 ppm due to the proton of the benzene ring and 2.5–2.6 ppm due to the benzyl group decreased and disappeared in the hardened dry (HD) film, and the peak at 4.9 ppm due to the hydroxyl of catechol also decreased. It was considered due to the molecular movement decreases as the nucleus–nucleus and nucleus–side chain reaction occurred during the enzyme polymerization.

Because the changing peak at 2.0 ppm due to the methylene group on the side of the double bond, 2.8 ppm due to the methylene group between the double bonds, and 5.3–5.5 ppm due to the double bond (*cis*) are not observed during the synthesis 4-alkenylcatechol (Figure 9.9), it can be concluded that the side chain and/or double bond of the side chain undergoes no reaction during the enzyme polymerization. However, because the 5.3–5.5 ppm peak due to the double bond (*cis*) of the side chain in the natural urushiol was decreased as dryness progressed (Figure 9.10), it was considered that the automatic oxidation reaction of the side chain occurred at approximately the same time during the enzyme polymerization.

The IR of synthesized 4-alkenylcatechol and native urushiol films are shown in Figures 9.13 and 9.14, respectively.

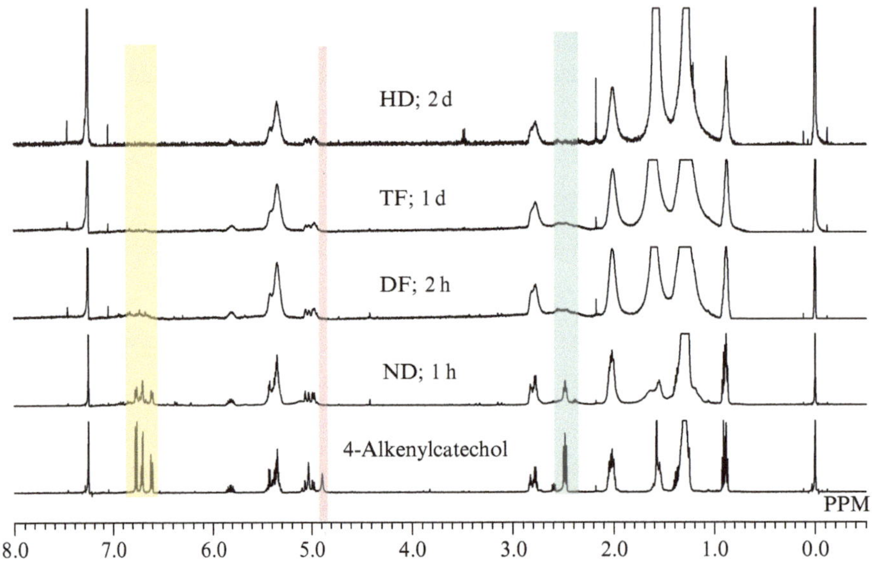

FIGURE 9.11

^{1}H NMR spectra of C15: 4-alkenylcatechol film in drying time course.

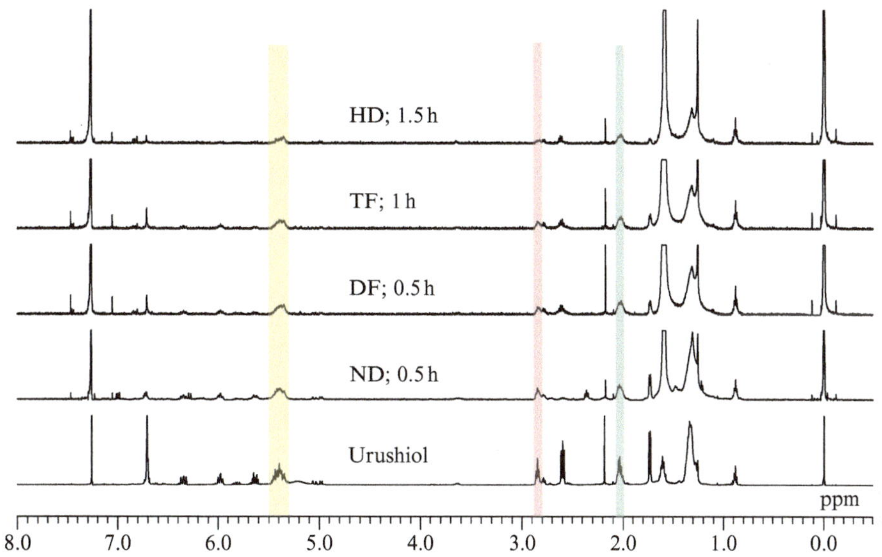

FIGURE 9.12

^{1}H NMR spectra of natural urushiol film during drying.

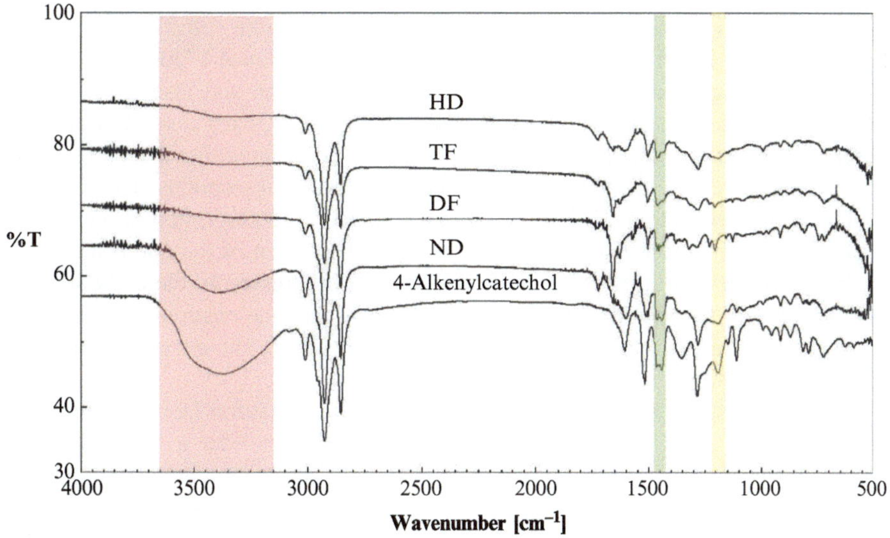

FIGURE 9.13

IR spectra of synthesis 4-alkenylcatechol film during drying.

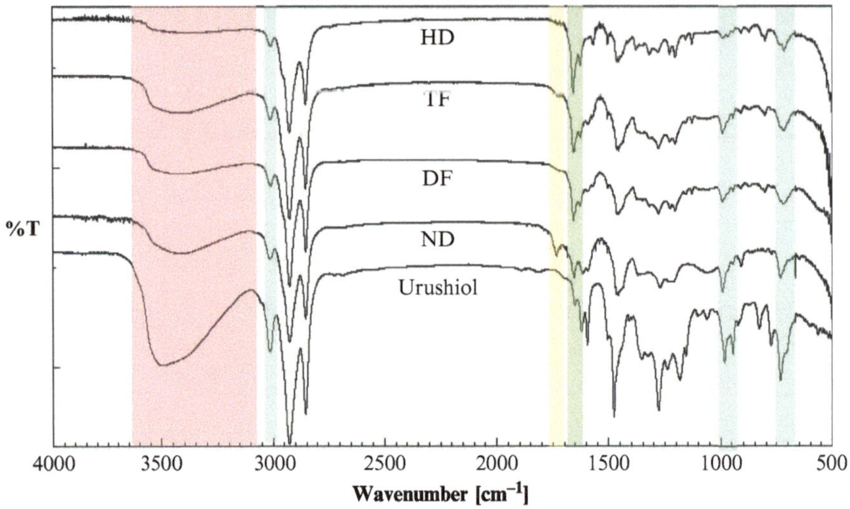

FIGURE 9.14

IR spectra of native urushiol film during drying.

The absorption at 3400 cm^{-1} due to a hydroxyl group decreased as the dryness progressed from ND to HD. The absorption at 1650 cm^{-1} for 4-alkenylcatechol and urushiol (Figures 9.13 and 9.14) and at 1720 cm^{-1} for 4-alkenylcatechol (Figure 9.14) due to the α,β-unsaturated carbonyl group increased and suggested that the catechol nucleus was oxidized by laccase and changed into carbonyl quinone. In detail, the absorption at 1720 cm^{-1} appeared in the ND state, and then decreased as the drying progressed, while absorption at 1650 cm^{-1} increased. It can be considered that a different type of carbonyl group is generated when the absorption at 1720 cm^{-1} decreased and 1650 cm^{-1} increased. In addition, a different absorption was also observed due to the difference in position in the side chain, and the 1720 cm^{-1} absorption was not observed for the side chain in the 3-position. It was considered that the yellow color of 4-alkenylcatechol films is related to the carbonyl group absorptions at 1720 cm^{-1}.

Due to the absorption of olefins at about 3010 cm^{-1}, the absorption of olefin (*cis*) at about 720 cm^{-1} for 4-alkenylcatechol, and the absorption of the conjugate olefin at about 990 cm^{-1} for urushiol decreased as drying progressed in the early stage of polymerization, the automatic oxidation reaction of the side chain and the reaction between the side chain and the catechol nucleus occurred. The considered reaction mechanism is shown in Scheme 9.23.

The colors of the synthesized 4-alkenylcatechol and natural lacquer films were quite different, and UV measurement was carried out in order to investigate the reason. It can be observed that urushiol showed wide absorption at 400 nm or more in the visible optical region during the drying process (Figure 9.15), and the film is brown to near black. For 4-alkenylcatechol, the absorption is only observed between 400 and 500 nm (Figure 9.16). Because the complementary of this low wavelength area is yellow, the synthesized 4-alkenylcatechol films are yellow with a little red.

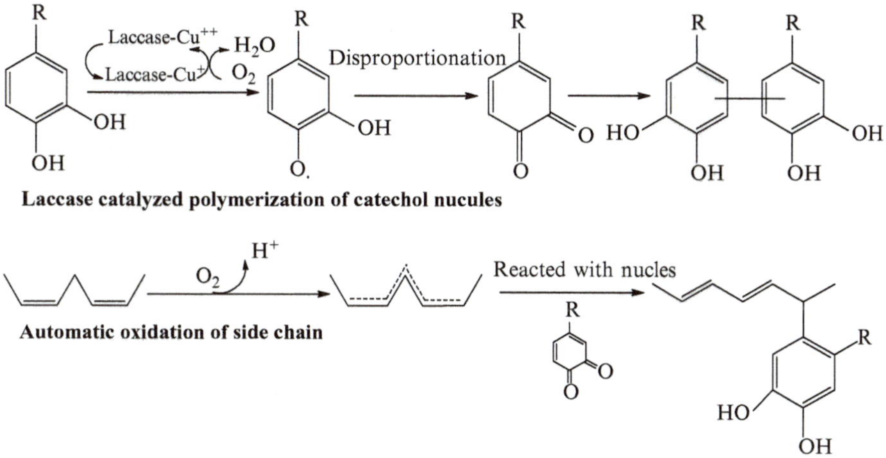

SCHEME 9.23

Postulated reaction mechanism of 4-alkenylcatechol.

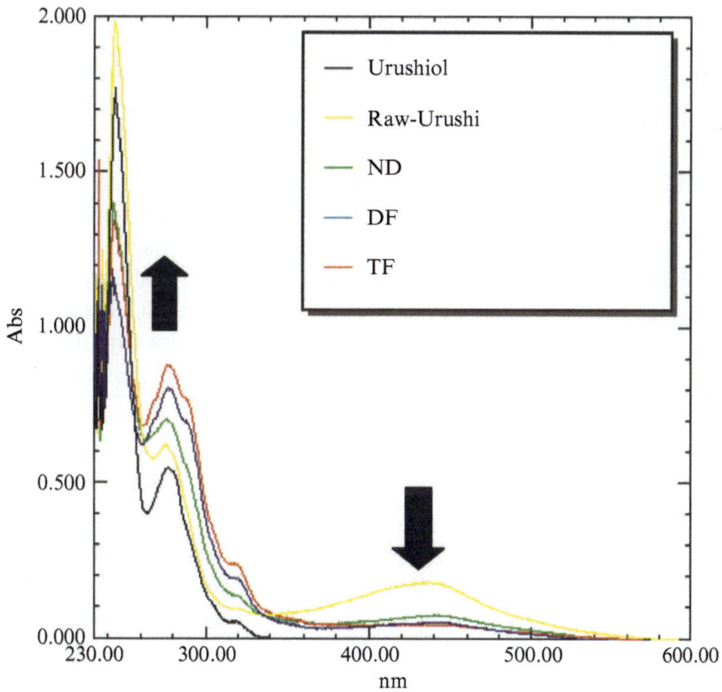

FIGURE 9.15

UV spectra of urushiol during drying time course.

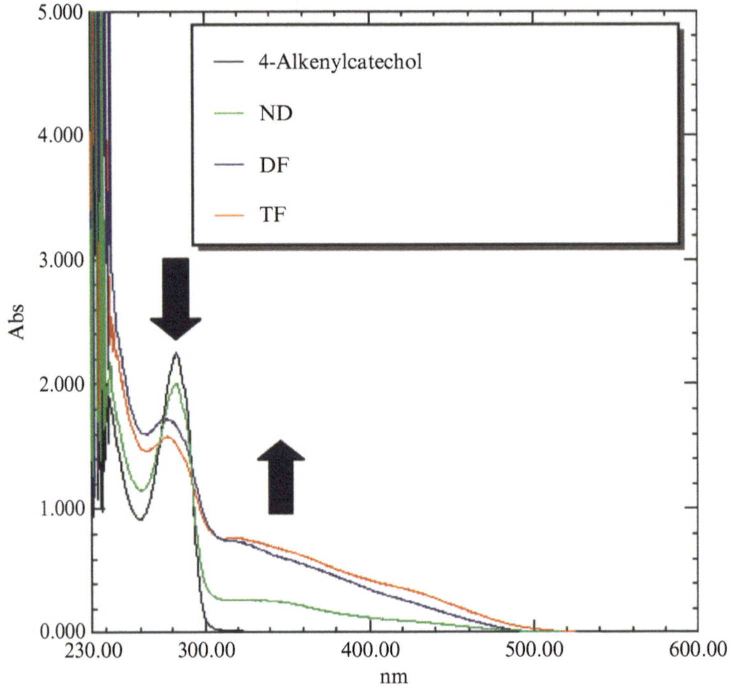

FIGURE 9.16

UV spectra of 4-alkenylcatechol during drying time course.

In the ultraviolet region, as for 4-alkenylcatechol, the benzene ring of catechol absorption at 280 nm decreases, and the absorptions of the biphenyl at 245 nm and quinone at 320 nm increase as the drying progressed. The increased absorption of quinone at 320 nm was observed in urushiol, but scarcely in 4-alkenylcatechol. From these results, it can be considered that the quinone of 4-alkenylcatechol with the side chain at the 4-position is stronger than the quinone of urushiol with the side chain in the 3-position, and the yellow color due to this quinone remains in the synthesized 4-alkenylcatechol film as the drying occurs.

REFERENCES

[1] R. Majima, J. Tahara, Über den Hauptbestandteil des Japanlacks. VI. Mitteilung: Über die Synthese des Hydro-urushiols, Ber. Deut. Chem. Ges. 21 (1915) 1606.

[2] H.J. Backer, N.H. Haack, Le principe toxique des fruits de renghas (*Semecarpus heterophylla* Bl.), Rec. Trav. Chim. 57 (1938) 225.

[3] H.S. Mason, The synthesis of 3-n-pentadecylcatechol (hydrourushiol), J. Am. Chem. Soc. 67 (1945) 1538.

[4] C.R. Dawson, D. Wasserman, H. Keil, 3-n-Pentadecylcatechol, J. Am. Chem. Soc. 68 (1946) 534.

[5] J.S. Byck, C.R. Dawson, Synthesis of compounds structurally related to poison ivy urushiol II. 4-Methyl-, 5-methyl-, 6-methyl-, and 4,5,6-trimethyl-3-pentadecylcatechol, J. Org. Chem. 33 (1968) 2451.

[6] A.P. Kurtz, C.R. Dawson, Synthesis of compounds structurally related to poison ivy urushiol. 3. 3-n-Pentadecylcatechol and 3-n-alkylcatechols of varying side-chain length, J. Med. Chem. 14 (1971) 729.

[7] C.R. Dawson, G.P. Ng, An improved synthesis of 3-*N*-pentadecyl catechol (3-PDC), Org. Prep. Proc. Int. 10 (1978) 167.

[8] H. Halim, H.D. Locksley, J.J. Memon, Vesicant principles of poison ivy and related plants: synthesis of the urushiols, 1,2-dihydroxy-3-[(*Z*)-pentadec-8-enyl]benzene and 1,2-dihydroxy-3-pentadecylbenzene, J. Chem. Soc. Perkin I (1980) 2331–2337.

[9] B. Loev, C.R. Dawson, Alkenylphenols related to the poison ivy principle. An improved method of synthesis involving the Na-butanol cleavage of benzyl ethers, J. Am. Chem. Soc. 78 (1956) 6095.

[10] J.H.P. Tyman, Synthetic and natural phenols, in: Studies in Organic Chemistry, vol. 52, Elsevier, 1996, pp. 501–502.

[11] J.H.P. Tyman, B.G. Schofield, C.H. Khor, A synthesis of the phenolic lipid, 3-[(Z)-pentadec-8-enyl] catechol, (15:1)-urushiol, Chem. Phy. Lip. 120 (2002) 101.

[12] H.J. Bestmann, N. Hessenauer, Synthesis of mono- and diolefinic components of urushiol, Synlett 555 (1990).

[13] M.V. Sargent, S. Wangchareontrakui, The synthesis and identification of alkenyl and alkadienyl catechols of Burmese lac, J. Chem. Soc. Perkin Trans. I 431 (1989).

[14] T. Miyakoshi, H. Kobuchi, N. Nimura, Y. Yoshihiro, Synthesis of urushiols with pentadecatrienyl side chain, two constituents of the sap of a lac tree, *Rhus vernicifera*, Bull. Chem. Soc. Jpn. 64 (1991) 2560.

[15] N. Niimura, Y. Kamiya, T. Sato, I. Katano, T. Miyakoshi, Synthesis of 3-[(8Z,11E,13Z)-8,11,13-pentadecatrienyl]catechol and analysis of urushiol-triene in *Rhus vernicifera* lacquer sap, J. Jpn. Oil Chem. 47 (1998) 171.

[16] Y. Kamiya, Y. Niimura, T. Miyakoshi, Evaluation of synthesized lacquer films using pyrolysis-gas chromatography/mass spectrometry, Bull. Chem. Soc. Jpn. 73 (2000) 2621.

[17] Y. Kamiya, T. Miyakoshi, Synthesis of urushiol components and analysis of urushi sap from *Rhus vernicifera*, J. Oleo Sci. 50 (2001) 865.

[18] Y. Kamiya, W. Saito, T. Miyakoshi, Synthesis and identification of laccol components from *Rhus succedanea* lacquer sap, J. Oleo Sci. 51 (2002) 473.

[19] T.V. Hansen, L. Skattebol, One-pot synthesis of substituted catechols from the corresponding phenols, Tetrahedron Lett. 46 (2005) 3357.

[20] W.J. Dejarlais, E.A. Emken, Syntheses of tetra- and hexadeuterated octadecenoates, Lipids 11 (1976) 594–598.

[21] N. Cohen, B.L. Banner, R.J. Lopresti, F. Wong, M. Rosenberger, Y.Y. Liu, E. Thorn, A.A. Liebman, Enantiospecific syntheses of leukotrienes C4, D4, and E4, and [14,15-3H2] leukotriene E4 dimethyl ester, J. Am. Chem. Soc. 105 (1983) 3661–3672.

[22] T. Miyakoshi, K. Nagase, T. Yoshida, Progress of Lacquer Chemistry, IPC, Japan, 1999.

CHAPTER

Modification of lacquer

10

10.1 FAST-DRYING LACQUER

Lacquer sap is polymerized into a tough and brilliant film when catalyzed by laccase. During the drying process, no organic solvent evaporates, only water. Because of the self-drying system, natural lacquer can be called a solventless and high-solid content component-type coating material. However, because the drying process of lacquer is based on enzymatic polymerization, it requires strict management of temperature and relative humidity, and a longtime to reach full dryness.

We have developed a small kneading mixer capable of repeatedly kneading lacquer sap to accelerate the enzymatic polymerization in raw lacquer sap, as shown in Figure 10.1 [1]. The repeated kneading process is called "kurome" in Japanese. It is a process that accelerates polymerization of lacquer monomer to form dimers, trimers, tetramers, ..., oligomers, and polymers. In detail, raw lacquer sap is stirred in an open vessel (Figure 10.1) at room temperature for about 1.5 h, and then the temperature is increased from 20 °C to 40 °C for 2-4 h until the water concentration is reduced to 3-5%, for kurome-0 (K-0). Water is added to K-0 to increase the moisture to about 20-25%, comparable to the water concentration of the raw lacquer, and the kurome process is repeated. By repeatedly adding water and reducing it, K-1, K-2, K-3, and K-4 are obtained. Here, K-0 is refined only once, not repeated. K-1 repeats kurome once, K-2 repeats kurome twice, and so on [2]. After kurome treatment, the molecular weight distribution of the raw and repeatedly reduced kurome lacquers were measured at 40 °C by aqueous-phase gel permeation chromatography (GPC; column: Tosoh, TSK-gel, α-3000, α-4000, and α-M, Φ7.8 mm × 300 mm × 3; eluent, DMF with 0.01 M LiBr) running on an HPLC system with an RI detector using polystyrene standards with molecular weights of 5.00×10^2, 2.63×10^3, and 1.81×10^4. The results are summarized in Table 10.1. It showed that the monomer decreased in the repeated kurome lacquers, while the oligomers and polymers increased from K-0 to K-4 suggested that enzyme polymerization was occurring during the repeated kurome process.

The antioxidation of raw and kurome lacquer was also examined using thermogravimetry and derivative thermogravimetry (DTG-TG) (Shimadzu, Japan). It is known that oxidation of methyl linoleate occurs easily. In this study, oxidation of pure methyl linoleate required 1.50 h. After adding raw and series kurome lacquers (as an antioxidation reagent) to the methyl linoleate, the induction times of the oxidation of raw to K-4 lacquers are 13.03, 11.03, 9.65, 8.33, 6.23, and 4.19 h, respectively,

Lacquer Chemistry and Applications. http://dx.doi.org/10.1016/B978-0-12-803589-4.00010-9

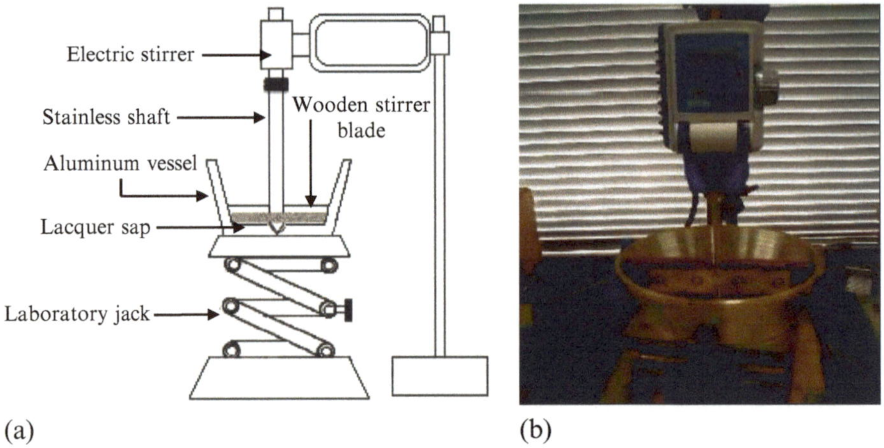

FIGURE 10.1

Image (a) and photo (b) of kneading mixer.

Table 10.1 Molecular weight distributions of raw and repeated kurome lacquers

Name of lacquer	Polymer	Oligomer	Monomer
Raw	0.00	16.28	83.72
K-0	0.64	33.30	66.06
K-1	0.82	39.73	59.45
K-2	1.13	40.08	58.79
K-3	1.56	40.50	57.93
K-4	2.19	42.85	55.03

Polymer: 10,000 ≥ MW > 3000, oligomer: 3200 ≥ MW ≥ 640, monomer: MW = 316-320.

which decreased according to the kurome treatments (Figure 10.2). These results suggest that the antioxidation of the lacquer decreased and that autoxidation of the side chain increased from the raw to K-4 lacquers, naturally, and the drying time are also greatly reduced from K-0 to K-4.

10.2 HYBRID LACQUER

The liquid components of lacquer sap, urushiol, laccol, and thitsiol, are catalyzed by laccase to form a durable film. During the enzyme-catalyzed polymerization, the activity of laccase and the antioxidation power of the phenolic structure play significant roles. High laccase activity can strongly accelerate the polymerization reaction, thereby greatly reducing the drying time of lacquer. On the other hand, the low antioxidation power of phenolic components can also contribute significantly to the

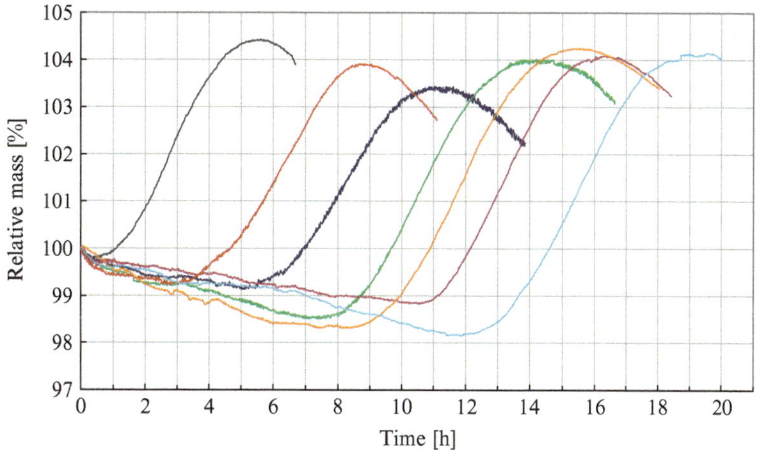

FIGURE 10.2

DTG spectra of (A) methyl linoleate (ML), (B) ML+raw, (C) ML+K-0, (D) ML+K-1, (E) ML+K-2, (F) ML+K-3, and (G) ML+K-4 lacquer.

olefin in the side chain of the autoxidation reaction. Although we cannot increase the activity of laccase, we can reduce the antioxidant capacity of urushiol by adding chemical reagents, such as silane or silicone, to react with the hydroxyl group of urushiol to decrease the concentration of urushiol monomers.

Hybrid lacquers based on raw [3] and kurome [4,5] lacquers were investigated in our laboratory for a longtime. Taking into account the fact that an organic silicone compound that has chlorine [6] and a mercapto [7] radical could possibly weaken the activity of the laccase, we selected many kinds of silane without chlorine and/or mercapto residues to prepare hybrid lacquers. The reaction mechanism between urushiol and organic silane also has been examined [8].

After analysis of the IR, NMR, and TGA/DSC measurements, the results showed that amino and an epoxy silane with the structures of *N*-(β-amino ethyl)-γ-amino propyl trimethoxy silane and β-(3,4-epoxy cyclohexyl ethyl trimethoxy silane had a fast alcoholysis reaction with urushiol and were understood to promote polymerization remarkably. Moreover, it is possible that the amino residue can cause the ring opening of epoxy residue and produce a mesh network structure with mutual penetration. Because the alcoholysis reaction occurred between the alkoxy residue of the organic polysiloxane and the hydroxyl group of urushiol, it can be surmised that a high polymer lacquer film was formed from the three-dimensional bridge construction by repetition of the Si-O combination. The reaction image of alcoholysis between urushiol and organic silane is shown in Scheme 10.1, and the reaction mechanism of urushiol with amino group in organic silane is shown in Scheme 10.2.

From the results of hybrid lacquer studies, we concluded that the reaction occurred at the hydroxyl group of the urushiol when adding organic silane to lacquer sap as shown in Scheme 10.3.

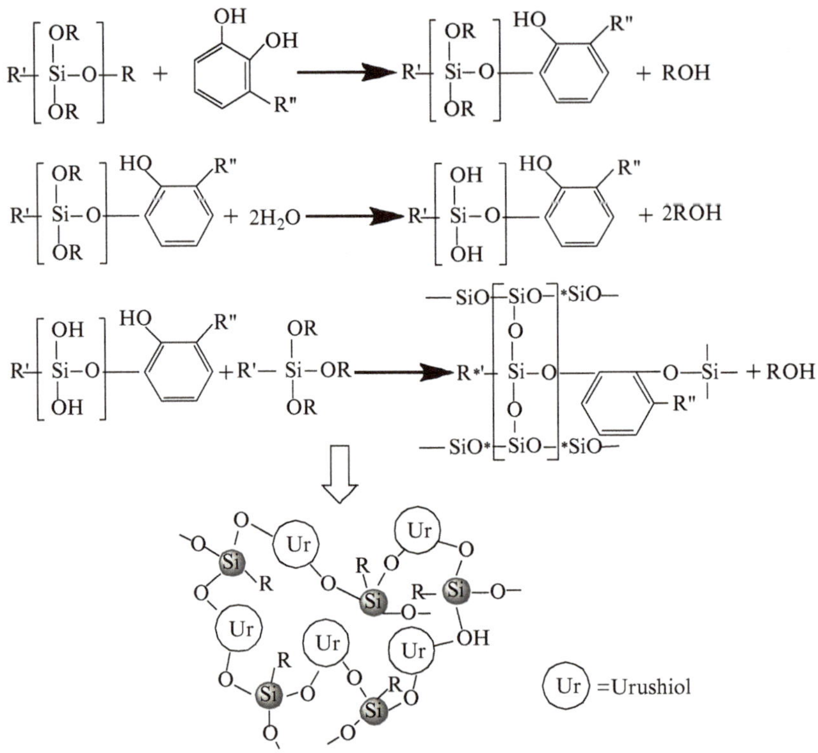

SCHEME 10.1

Alcoholysis between urushiol and organic silane.

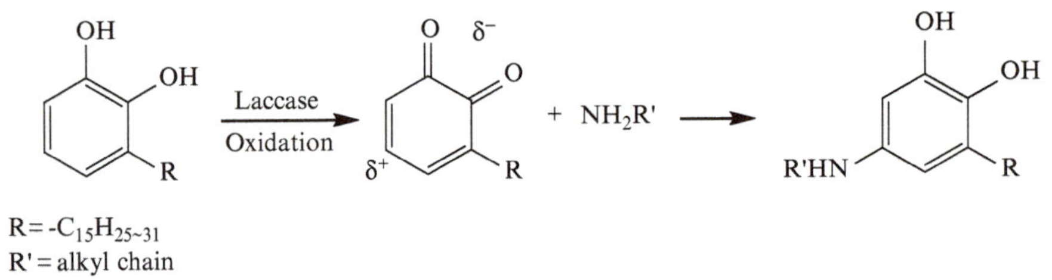

SCHEME 10.2

Reaction mechanism of urushiol with amino groups.

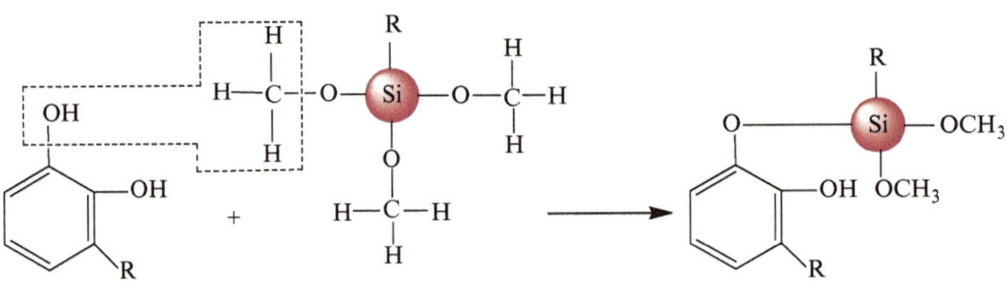

SCHEME 10.3

Reaction between urushiol and silane.

The silyloxy residue was generated by reaction of the sol-gel with the free water in the lacquer sap, and then the hydroxyl group of urushiol was attacked to generate urushiol radical. This reaction was repeated and increased the polymer ratio to form the lacquer film. The other components in lacquer sap, for example, polysaccharides and glycoproteins, did not react with organic silane in our tests. In general, a lacquer film forms via the oxidation of urushiol catalyzed by laccase, which then is followed by a coupling reaction as well as an autoxidation reaction on the long aliphatic unsaturated side chain. Lacquer has been used only in crafts until now because it requires a fixed temperature and about 70-80% relative humidity for the enzyme activities, and therefore, requires a special chamber for drying. Thus, management of the drying process is difficult. However, when lacquer was hybridized with organic silane, a reaction between urushiol and the silane compound occurred. It is expected that the lacquer hybridized with organic silane will be used not only for crafts, but also in the open air by industries due to its quick drying property based on the urushiol-silane and sol-gel reaction. Recently we reported one application of hybrid lacquer as a chrome-free corrosion-protective coating [9], and the results showed that the urushiol-silicate compound offers almost the same corrosion protection as the conventionally used chrome treatment.

10.3 NANOLACQUER

It is well known that the smaller the diameter of a particle, the greater is its specific surface area. If the particle diameter of a water drop in a lacquer sap emulsion becomes smaller, the area of contact between urushiol and laccase will increase, the enzymatic reaction will be catalyzed more easily, and a faster drying lacquer will be obtained. Based on this idea, a kneading mixer created in our laboratory was used to treat the raw lacquer (Figure 10.1). The size of water drops were analyzed by a nanoparticle size analyzer SALD-7100 (Shimadzu, Japan).

When the lacquer sample was stirred at 240 rpm for 60 min, the best particle dispersion was obtained, and the average particle diameter was 0.194 mm; for other conditions, the average particle sizes were approximately (A) 0.337 μm, (B) 0.233 μm, and (C) 0.279 μm, as shown in Figure 10.3 [10].

A segment mixer KF-6 with more power produced by Toyoseiki Co., Tokyo, Japan, also was used to treat natural raw lacquer sap, and the result is shown in Figure 10.4. It can be observed that the average diameter of water drops is about 0.1 μm, and some of them under 100 nm achieve nanolevel according to the nano-definition; this nanolacquer was stable during 6 months storage according to our test. The decrease in particle size greatly increased the color and gloss of the nanolacquers and also the UV resistance.

In general, colored lacquer is prepared by adding organic and/or inorganic pigment to lacquer sap. However, some pigments, such as cinnabar (HgS) for red color and for realgar (As_4S_4) for yellow, are toxic, some pigments, such as red lead (Pb_3O_4) for red color, interfere with lacquer drying, and some pigments, such as titanium dioxide (TiO_2)

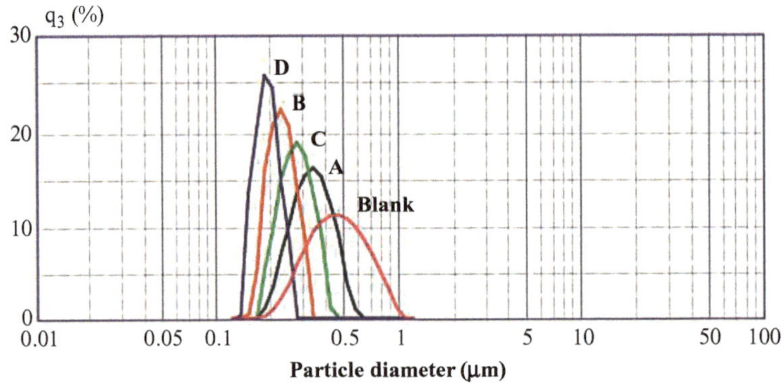

FIGURE 10.3

Particle size distribution of refined lacquer saps. The stirring speed and time were (A) 180 rpm, 30 min, (B) 180 rpm, 60 min, (C) 240 rpm, 30 min, and (D) 240 rpm, 60 min.

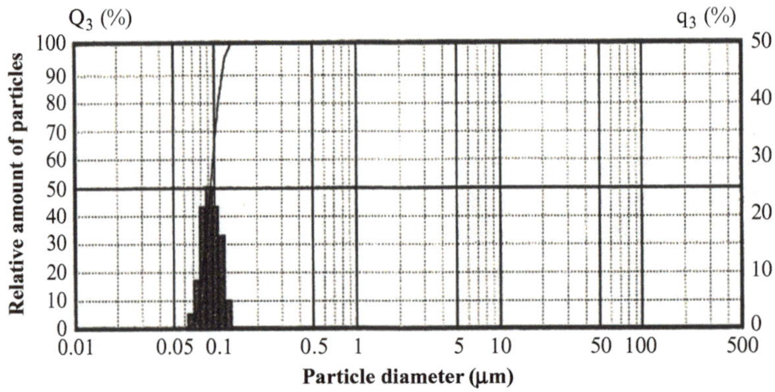

FIGURE 10.4

Size of water drops estimated by SALD-7100.

for white color, could cause a photochemical reaction in the lacquer film [11]. An organic colorant or dye, either from animals and plants, or a synthetic dye almost the same color as an organic compound, has a conjugate double bond. The double bond is easily decomposed by optical irradiation; therefore, the color tends to fade. This is the reason why traditional-colored lacquer does not withstand light well. A new method of dyeing by using precious metal ultrafine particles has been developed [12]. We have used nano-Au-colloid (~15 nm) and Ag-colloid (5-8 nm) obtained from Nippon Paint Co., Ltd., Japan, to prepare a newly designed nanolacquer [13].

The precious metal colloids showed a homogeneous dispersion in lacquer sap (Figure 10.5). The Au-colloid produces a homogeneous red tone similar to the standard paint color S41-142, and Ag-colloid lacquer film a homogeneous yellow tone similar to the standard paint color S44-257, as shown in Figure 10.6.

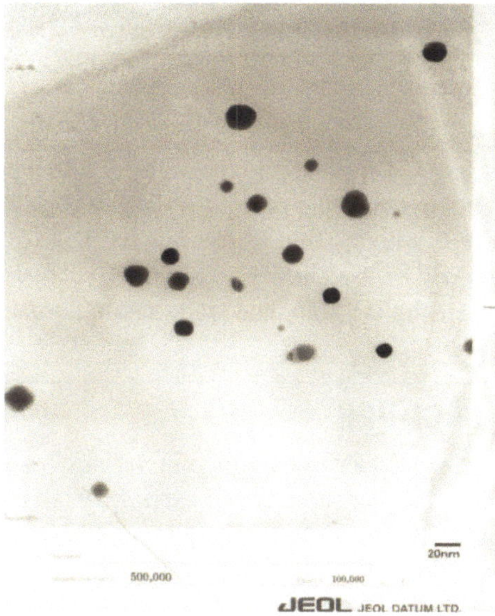

FIGURE 10.5

Dispersion of Au-colloids in lacquer sap.

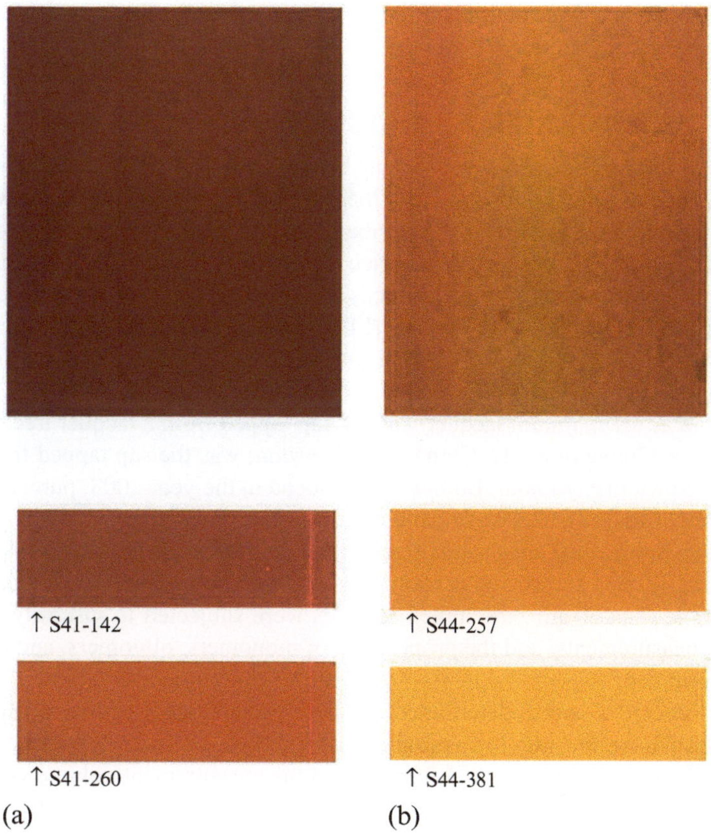

FIGURE 10.6

(a) Lacquer/Au and (b) lacquer/Ag films. The photos below are standard paint colors (Japan Paint Manufacturers Association, 1993 version).

The lacquer is first microdispersed and then mixed with a precious metal colloid showing the preeminent characteristic of light resistance. It is expected that the fast-drying lacquer hybridized with organic silane, polyurethane [14], epoxy, and nano-precious metal colloids will be used not only for crafts, but also tested in the open air by industries due to its strong UV light and water resistances.

10.4 BLENDED LACQUER

Lacquer saps tapped from lacquer trees grown in various Asian countries were described in Chapter 1. The lacquer sap from *G. usitata*, which has thitsiol as the main component, has a slower drying time and requires more severe drying conditions than sap from *T. vernicifluum*, which has urushiol as the main component, and produces a smoother and glossier film surface. However, the lacquer sap from *G. usitata* is about $20-30/kg in the international market, which is cheaper than that from *T. vernicifluum* at about $100-120/kg in the market. Therefore, how to effectively use this inexpensive lacquer sap in the coating industry becomes a very important subject for lacquer science. We have made many attempts to effectively use this cheaper material and developed many methods.

10.4.1 URUSHIOL-BLENDED THITSIOL

We have developed a series of fast-drying lacquers using the refining (kurome), hybridized, and modified methods. Using these techniques to improve the slow drying time of thitsiol, in this section, we describe blending urushiol with thitsiol to make a new blended lacquer. Comparing the viscosity, drying property, film characteristics, and analyzing the results from two-dimensional nuclear magnetic resonance (NMR) and field desorption mass spectrometry (FD-MS) elucidated the reaction mechanism between urushiol and thitsiol. We wish to further explore the use of thitsiol in the coating industry based on the results described in this section in the future.

The raw *T. vernicifluum* lacquer was the sap tapped from a lacquer tree grown in Chengkou of Chongqing city, China, and *G. usitata* was the sap tapped from a lacquer tree grown in Myanmar. Both were produced in the year 2008, purchased from Tohityu Co., Ltd. in Osaka, Japan, and used directly.

The two raw lacquer saps were mixed in various ratios, as shown in Table 10.2, in plastic sample cups and stirred at room temperature at least for 10 min. After standing for more than 30 min, the mixed samples were subjected to viscosity and drying time measurements, and the proportions of monomers, oligomers, and polymers were determined.

The viscosities were determined at room temperature using a Brookfield Engineering Laboratories, Inc., programmable DV-II+ viscometer and a CPE-51 spindle. The rotation speed was 5-20 rpm, and the measurement volume was 0.5 ml. The results are summarized in Table 10.3. Because thitsiol contains more oligomers and polymers than urushiol, the viscosities of the blended lacquers increased with the addition of thitsiol, in good agreement with the increase in the proportions of

Table 10.2 Ratios of blended Chinese and Thai lacquers

Blended lacquer	*T. vernicifluum* (wt%)	*G. usitata* (wt%)
BL-1	100	0
BL-2	100	10
BL-3	100	30
BL-4	100	50
BL-5	100	100
BL-6	0	100

Table 10.3 Molecular weight distribution and viscosities of blended lacquers

Blended lacquer name	Monomer	Oligomer	Polymer	Viscosities (mPa s^{-1})
BL-1	85.3	14.7	0.0	1251
BL-2	83.3	16.5	0.2	1635
BL-3	76.2	22.7	1.1	2238
BL-4	69.2	28.2	2.6	3678
BL-5	64.4	32.5	3.1	4857
BL-6	52.6	40.3	7.1	7047

oligomers and polymers in the blended lacquers due to the thitsiol. All viscosities of the blended lacquers are suitable for coating material in 1000 to 7000 mPa s^{-1}.

The proportions of monomers, oligomers, and polymers in the blended lacquers were determined at 40 °C by aqueous-phase gel permeation chromatography as described in Section 10.1. The polystyrene standards used had molecular weights of 1.90×10^5, 4.39×10^4, 5.40×10^3, and 5.00×10^2. The elution rate and pressure of the DMF eluent were 0.8 ml/min and 48 kgf/cm^2, respectively. The calculated data are summarized in Table 10.3. The results showed that as the ratio of thitsiol increased in the blended lacquers, monomers decreased and oligomers and polymers increased. The increase in the percentages of oligomers and polymers corresponded to the increased ratio of thitsiol, suggesting that a reaction between urushiol and thitsiol had occurred.

The surface morphology, molecular motion, and glass transition behavior of polymer films are quite important because many properties such as chemical and abrasion resistance and wettability are determined predominantly by the structure within a few nanometers of the surface [15,16]. In order to reveal the surface properties of the blended lacquer films, the glass transition temperature (T_g) was analyzed by a rigid-body pendulum physical property-testing (RPT) instrument (A&D, Japan). The RPT oven was programmed to rise from 25 °C to 250 °C at a rate of 2 °C min^{-1}. All blended lacquer samples were dried under 80% relative humidity at 30 °C for 7 days. The results in Figure 10.7 show that the T_g decreased with the increase in the percentage of thitsiol. Compared with the T_g of 119 °C for

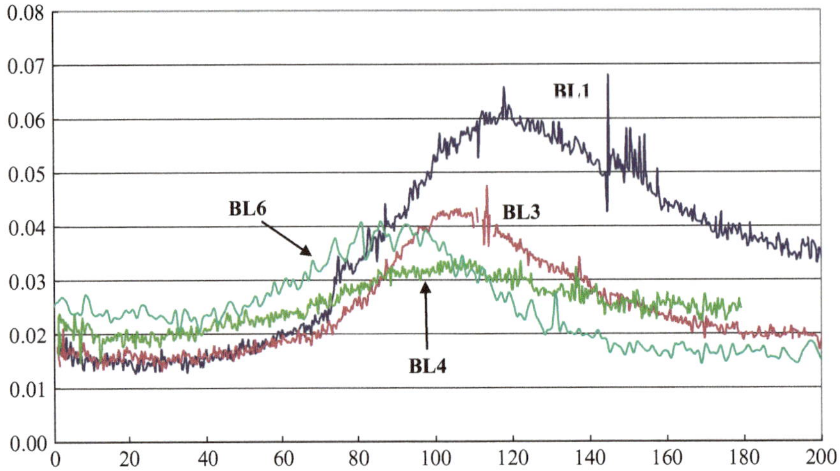

FIGURE 10.7

T_g of blended lacquer films with RPT measurement.

the *T. vernicifluum* lacquer film, the T_g of the blended lacquer films were lower, that is, 108 °C for 30 wt%, 104 °C for 50 wt%, but higher than the 79 °C of the *G. usitata* lacquer film. All T_g of the blended lacquer films were higher than that of the thitsiol lacquer film, suggesting that the cross-linkage polymerization in blended lacquers is faster than that in thitsiol lacquer. All properties of the lacquer films, such as color, gloss, and T_g, changed depending on the drying time and drying conditions because the side chains of urushiol and/or thitsiol undergo continuous autoxidation during the drying process.

The drying properties of the blended lacquers were measured with an automatic drying time recorder (Tai Yu Equipment Co., Ltd., Osaka, Japan) under 80% relative humidity at 30 °C. The thickness of the tested lacquer films was 76 μm. The lacquer drying process could be divided into three stages of DF, TF, and HD as described in Section 9.3. The time until the mark of the needle appears on the glass board is equivalent to DF, the time until the mark of the needle appears on the lacquer film surface is equivalent to TF, and the time until the mark of the needle on the lacquer film surface disappears completely is equivalent to HD. The results are summarized in Table 10.4. It can be observed that the drying times of 50 wt% (BL4) and 100 wt% (BL5) additions of thitsiol to the urushiol have the same DF after 2 h, HD after 5 h, and HB hardness after 5 days. The blended lacquers with 10 wt% (BL2) and 30 wt% (BL3) thitsiol achieved DF after 1 h and 2H hardness after 5 days, the same as *T. vernicifluum* lacquer (BL1). The reason will be discussed in the following section. There was no change in the drying time of the *G. usitata* lacquer (BL6) even when laccase separated from *T. vernicifluum* lacquer was added, implying that the polymerization of thitsiol is activated by the active intermediates of urushiol. This conjecture was confirmed by the results from two-dimensional NMR and FD-MS measurements described in the following section.

Table 10.4 Drying properties of blended lacquers under 80% relative humidity at 30 °C

	Drying properties									
Name	**1 h**	**2 h**	**3 h**	**4 h**	**5 h**	**1 day**	**2 day**	**3 day**	**4 day**	**5 day**
BL-1	DF	TF	HD	HD	6B	F	F	H	2H	2H
BL-2	DF	TF	HD	HD	6B	F	F	H	2H	2H
BL-3	DF	DF	TF	HD	6B	F	F	F	2H	2H
BL-4	ND	DF	TF	TF	HD	B	B	B	B	HB
BL-5	ND	DF	TF	TF	HD	2B	B	B	B	HB
BL-6	ND	ND	ND	ND	ND	TF	HB	6B	6B	6B

Abbreviations: DF, dust-free dry; TF, touch-free dry; HD, hardened dry.

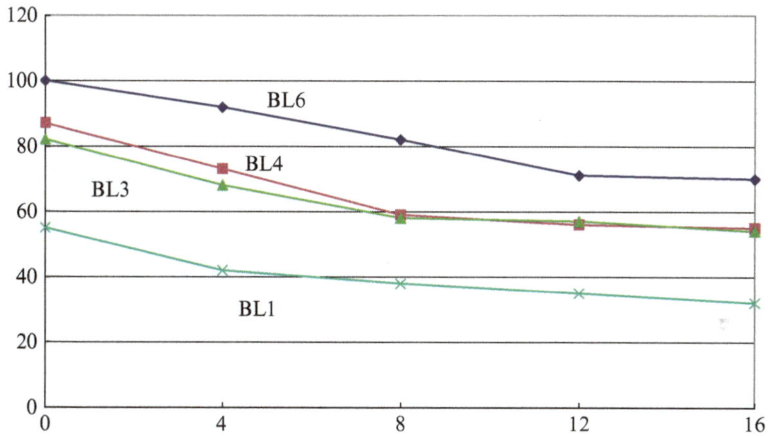

FIGURE 10.8

Gloss maintenance of blended lacquer films.

UV irradiation from one side of the square glass plate film was measured using a Handy UV800 system (ORC, Tokyo, Japan), equipped with a high-pressure mercury lamp HHL-800/C-FL (80 W cm^{-1}, emission light length 100 mm). The distance from the source of UV irradiation in the experiment was 82.5 mm. BL3 and BL4 samples were subjected to the ultraviolet light resistance test, and the results were compared with those of BL1 and BL6. All samples were applied to glass sheets (40 mm × 20 mm × 1 mm) at 76 μm thickness and dried in 80% relative humidity at 30 °C for over 1 month. The decrease in the gloss of the blended lacquer films is shown in Figure 10.8. The glosses of the blended lacquer films were lower than those of the pure Thai lacquer film (BL6), but higher than those of the pure China lacquer film (BL1). The gloss maintenance was calculated as 70% for BL1, 63% for BL3, 66% for BL4, and 53% for BL6, respectively (Figure 10.8). Because a reaction between urushiol and thitsiol occurred to form polymer networks, it can be concluded that maintenance of the gloss of blended lacquer film may be a contribution of the thitsiol component.

In order to examine the behavior of the enzymatic dehydrogenative polymerization of blended lacquers during the drying process, the time course of IR spectra was determined using a JASCO Fourier transform infrared 460$^+$ spectrometer (Tokyo, Japan). Because it was possible for the moisture in the lacquer exudates to melt a KBr or NaCl crystal, a polyvinylidene chloride film (Saran Wrap, Asahi Kasei Co., Tokyo, Japan) were used to support the IR samples. The lacquer saps were spread thinly on the polyvinylidene chloride film with a spatula, and the same kind of polyvinylidene chloride film was used as a blank. Spectra of samples were acquired at 2 cm^{-1} resolution ranging from 500 to 4000 cm^{-1}.

All lacquer samples were dried in 80% RH at 30 °C for 0, 1, 3, 6, 24, and 72 h. Figure 10.9 shows the IR spectra of blended lacquers that were dried over the time course. The peaks at 3400 cm^{-1} and 1400 cm^{-1} due to the absorption of the hydroxyl groups decreased, suggesting that laccase-catalyzed polymerization had occurred. However, the absorption peaks at 1470 cm^{-1} and 1080 cm^{-1} of quinone-olefin and/or dibenzofuran did not appear, and the absorptions at 3013, 1280, 732, and 698 cm^{-1} were decreased, indicating that monomeric urushiol decreased as the laccase-catalyzed polymerization advanced.

The enzyme-catalyzed reaction products of the blended lacquers were separated with acetone (50 ml) at 3.5 h, and the TF and HD stage were determined after drying in a chamber with 80% relative humidity at 30 °C. The acetone solution

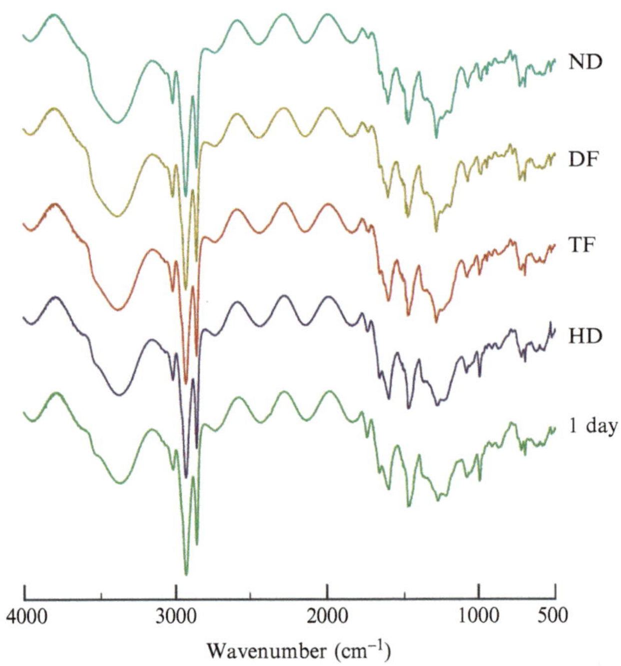

FIGURE 10.9

IR spectra of blended lacquer at ND, DF, TF, HD, and 1 day.

was filtered off to obtain the acetone-insoluble fraction, namely acetone powder, which includes the enzyme laccase and polysaccharides. The acetone-soluble fraction was evaporated under vacuum to obtain the residue. This residue was analyzed by an HPLC LC-908 system as described in Section 3.1. The sample (5 mg) was dissolved in 2 ml of $CHCl_3$. The resulting solution was injected using a 5 ml syringe into the column through a pretreatment PTFE cartridge containing an H-13-5 filter disc into the column, and then the column was eluted with $CHCl_3$ at a rate of 2.0 ml min^{-1}; polystyrenes with molecular weights of 200, 400, 600, and 1000 were used as standards.

The blended lacquers were dried in 80% RH at 30 °C, and then used as NMR samples in the three stages of DF, TF, and HD. All NMR spectra including two-dimensional NMR such as double-quantum-filtered correlated spectroscopy (COSY), heteronuclear multiple quantum coherence (HMQC), and heteronuclear multiple bond coherence (HMBC) were recorded at 500 MHz on a JEOL JNM-ECA500 spectrometer by using the phase-sensitive mode and a field-gradient probe. DMSO-d_6 was used as a solvent for tetramethylsilane (TMS) as the internal standard. JEOL standard software was applied to the two-dimensional spectra.

The structures of oligomers produced by the laccase-catalyzed dehydrogenative polymerization of blended lacquers were analyzed over the time course by proton-NMR and two-dimensional NMR measurements. The samples were dried in 80% RH at 30 °C, and then taken as NMR samples in the three stages of DF, TF, and HD. Figure 10.10 shows the ^1H NMR spectra of blended lacquers in three stages (DF, TF, and HD) due to

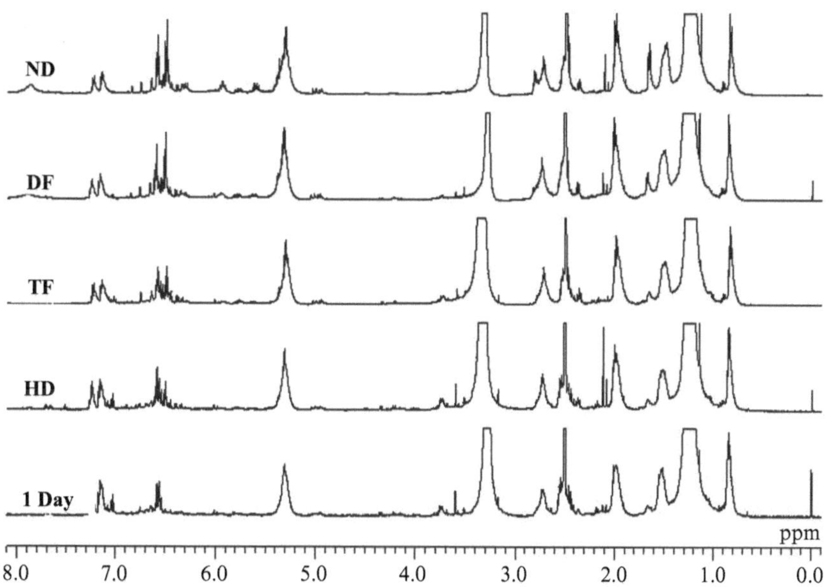

FIGURE 10.10

^1H NMR spectra of the blended lacquers.

the enzymatic dehydrogenative polymerization; "ND" stands for nondry (wet blended lacquer); and "1 day" stands for lacquers dried for 1 day. In the ^{1}H NMR spectra, signals in the δ_H-range of 5-6 showed that olefin protons of the urushiol side chain decreased rapidly due to the occurrence of laccase-catalyzed dehydrogenative polymerization that occurred between urushiol and thitsiol. It can be observed that the enzymatic dehydrogenative polymerization of urushiol is higher than that of thitsiol. The difference in the number of C–C double bonds is about 4.6H in the side chains of urushiol and about 3.2H in those of thitsiol. Figure 10.10 also shows that the signals for protons at δ_H-range of 6.37-6.67, corresponding to biphenyl units increased in the integrated area ratio due to the coupling of urushiol and thitsiol, suggesting that radical coupling products of nucleus-nucleus (Ph-Ph) and nucleus-side chain (Ph-C) were generated [17]. Consequently, hydroxyl groups had decreased with the progress of the laccase-catalyzed dehydrogenative polymerization.

In order to characterize the chemical reaction between urushiol and thitsiol, two-dimensional NMR measurement was carried out. Figure 10.11 is one of the HMBC spectra of blended lacquer dried to the stage of DF; all correlation signals were assigned and the data is summarized in Table 10.5. From Table 10.5, it can be concluded that an urushiol-thitsiol dimmer was generated. After assigning all two-dimensional NMR spectra, we estimated that the urushiol radicals were first generated in the early stages of polymerization due to enzyme-catalyzed reaction, and then semiquinone radicals were generated as the radicals shifted. Some semiquinone radicals coupled to form biphenyl and/or dibenzofuran types of urushiol dimers. On the other hand,

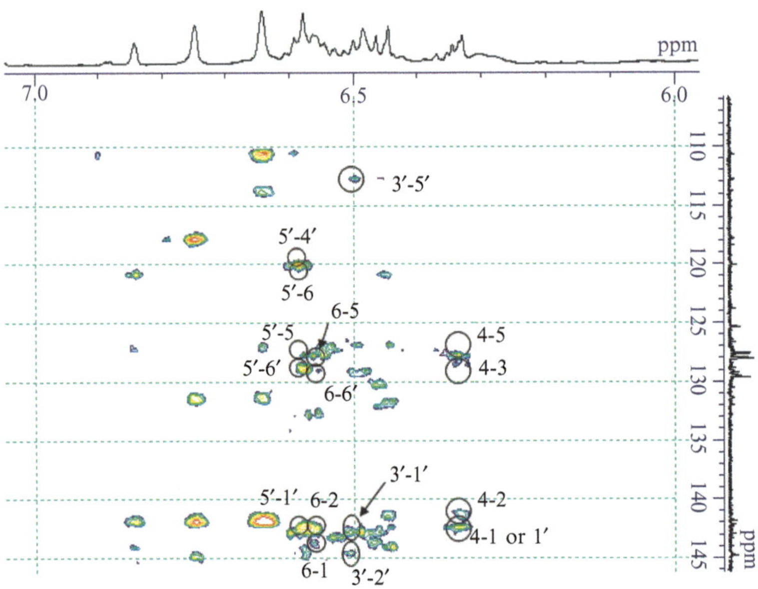

FIGURE 10.11

HMBC spectrum of blended lacquers (DMSO-d_6 as solvent, 25°C).

Table 10.5 Assignments of correlations in HMBC of blended lacquer

Peaks (ppm)	Coupling	Peaks (ppm)	Coupling
6.34-29.74	4-7	6.56-141.39	6-2
6.34-127.87	4-5	6.56-142.504	6-1
6.34-128.56	4-3	6.58-120.08	5′-4′
6.43-141.39	4-2	6.58-120.24	5′-6
6.43-142.50	4-1 or 4-1′	6.58-129.18	5′-6′
6.50-112.73	3′-5′	6.58-142.50	5′-1′
6.50-142.62	3′-1′	6.58-127.87	5′-5
6.50-144.58	3′-2′	2.36-118.67	3′-7′
6.56-127.87	6-5	2.53-121.07	4-7
6.56-129.18	6-6′		

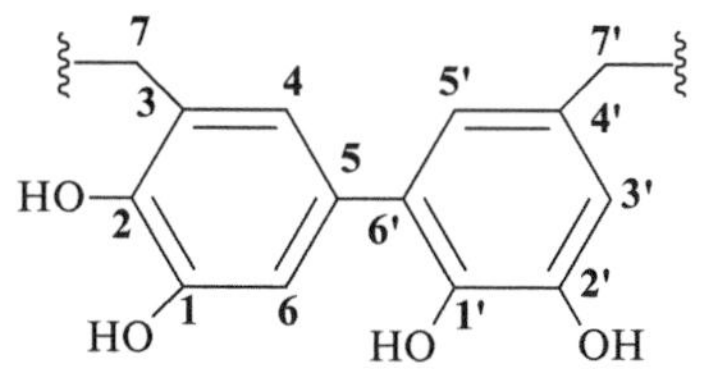

FIGURE 10.12

Structure of urushiol-thitsiol dimer.

some urushiol semiquinone radicals as initiators attack thitsiol to form thitsiol radicals, so that three kinds of dimers, urushiol-urushiol, urushiol-thitsiol (Figure 10.12), and thitsiol-thitsiol, were generated and detected in NMR measurements. Based on these results, the reaction mechanism will be discussed in following section.

Field desorption mass spectrometry (FD-MS) measurement was carried out using a JEOL JMS-700 QQ-type spectrometer system at the acceleration voltage of 6 kV with a mass range from *m/z*=45-1000 with Me_2CO as the analysis solvent.

The FD-MS spectrum obtained from the ether extract of the laccase-catalyzed dehydrogenative polymerization of urushiol (BL1) exhibited a mixture of dimethyl ethers of monomers and dimers. The peaks at *m/z*=342 (C_{15}-triene type), 344 (C_{15}-diene type), 346 (C_{15}-monoene type), and 348 (C-15 saturated type) were *O*-dimethyl ethers of unreacted monomers. However, the prominent peaks in the *m/z* range of 684-692 correspond to the *O*-tetramethylated ethers of dimeric products (Figure 10.13A) that were produced by the laccase-catalyzed dehydrogenative polymerization of urushiol.

Similarly, the FD-MS spectrum of thitsiol (BL6) also exhibited a mixture of *O*-dimethyl ethers of monomers and dimers. The peaks at *m/z*=370 (C_{17}-triene type), 372 (C_{17}-diene type), 374 (C_{17}-monoene type), 376 (C_{17}-saturated type), and 382 [C_6H_5-$(CH_2)_{12}C_6H_3$-$(OH)_2$ type] correspond to thitsiol and 12-phenyldodecylcatechol *O*-dimethyl ethers of unreacted monomers. The prominent peaks in the *m/z* range

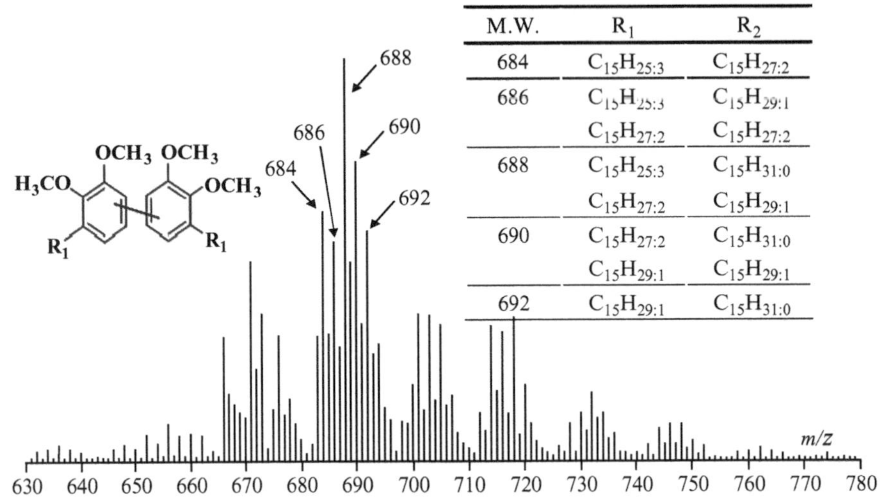

M.W.	R_1	R_2
684	$C_{15}H_{25:3}$	$C_{15}H_{27:2}$
686	$C_{15}H_{25:3}$	$C_{15}H_{29:1}$
	$C_{15}H_{27:2}$	$C_{15}H_{27:2}$
688	$C_{15}H_{25:3}$	$C_{15}H_{31:0}$
	$C_{15}H_{27:2}$	$C_{15}H_{29:1}$
690	$C_{15}H_{27:2}$	$C_{15}H_{31:0}$
	$C_{15}H_{29:1}$	$C_{15}H_{29:1}$
692	$C_{15}H_{29:1}$	$C_{15}H_{31:0}$

(a)

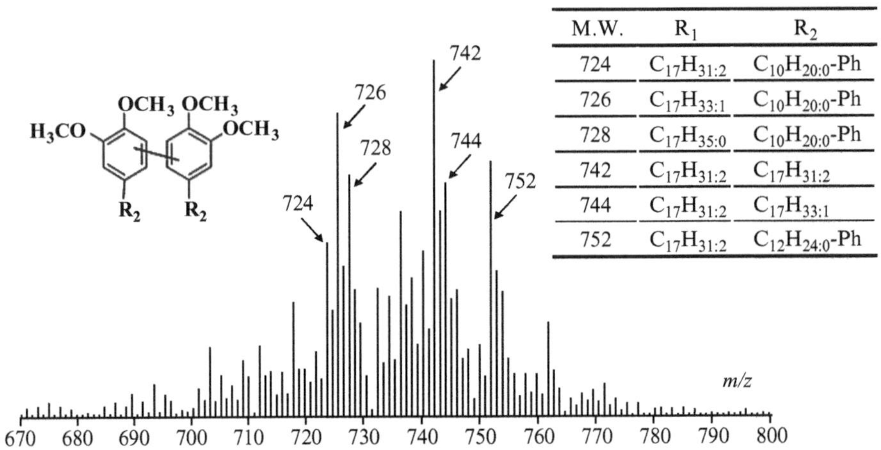

M.W.	R_1	R_2
724	$C_{17}H_{31:2}$	$C_{10}H_{20:0}$-Ph
726	$C_{17}H_{33:1}$	$C_{10}H_{20:0}$-Ph
728	$C_{17}H_{35:0}$	$C_{10}H_{20:0}$-Ph
742	$C_{17}H_{31:2}$	$C_{17}H_{31:2}$
744	$C_{17}H_{31:2}$	$C_{17}H_{33:1}$
752	$C_{17}H_{31:2}$	$C_{12}H_{24:0}$-Ph

(b)

FIGURE 10.13

FD-MS results of dimeric reaction of (a) urushiol and (b) thitsiol.

of 724-752 correspond to the *O*-di- and *O*-tetramethylated ethers of thitsiol dimeric products (Figure 10.13B). These dimeric ion peaks clearly indicate that the dimeric products were produced by coupling of thitsiol radicals due to the laccase-catalyzed dehydrogenative polymerization.

Furthermore, the FD-MS spectrum of BL5 exhibits a mixture of monomers and dimers of thitsiol and urushiol *O*-dimethyl ethers. The peaks at $m/z = 342$ (C15-triene type), 344 (C_{15}-diene type), 346 (C_{15}-monoene type), 348 (C_{15} saturated type), corresponding to urushiol *O*-dimethyl ethers and *m/z*=370 (C_{17}-triene type), 372 (C_{17}-diene type), 374 (C_{17}-monoene type), 376 (C_{17}-saturated type), 382 [C_6H_5-$(CH_2)_{12}C_6H_3$-$(OH)_2$ type], were *O*-dimethylated ethers of thitsiols and 12-phenyldodecylcatechol. The peaks at the *m/z* range of 684-692 correspond to *O*-dimethyl ethers of urushiol dimers, while the peaks in the *m/z* range of 724-752 correspond to *O*-dimethyl ethers of thitsiol dimers.

In addition, the spectrum exhibits several new molecular ions peaks in the *m/z* range of 712-718, which correspond to the *O*-dimethylated ethers of dimers produced by cross coupling among urushiols, thitsiols, and 10-phenyldodecanylcatechols including 12-phenyldodecylcatechol (Figure 10.14). These new dimeric molecular ions clearly indicate the occurrence of cross coupling among radical species produced from laccase-catalyzed urushiol, thitsiol, and 10-phenyldodecanylcatechols by enzymatic single-electron dehydrogenation to produce the corresponding compounds.

From the above analysis, we considered the reaction mechanism as follows. Radical formation of urushiol with the enzyme laccase occurred rapidly under 30 °C in an 80% relative humidity environment because urushiol is easily dried to form a film by enzymatic polymerization, but radical formation of thitsiol by the laccase was very slow in the same conditions. However, the blended lacquer with 30 wt% of thitsiol dried almost as fast as urushiol in the same drying conditions. The reaction mechanism in the blended lacquers can be described as follows: the laccase-catalyzed dehydrogentive formation of phenoxyl radicals from urushiol is first catalyzed by enzymatic oxidation, then thitsiol radicals are produced from thitsiols with phenoxyl radicals of urushiol by a radical transfer reaction [18] and then a coupling reaction between these two radical species occurs to form dimers. In the propagation step, the dimeric products would further undergo both radical transfer reactions and dehydrogentive oxidation by laccase under the presence of dioxygen in the atmosphere. The resulting radical species including monomer radical species would then couple among themselves to give the higher molecular weight polymer, as shown in Scheme 10.4.

Scheme 10.4 suggests that phenoxyl radicals of urushiols are formed by laccase-catalyzed oxidation in the presence of dioxygen in the atmosphere; then, thitsiol radicals are formed by the radical transfer of the radical species of urushiols in the presence of laccase. Subsequently, the radical coupling takes place among these radical species, leading to the formation of urushiol-thitsiol mixed polymers. These results clearly show that urushiol and thitsiol can be copolymerized via a radical transfer reaction by the enzymatic oxidative coupling.

In the blended lacquers of thitsiol and urushiol, we found that blends with about 30 wt% of thitsiol to urushiol dried quickly and had a good viscosity near that of the mother lacquer urushiol. The blended lacquer film also has a higher gloss and UV light resistance compared with the China lacquer film. The reaction mechanism that occurred between urushiol and thitsiol due to laccase under atmospheric oxygen is discussed in the section on NMR and FD-MS measurements. The results clearly showed that urushiol and thitsiol can be copolymerized via a radical transfer reaction by the enzymatic catalysis. Because many kinds of antioxidants with hydroxyl groups such as catechols and phenols exist in natural products, it can be speculated that the radical transfer and radical coupling is a very common reaction.

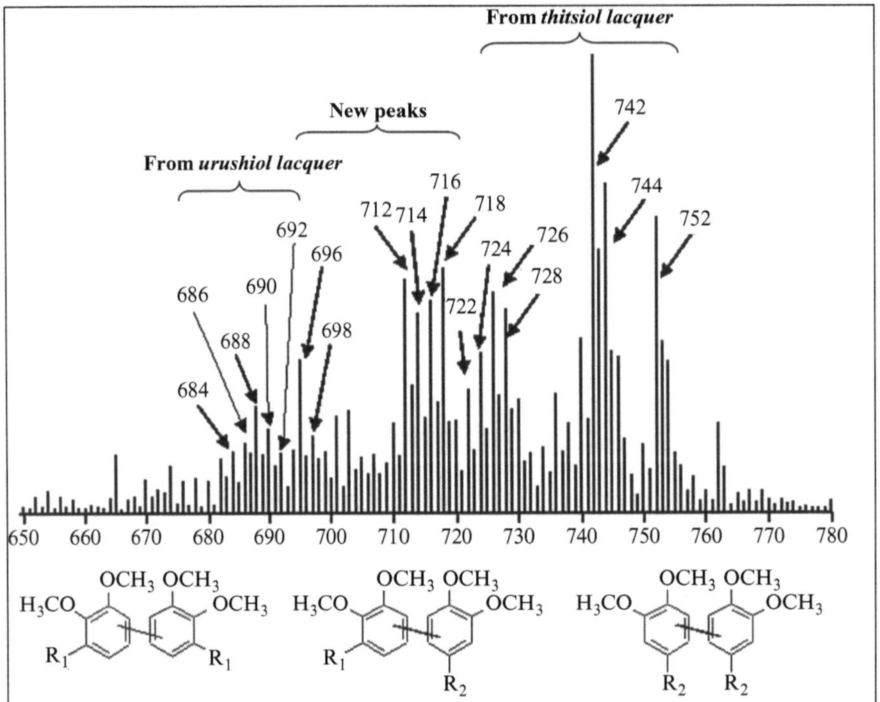

From *R. vernifera*			New peaks			From *G. usitata*		
M.W.	R_1	R_2	M.W.	R_1	R_2	M.W.	R_1	R_2
684	$C_{15}H_{25:3}$	$C_{15}H_{27:2}$	694	$C_{15}H_{25:3}$	$C_{10}H_{20:0}$-Ph	724	$C_{17}H_{31:2}$	$C_{10}H_{20:0}$-Ph
686	$C_{15}H_{25:3}$	$C_{15}H_{29:1}$	696	$C_{15}H_{27:2}$	$C_{10}H_{20:0}$-Ph	726	$C_{17}H_{33:1}$	$C_{10}H_{20:0}$-Ph
	$C_{15}H_{27:2}$	$C_{15}H_{27:2}$	698	$C_{15}H_{29:1}$	$C_{10}H_{20:0}$-Ph	728	$C_{17}H_{35:0}$	$C_{10}H_{20:0}$-Ph
688	$C_{15}H_{25:3}$	$C_{15}H_{31:0}$	712	$C_{15}H_{25:3}$	$C_{17}H_{31:2}$	742	$C_{17}H_{31:2}$	$C_{17}H_{31:2}$
	$C_{15}H_{27:2}$	$C_{15}H_{29:1}$	714	$C_{15}H_{25:3}$	$C_{17}H_{33:1}$	744	$C_{17}H_{31:2}$	$C_{17}H_{33:1}$
690	$C_{15}H_{27:2}$	$C_{15}H_{31:0}$		$C_{15}H_{27:2}$	$C_{17}H_{31:2}$	752	$C_{17}H_{31:2}$	$C_{12}H_{24:0}$-Ph
	$C_{15}H_{29:1}$	$C_{15}H_{29:1}$		$C_{15}H_{25:3}$	$C_{17}H_{35:0}$			
692	$C_{15}H_{29:1}$	$C_{15}H_{31:0}$	716	$C_{15}H_{27:2}$	$C_{17}H_{33:1}$			
				$C_{15}H_{29:1}$	$C_{17}H_{31:2}$			
				$C_{15}H_{27:2}$	$C_{17}H_{35:0}$			
			718	$C_{15}H_{29:1}$	$C_{17}H_{33:1}$			
				$C_{15}H_{31:0}$	$C_{17}H_{31:2}$			

FIGURE 10.14

FD-MS results of dimeric reaction of blended lacquers.

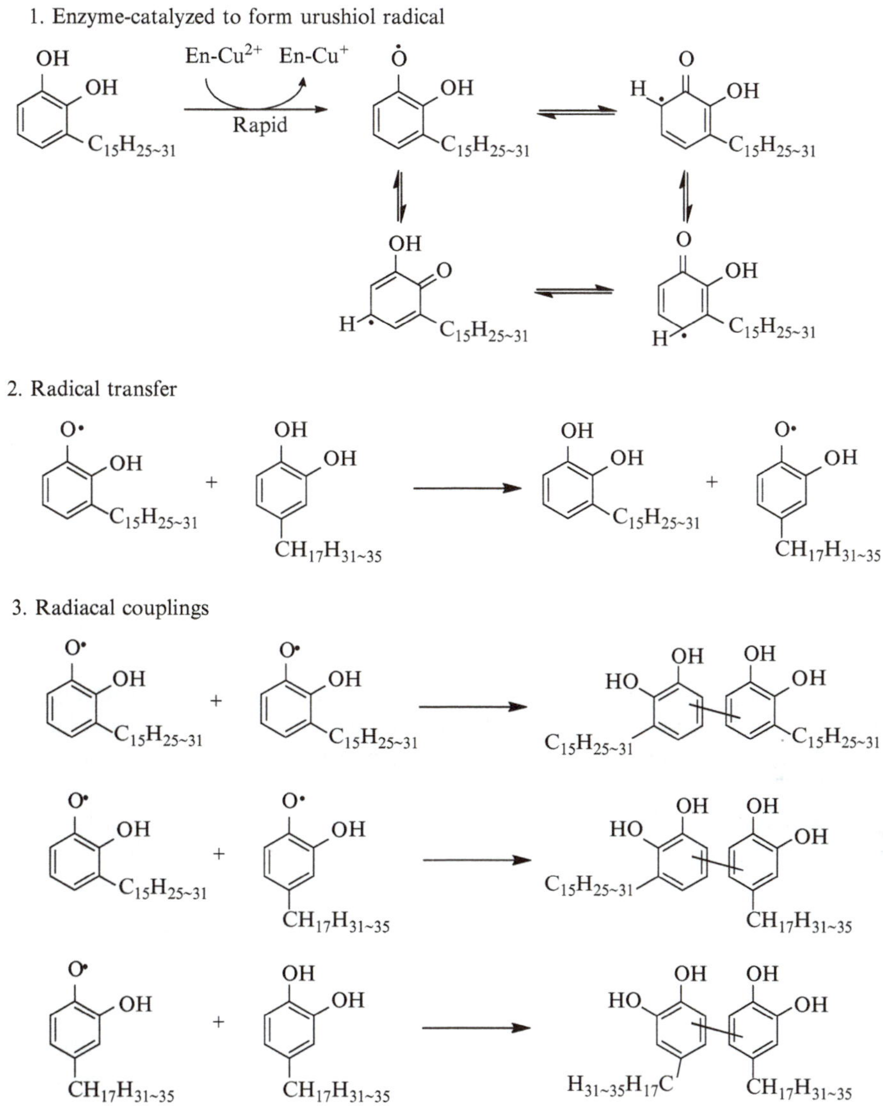

SCHEME 10.4

Mechanism of reaction between urushiol and thitsiol.

10.4.2 LACCOL-BLENDED URUSHIOL

It has been known that lacquer sap tapped from *T. vernicifluum* lacquer tree with urushiol as the lipid component has a quick drying time and high film quality, and the sap tapped from *T. succedanea* with laccol as the lipid component dries very slowly and the film quality is not so good. It can be due to the relatively lower laccase concentration and lower laccase activity, and laccol lacquer requires a severe drying condition, high temperature (~30 °C), and high relative humidity (>80%). On the other hand, laccol lacquer is cheaper than urushiol lacquer due to high production and poor drying properties. Thus, improving the drying properties of laccol lacquer is an important research topic.

Microdispersion by kneading for refining, a hybrid of organosilicone compounds [4] and adding precious metal nanoparticles [19] is known to improve the drying speed and other performance aspects of lacquer coating materials. Urushiol blended with thitsiol was described in Section 10.4.1. In this section, we describe a laccol lacquer blended with urushiol lacquer in certain proportions. The properties of the drying mechanism of enzymatic polymerization of the blended lacquer were examined by two-dimensional nuclear magnetic resonance (2D-NMR) and field desorption mass spectrometry (FD-MS).

Laccol lacquer was sap tapped from a *T. succedanea* lacquer tree grown in the Tho Van commune Tam Nong, Pho Tho, Vietnam. Urushiol lacquer sap was tapped from a *T. vernicifluum* lacquer tree grown in Chengkou of Chongqing city, China. The two raw lacquer saps were mixed in various ratios as summarized in Table 10.6. After stirring for 5 min at room temperature, the blended lacquers were put aside for about 30 min to let the bubbles produced by mixing disperse, and then they were uniformly coated on square glass plates (70 mm × 70 mm × 1.3 mm) with a 76 μm wet thickness, and dried in a humidity-controlled chamber with 80% relative humidity at 25 °C. After 2 days, the samples were removed from the chamber and stored at room temperature.

The drying time of blended lacquers was measured with an automatic drying time recorder (Tai Yu Equipment, Osaka, Japan), and the thickness of the tested lacquer films on the glass plate was 76 μm, the same method as used for thitsiol blended with urushiol.

The hardness and color of lacquer films were measured by pencil hardness tester (Yoshimituseiki, Japan) and Color-Guide instrument (BYK-Gardner, Germany), respectively. The glass transition temperature of the film surface was measured by a rigid-body pendulum physical property-testing (RPT) instrument (A&D, Japan). The RPT oven was programmed to rise from −10 to 200 °C at a rate of 3.5 °C/min.

One blended lacquer, BL5 (Table 10.6) was selected to examine the polymerization mechanism. BL5 was first kneaded in a mixer at a constant speed of 60 rpm for

Table 10.6 Ratio of blended laccol and urushiol lacquers

Blend lacquers	Laccol lacquer (wt%)	Urushiol lacquer (wt%)
BL1	100	0
BL2	100	10
BL3	100	30
BL4	100	50
BL5	100	100
BL6	0	100

1 h to promote the polymerization reaction, and then the molecular weight distribution was measured by GPC. The BL5 lipid component was extracted with acetone, and lipid dimers were separated and purified by HPLC as described in Chapter 3.

NMR spectra were recorded on a JEOL JNM-ECA500 (500 MHz), using DMSO-d_6 containing 0.01% tetramethylsilane (TMS) as a solvent. Five milliliters 10 mM HCl was added to samples to obtain a strong correlation signal-OH in 2D-NMR [20]. The system was controlled by an IBM RS = 6000 43P model 150 host workstation and AIX V 4.3.2 (IBM UNIX) operating system. Standard JEOL software was applied to the two-dimensional spectra.

Dimers were methylated as follows: 4 equivalents of dimethyl sulfate and 5 equivalents of K_2CO_3 were added to 10 ml acetone in which 70 mg laccol dimers was dissolved. After refluxing at 60 °C for 48 h, methylated laccol dimers were obtained.

Gas chromatography-mass spectrometry (GC-MS) measurement was carried out using an Agilent 6890 N/5975 GC/MS system with a FL UA-1 column. The GC oven temperature was programed to hold 1 min at 80 °C initially, increase by 4 °C/min to 180 °C, hold at 180 °C for 10 min, raise the rate at 10 °C/min to 320 °C, and hold for 10 min. The injection and interface temperature was 280 °C, the split ratio was 20:1, and the flow rate of helium gas was 1 ml/min. The mass spectrometry ionization energy was 70 eV (EI-mode).

FD-MS measurement was carried out in a JEOL JMS-T100GCV-type spectrometer system at the acceleration voltage of 7 kV with a mass range of $m/z = 300\text{-}900$. Samples were dissolved in acetone, and cholesterol was used as the internal standard. The drying times of blended lacquers are summarized in Table 10.7.

The drying time of blended lacquers decreased from BL1 to BL6 with increased ratios of urushiol lacquer sap. The hardness of the BL3 lacquer film, which was a mixture of 100 wt% laccol lacquer and 30 wt% urushiol lacquer, was 3B after 24 h and equaled that of urushiol lacquer (BL6); however, the hardness of laccol lacquer (BL1) was under 6B after 24 h, suggesting that the enzymatic oxidative polymerization of BL3 was completed almost like BL6; that is, laccol lacquer blended with urushiol lacquer had a polymerization reaction different from laccol lacquer sap alone. The color became lighter from BL3 to BL5, and the lightness of BL5 was almost as the same as that of BL6 (Table 10.8). This result also suggested that the drying time of blended lacquer became shorter with increasing proportion of urushiol.

The RPT results of BL1, BL5, and BL6 are shown in Figure 10.15. Laccol lacquer (BL1) had the lowest glass transition temperature of any drying time, about 77 °C after drying for 30 days. Meanwhile, the surface glass transition temperature of BL5 was 112 °C and that of BL6 was 122 °C, respectively, after 30 days drying.

The surface glass transition temperature increased with increasing ratio of urushiol lacquer. It may be due to the high activity of urushiol laccase and the fast polymerization between laccol and urushiol, which increased the cross-linking in the blended film. Because the percentage of double bonds in the side chains of urushiol

Table 10.7 Drying time and film hardness of blended lacquers

Samples	Drying time[a] (h)			Hardness[b] for drying days								
	DF	TF	HD	1	2	3	4	6	7	14	21	28
BL1	8.7	11.3	24.0	<6B	<6B	<6B	<6B	<6B	<6B	<6B	6B	5B
BL2	6.0	7.1	11.3	6B	6B	6B	5B	5B	4B	3B	2B	2B
BL3	3.8	7.0	7.1	3B	3B	3B	3B	3B	B	B	F	F
BL4	2.3	3.0	4.3	3B	3B	3B	3B	B	B	B	F	F
BL5	1.9	2.4	3.8	3B	3B	2B	B	HB	F	F	F	H
BL6	1.0	1.7	2.9	3B	3B	2B	B	F	F	F	F	H

[a] *Drying condition: 25 °C, 80%RH, wet thickness 76 μm. DF, dust-free dry; TF, touch-free dry; HD, harden dry.*
[b] *Based on pencil hardness test method.*

Table 10.8 Lightness, gloss, and color of blended laccol and urushiol lacquers

Lightness and coloration[1]			Gloss
*L**	*a**	*b**	
25.4	41.2	40.4	51.4
18.7	37.0	27.5	35.9
13.3	30.3	16.5	41.8
12.4	31.9	17.6	51.8
11.8	30.9	16.5	53.4
10.7	17.6	7.1	32.2

*L**: Lightness, *a**: +red/-green, *b**: +yellow/-blue, measured on spectro-guide 45/0 (Gardner).

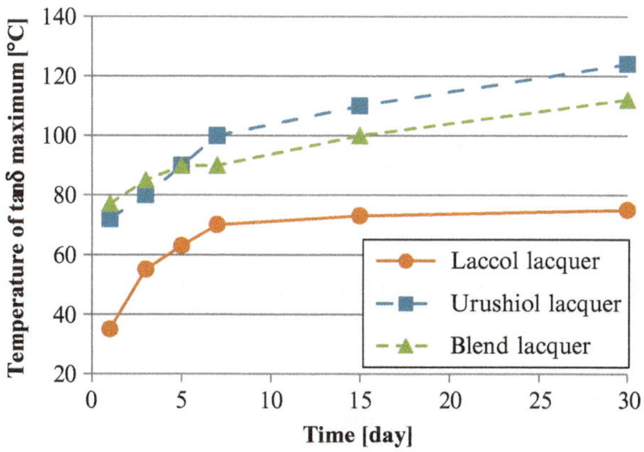

FIGURE 10.15

Glass transition temperature of film surface.

Table 10.9 Double bond ratio at side chain of urushiol and laccol

Kind of lacquer sap	Side-chain ratio [%]							
	$C_{15:0}$	$C_{15:1}$	$C_{15:2}$	$C_{15:3}$	$C_{17:0}$	$C_{17:1}$	$C_{17:2}$	$C_{17:3}$
Laccol lacquer	1.7	0.6	-	-	3.3	37.3	19.3	37.9
Urushiol lacquer	3.0	11.7	5.2	75.7	0.2	1.9	1.3	1.0

is higher than that of laccol (Table 10.9), the autoxidation polymerization occurred more easily in the urushiol side chain than in the laccol side chain.

After kneading at 60 rpm for 1 h, the blended lacquer BL5 was 45.1%, 51.6%, and 3.4% monomers, oligomers, and polymers, respectively. The dimers were separated by HPLC. After purification, BL5 was analyzed by NMR including two-dimensional NMR and FD-MS measurements.

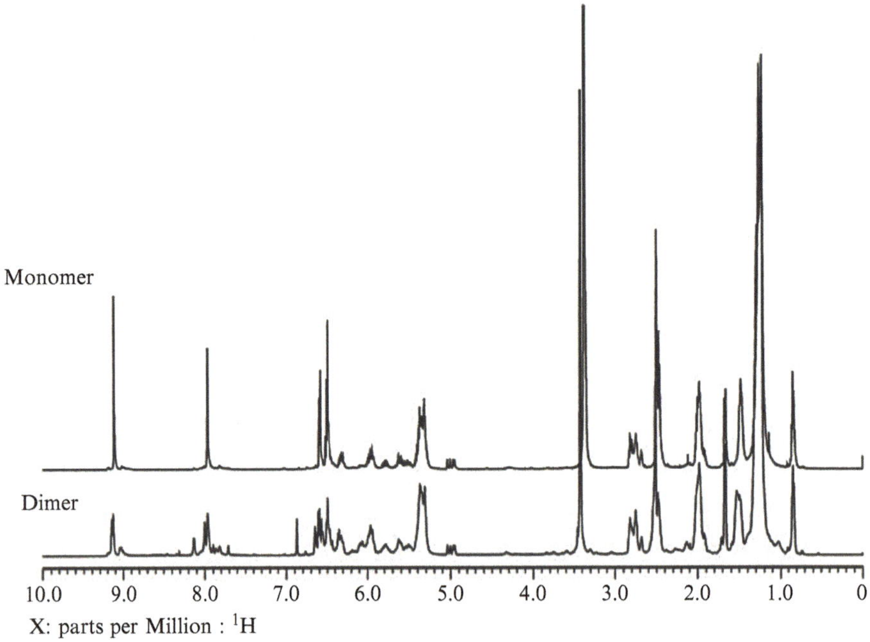

FIGURE 10.16

^{1}H NMR spectra of monomer and dimer from BL5.

^{1}H NMR spectra of the BL5 monomer and dimer are shown in Figure 10.16. Both of them showed peaks at δ6.4-6.7 ppm, which are due to the catechol ring, and the dimer showed a new peak in the low magnetic field (δ6.9, δ8.15).

Figures 10.17, 10.18, and 10.19 are the 2D-NMR of the BL5 dimer. Although the structure of catechol coupling at position 4 was verified by the HMBC spectrum, a correlation between catechol nucleus protons and other nucleus carbons was not observed, suggesting that this dimer has a symmetrical biphenyl-type structure. This structure usually exists in laccol dimer [20] and/or urushiol dimer [2]. The details of results of 2D-NMR are summarized in Table 10.10, and they show that BL5 dimer has a main 4-4′ structure.

In order to confirm the structure observed by NMR, 3-methylcatechol, which is an abbreviated structure of the lacquer lipid components, was catalyzed by laccase from laccol and urushiol lacquer saps, respectively. The derived dimers were separated by HPLC and analyzed by GC/MS. The result is shown in Figure 10.20. There are three peaks, namely A, B, and C in both total ion chromatograms at the retention times of 32.4, 42.3, and 43.1 min, respectively.

Analyses of the mass spectra showed that peak A has a biphenyl-type structure, while peaks B and C have a dibenzofuran-type structure, which may be due to the catechol nucleus initially coupling at position 6. Due to the radical transfer, it

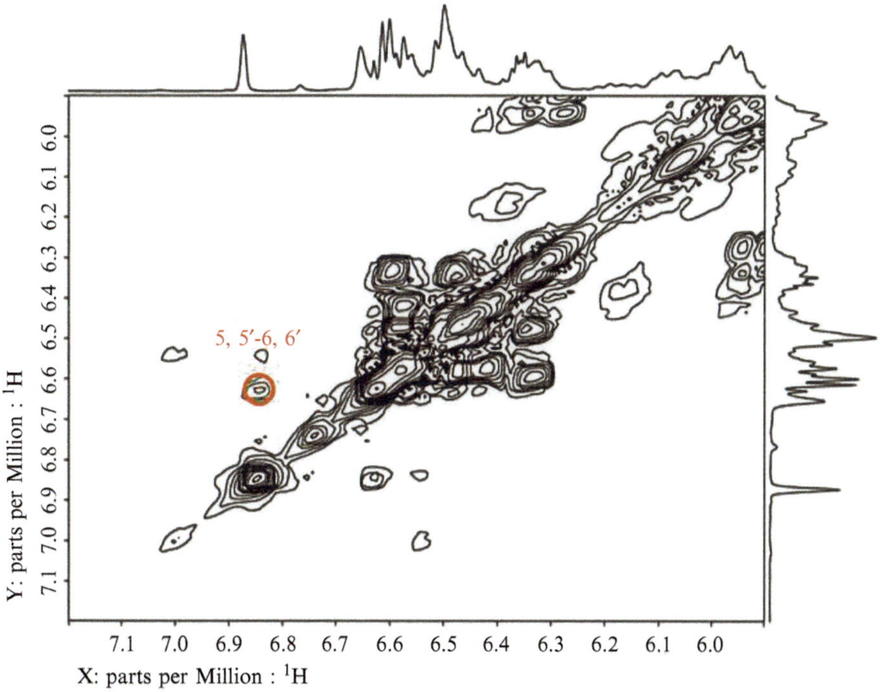

FIGURE 10.17

COSY spectrum of BL5 dimer.

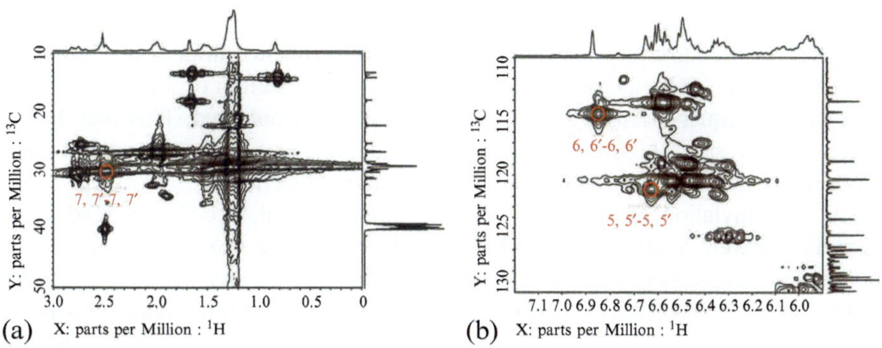

FIGURE 10.18

HMQC spectra of BL5 dimer: (a) ^{1}H: 0.0-3.0 ppm; ^{13}C: 10.0-50.0 ppm, (b) ^{1}H: 6.0-7.1 ppm; ^{13}C: 110.0-130.0 ppm.

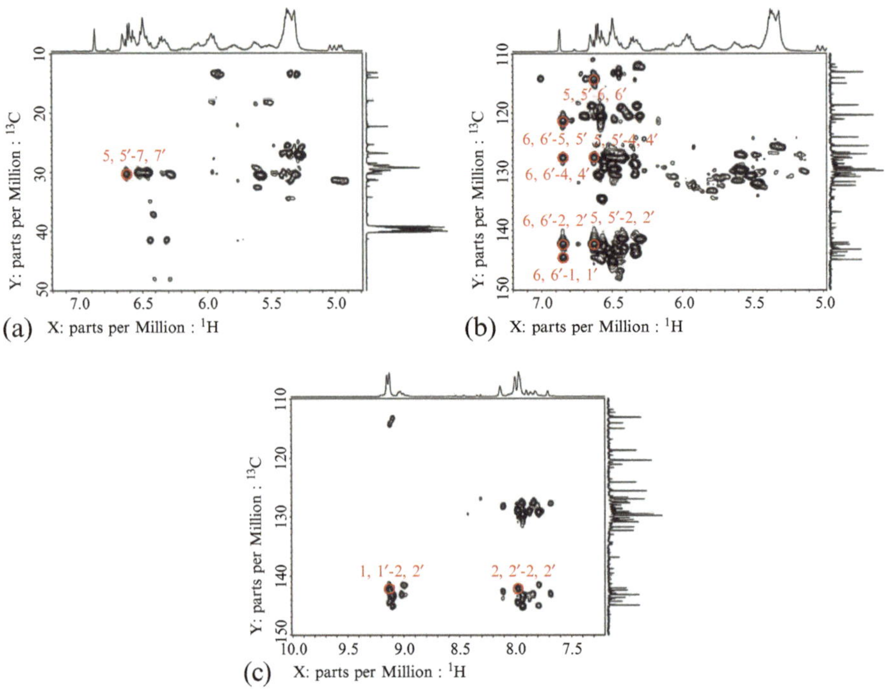

FIGURE 10.19

HMBC spectra of BL5 dimer: (a) ^{1}H: 5.0-7.0 ppm; ^{13}C: 10.0-50.0 ppm, (b) ^{1}H: 5.0-7.0 ppm; ^{13}C: 110.0-150.0 ppm, (c) ^{1}H: 7.5-10.0 ppm; ^{13}C: 110.0-150.0 ppm.

forms 4-4′, 4-6′ (4′-6), and 6-6 (6-6′) dimer structures. The dimer of peak A is 2, 2′-dimethyl-[1, 1-biphenyl]-3,3′,4,4′-tetraol, which is the symmetrical type combined at position 4, agrees with the structure of dimers observed by 2D-NMR. Dimers of peaks B and C are dibenzofuran structures coupling at 4-6 and 6-6, respectively, as shown in Scheme 10.5.

It has reported that laccol lacquer does not form a dibenzofuran dimer during the initial enzymatic polymerization [20]. However, in the reaction of 3-methylcatechol with laccol lacquer laccase, a dibenzofuran dimer was obtained. This may be because the low steric hindrance of the small side chain of 3-methylcatechol made low-activity laccol laccase to accelerate the oxidation reaction.

After methylation, all lacquer dimers including urushiol, laccol, and blended lacquer dimers were analyzed by FD-MS, and the results are shown in Figure 10.21. In the dimer mass spectra, peaks m/z=682-694 are the side chain of urushiol (C15) and m/z=738-750 are the side chain of laccol (C17). On the other hand, new peaks m/z=710-722 between urushiol and laccol dimers appeared in the blended lacquer. These new peaks are due to dimer coupling between the side chains C15 and C17, suggesting that a polymerization reaction between urushiol and laccol occurred in the blended lacquer.

Table 10.10 Assignment of blend lacquer lipid dimer

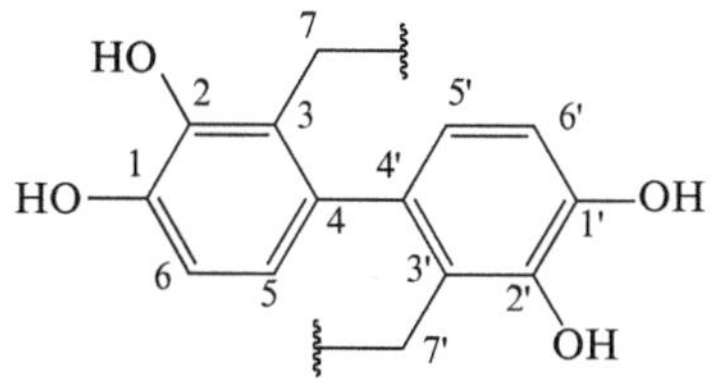

Method	Peaks	Peaks coupling	
HMQC	δ_H [ppm]	-	δ_C [ppm]
2.46	-	30.08	7,7′-7,7′
6.63	-	121.38	5,5′-5,5′
6.63	-	114.34	6,6′-6,6′
HMBC	δ_H [ppm]	-	δ_C [ppm]
6.63	-	30.08	5,5′-7,7′
6.63	-	114.34	5,5′-6,6′
6.63	-	127.63	5,5′-4,4′
6.63	-	142.29	5,5′-2,2′
6.85	-	121.40	6,6′-5,5′
6.85	-	127.63	6,6′-4,4′
6.85	-	142.29	6,6′-2,2′
6.85	-	144.44	6,6′-1,1′
7.98	-	142.29	2,2′-2,2′
9.13	-	142.29	1, 1′-2,2′
COSY	δ_H [ppm]	-	δ_H [ppm]
6.63	-	6.85	5,5′-6,6′

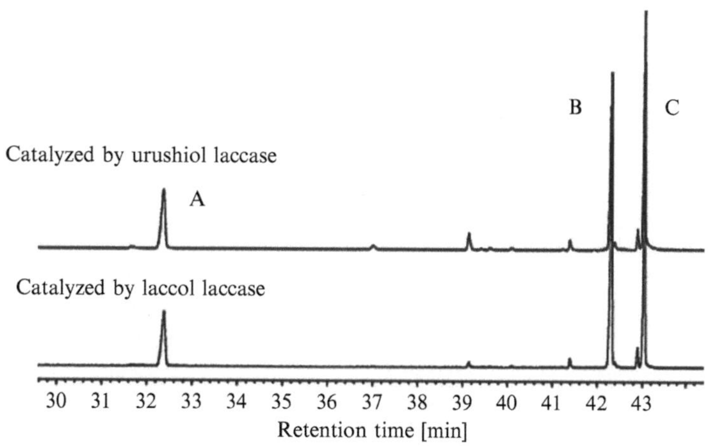

FIGURE 10.20

TIC of 3-methylcatechol dimer.

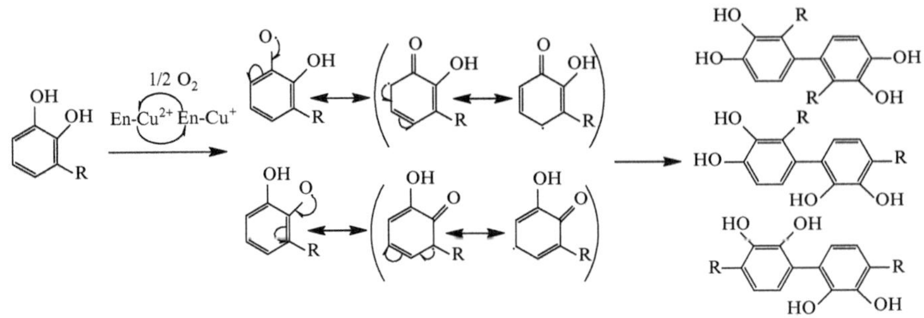

SCHEME 10.5

Enzymatic polymerization of blend lacquer.

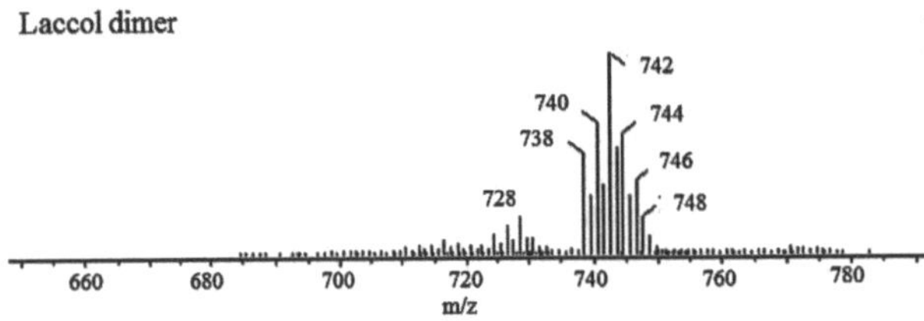

M.W.	Side chain type		
750	$C_{17}H_{35:0}$	-	$C_{17}H_{35:0}$
748	$C_{17}H_{35:0}$	-	$C_{17}H_{33:1}$
746	$C_{17}H_{35:0}$	-	$C_{17}H_{31:2}$
	$C_{17}H_{33:1}$	-	$C_{17}H_{33:1}$
744	$C_{17}H_{35:0}$	-	$C_{17}H_{29:3}$
	$C_{17}H_{33:1}$	-	$C_{17}H_{31:2}$
742	$C_{17}H_{33:1}$	-	$C_{17}H_{29:3}$
	$C_{17}H_{31:2}$	-	$C_{17}H_{31:2}$
740	$C_{17}H_{31:2}$	-	$C_{17}H_{29:3}$
738	$C_{17}H_{29:3}$	-	$C_{17}H_{29:3}$

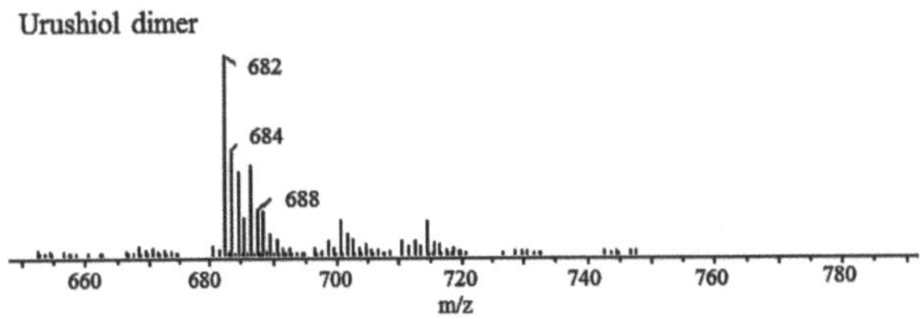

M.W.	Side chain type		
694	$C_{15}H_{31:0}$	-	$C_{15}H_{31:0}$
692	$C_{15}H_{31:0}$	-	$C_{15}H_{29:1}$
690	$C_{15}H_{31:0}$	-	$C_{15}H_{27:2}$
	$C_{15}H_{29:1}$	-	$C_{15}H_{29:1}$
688	$C_{15}H_{31:0}$	-	$C_{15}H_{25:3}$
	$C_{15}H_{29:1}$	-	$C_{15}H_{27:2}$
686	$C_{15}H_{29:1}$	-	$C_{15}H_{25:3}$
	$C_{15}H_{27:2}$	-	$C_{15}H_{27:2}$
684	$C_{15}H_{27:2}$	-	$C_{15}H_{25:3}$
682	$C_{15}H_{25:3}$	-	$C_{15}H_{25:3}$

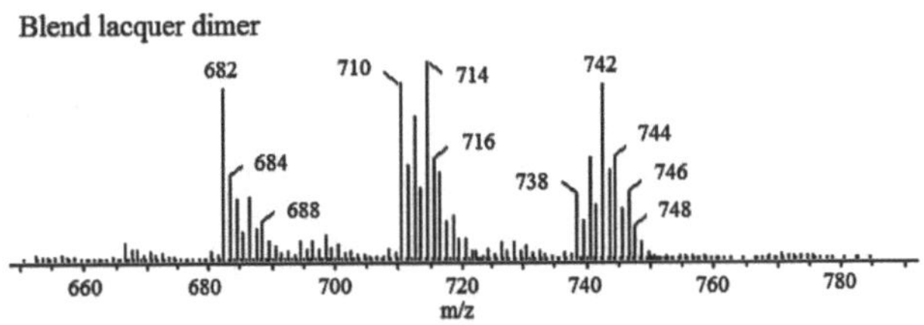

M.W.	Side chain type			M.W.	Side chain type		
722	$C_{17}H_{35:0}$	-	$C_{15}H_{31:0}$	716	$C_{17}H_{31:2}$	-	$C_{15}H_{29:1}$
720	$C_{17}H_{35:0}$	-	$C_{15}H_{29:1}$		$C_{17}H_{29:3}$	-	$C_{15}H_{31:0}$
	$C_{17}H_{33:1}$	-	$C_{15}H_{31:0}$	714	$C_{17}H_{33:1}$	-	$C_{15}H_{25:3}$
718	$C_{17}H_{35:0}$	-	$C_{15}H_{27:2}$		$C_{17}H_{31:2}$	-	$C_{15}H_{27:2}$
	$C_{17}H_{33:1}$	-	$C_{15}H_{29:1}$		$C_{17}H_{29:3}$	-	$C_{15}H_{29:1}$
	$C_{17}H_{31:2}$	-	$C_{15}H_{31:0}$	712	$C_{17}H_{31:2}$	-	$C_{15}H_{25:3}$
716	$C_{17}H_{35:0}$	-	$C_{15}H_{25:3}$		$C_{17}H_{29:3}$	-	$C_{15}H_{27:2}$
	$C_{17}H_{33:1}$	-	$C_{15}H_{27:2}$	710	$C_{17}H_{29:3}$	-	$C_{15}H_{25:3}$

FIGURE 10.21

FD-MS spectra of lacquer dimers.

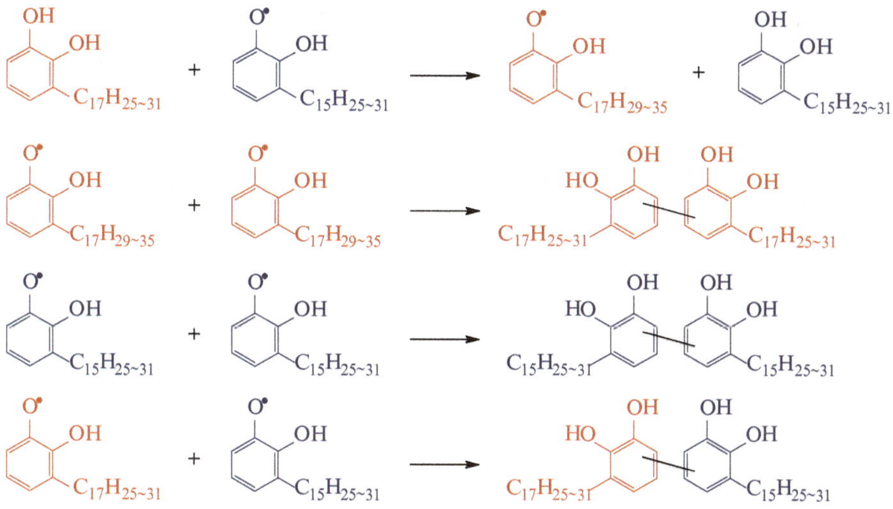

SCHEME 10.6

Reaction between urushiol and laccol.

Adding 30 wt% urushiol to 100 wt% laccol lacquer greatly improved the drying time and the film hardness (BL3). During the enzymatic polymerization, a radical transfer [18,21,22] and reaction between urushiol and laccol occurred, and then several new peaks appeared in the blended lacquer FD-MS spectrum. The reaction mechanism in the blended lacquer was considered to be an intermolecular reaction, as shown in Scheme 10.6.

Because there are no differences between the coupling position of laccol and 3-methylcatechol dimers in the NMR and GC/MS analysis results, that is, although the enzyme activity is different, higher urushiol laccase activity and lower laccol laccase activity, the substrate specificity is the same.

Blending urushiol lacquer sap with laccol lacquer sap can obviously improve the drying property and hardness of laccol lacquer. Blended lacquers with 30 wt% urushiol and 100 wt% laccol have almost the same drying rate as the enzymatic polymerization of urushiol lacquer. The film hardness is dependent on the number of the double bonds of urushiol, that is, the film hardness increased as the proportion of urushiol increased.

Structure analysis of dimers by 2D-NMR and FD-MS showed that radical transfers occurred between urushiol and laccol, so the dimer structure of 4,4′ biphenyl and 4-6′ (4′-6) and 6-6 (6-6′) dibenzofuran types, which occurred between urushiol and urushiol, laccol and laccol, and urushiol and laccol were obtained. Based on these results, it was concluded that a reaction between laccol and urushiol occurred.

10.4.3 URUSHIOL BLENDED WITH POLYURETHANE

Due to the environmental friendly nature and biocompatibility and the versatile structure property relationships of polyurethane [23–25], polyurethane products offer desirable properties such as high setting speeds, low viscosity, good flexibility, and thermal and ultraviolet stability [26], and their application is rapidly extending to fields such as coatings and paints, automotive, apparel, medical, and window treatment products [27,28]. Polyurethane is typically composed of isocyanate-capped macromolecular chains containing urethane groups in their backbones, usually with a 1-15 wt% free isocyanate concentration [29], obtained by reaction of a polyol with a molar excess of di- or polyisocyanate:

(1) R-N=C=O+HO-R′→R-NH-COOR′
(2) R-N=C=O+R-NH-COOR′→R-N(COOR)-COOR′

where in reaction formulas (1) and (2), R and R′, represent many different organic compounds such as alcohols or other hydroxy-containing compounds.

In this blend scheme, a commercial polyurethane product used as a coating was mixed with a refined lacquer (kurome lacquer) in order to investigate blends of these two coatings, both of which have good qualities, and the properties of blended lacquer were examined.

The refined lacquer is prepared according to Section 10.1. Polyurethane resin (R′=main phenyl group and minor alkyl group) and its hardener were purchased from Cashew Co. Ltd., Saitama, Japan, and used as received. The components of the main resin A and its hardener B are shown in Table 10.11.

The blends of refined lacquer and polyurethane (LPU) were mixed in various ratios in glass sample cups with or without a hardener by stirring 5 min by hand at room temperature (20-25 °C). In each case, the ratio of polyurethane to its hardener was 2:1. According to the concentrations of lacquer used in the experiments, the blends were named LPU 100 (0% polyurethane), LPU 90 (10% polyurethane), LPU 80 (20% polyurethane), LPU 70 (30% polyurethane), LPU 60 (40% polyurethane), LPU 50 (50% polyurethane), and LPU 0 (100% polyurethane). Refined lacquer and the blends were uniformly coated on glass sheets (23.5×245×2 mm) and square glass

Table 10.11 Components of commercial polyurethane resin

Components	Polyurethane resin A (%)	Polyurethane hardener B (%)
Xylene	14.4	3.3
Butylacetate	12.3	9.0
Propylene glycol monoethyl ether acetate	3.0	-
Ethylbenzene	11.8	2.7
Ethylacetate	-	18.0
Toluene	-	4.0

plates (70 mm × 70 mm × 1.3 mm), respectively, with a 76 μm thickness wet, and hardened in a humidity-controlled chamber at 20-25 °C. After 7 days, the sample-coated glass sheets were removed from the chamber and stored at room temperature for 6 months.

The drying time is measured using an automatic drying time recorder as described in Section 10.4. The hardness of pencil lead is described by letters and numbers determined according to the current national standard of GB/T6739-1996: the meanings of H and B designate how hard or soft the tested films are, respectively. A higher number with H or B expresses the hardness or softness of the tested films, respectively; F and HB indicate medium hardness, while F is a slightly harder film than HB. The pencil lead hardness was determined by a Yoshimitsu C-221 instrument (Tokyo, Japan). The results are summarized in Table 10.12.

From Table 10.12, it is interesting to note that the drying time necessary to reach DF dry obviously decreased, and that of TF dry changed little, while the change in drying time for HD changed to an asymmetric "U" curve. First, the drying times of HD at the ratio 9:1 and 8:2 were decreased, and then when the ratio was more than 7:3, the drying time of HD increased, although the DF drying time decreased. When the ratio of lacquer to polyurethane is less than 50%, the blends cannot be called lacquer coatings, but should be called lacquer-modified polyurethane coatings; thus, the results are not discussed here (the drying time also increased until the ratio was 2:8 and then decreased to 4 h for polyurethane). However, polyurethane was also examined as a control. After 3 months, the pencil hardness of the blended films LPU 100 to LPU 50 reached 8H while the polyurethane reached only 6H. It seemed that the drying rates of LPU 80 and LPU 70 with ratios of 8:2 and 7:3, respectively, were better than the others.

The color and gloss of LPU series films were detected with a spectro-guide BYK Gardner GmbH (Columbia, OH, USA) with illuminant/observer D65/10°, and the instrument geometry 45°/0°. UV irradiation from one side of the square glass plate film was measured using a Handy UV 800 (ORC, Tokyo, Japan) equipped

Table 10.12 Drying properties of various sample preparations (20-25 °C, 65-70% RH)

		Drying time (h:min)			
Names	**Ratio**	**DF**	**TF**	**HD**	**PH**
LPU 100	-	2:20	3:40	6:20	8H
LPU 90	9:1	2:20	3:35	5:05	8H
LPU 80	8:2	1:30	2:50	5:05	8H
LPU 70	7:3	1:15	3:25	6:25	8H
LPU 60	6:4	1:25	3:30	7:00	8H
LPU 50	5:5	1:05	3:25	7:55	8H
LPU	-	0:40	2:45	4:20	6H

LPU: lacquer/polyurethane blends; DF: dust free; TF: touch free; HD: harden dry; PH: pencil hardness.

with a high-pressure mercury lamp HHL-800/C-FL (80 W/cm, emission light length 100 mm). The exposure distance of UV irradiation was 82.5 mm.

The water resistance of LPU series films was tested as follows: LPU series films coated on square glass plates (70 mm × 70 mm × 1.3 mm) were dried for 3 months, dipped in boiling water for 10 min, taken out and dried, and dried in 50 °C oven for 1 h, then cooled in air to room temperature and examined.

The results are summarized in Table 10.13, Figure 10.22, and Figure 10.23. From Figure 10.22, it can be seen that the color of the films increased prominently. According to the CIE (Commission Internationale d'Éclairage) system, L^*, a^*, and b^* mean lightness, (+red)/(−green) and (+yellow)/(−blue), respectively.

As the concentration of polyurethane (C_{pu}) increased, the lightness of the LPU increased distinctly, and it seemed that there was a very good linear dependence between them, that is, $\Delta L^* = 1.7327 + 0.8159 \times C_{pu}$. The larger the number of a^* and b^*, the closer to red and yellow, respectively, the film is ΔE^* represents the total difference between sample and standard, and is equal to the following:

$$\Delta E^* = \sqrt{(\Delta L^*)^2 + (\Delta a^*)^2 + (\Delta b^*)^2}$$

From Figure 10.22, it is obvious that ΔE^* increased as the concentration of polyurethane increased, and there was a very good relationship between ΔE^* and lightness.

From Table 10.13, it is clear that the RRuv% (UV retention rate of the film gloss) increased as the concentration of polyurethane increased, but the change in the gloss of the film was a "U" curve like the drying time of HD. The UV resistances of the blended films, however, were all higher than that of the refined lacquer film LPU 100. Thus, it was clear that the addition of polyurethane improved the lacquer films. In terms of water resistance, LPU 70 is the best in the series with the highest resistance, 88.3%.

From Table 10.13 and Figure 10.23, it can be concluded that the resistance to weathering due to UV irradiation of the blended films increased as the concentration of polyurethane increased. When the blended film contained 50% polyurethane, the rate of gloss retention was close to a constant. It has been calculated that the Handy UV 800 produces UV irradiation in 8 h that is equivalent to 1 year of summer UV irradiation in Ryukyu, Okinawa, Japan [13]. Blending, therefore, increased the weathering resistance to UV irradiation and water resistance of LPU largely.

The surface morphology, molecular motion, and glass transition behavior of polymer films are quite important because many properties such as chemical and abrasion resistance and wettability are determined predominantly by the surface structure within a few nanometers [16,15,30]. In order to study the surface and bulk properties of the LPU series films, T_gs of RPT, T_g of DSC, and TGA are measured to further understand the film qualities and structures.

A rigid-body pendulum physical property-testing (RPT) instrument (AND, Japan) was used to determine the viscoelasticity property of the blended films. The RPT oven was programmed to rise at a rate of 2 °C/min from 25 to 250 °C [31].

Table 10.13 Changes in color and gloss of LPUs after boiling in water and UV irradiation

Names	Ratio	Color			Gloss				
		*L**	*a**	*b**	0 (h)	Boil	RRw (%)	IRuv	RRuv (%)
LPU 100	-	6.42	18.14	6.74	51.1	24.5	47.9	28.3	55.4
LPU 90	9:1	14.30	29.55	21.02	43.4	25.6	59.0	26.5	61.1
LPU 80	8:2	25.31	34.51	40.26	49.7	46.2	66.3	34.0	68.4
LPU 70	7:3	33.85	34.59	55.07	62.2	54.9	88.3	44.2	71.1
LPU 60	6:4	40.42	32.32	64.96	71.6	53.4	74.6	58.3	81.4
LPU 50	5:5	53.43	22.69	73.68	64.7	53.9	83.3	59.7	92.3
LPU	-	87.32	1.59	1.49	144.4	135.6	93.9	137.5	95.2

Boil: boiled for 10min in water.
RRw (%): retention rate of gloss after boiled 10min in water.
IRuv: irradiation by ultraviolet for 16h.
RRuv (%): retention rate of gloss after 16h irradiation by ultraviolet light.

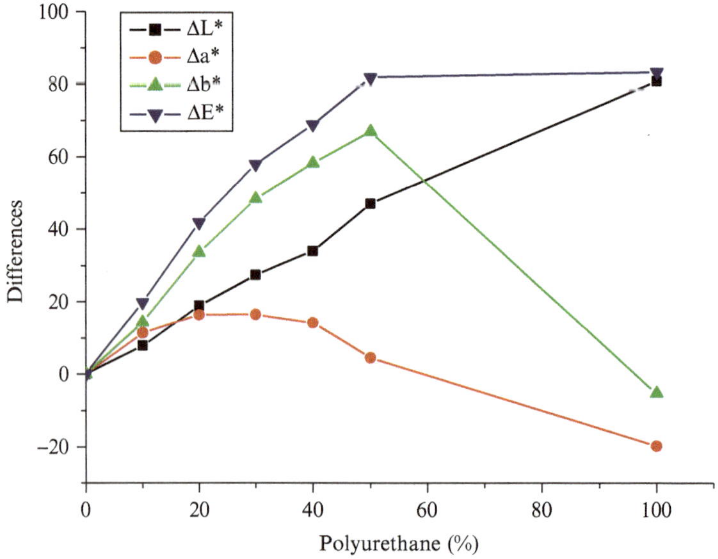

FIGURE 10.22

Color concentration of polyurethane of lacquer/polyurethane film.

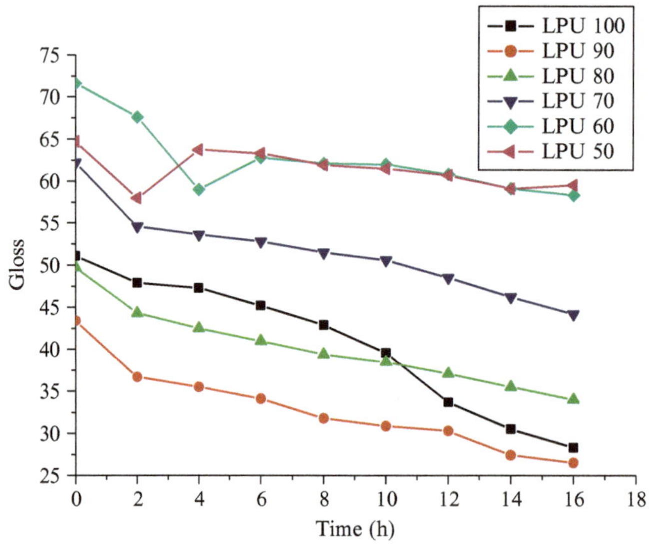

FIGURE 10.23

Gloss-time curves of LPU lacquers by UV irradiation.

A DSC-60 furnace (Shimadzu, Japan) was programmed at a constant heating rate rise of 10 °C/min in the temperature range from 30 to 300 °C. DTG-60 equipped with a TA-60 WS apparatus (Shimadzu, Japan) was also used to analyze the thermal behaviors of the blended films at a constant heating rate rise of 20 °C/min in the temperature range from 30 to 60 °C.

The results from RPT measurement are shown in Figure 10.24. The specimens were first dried for 7 days with RH 60% at 20 °C, then stored for 6 months in air at room temperature. The oven rise rate was 2 °C/min. Compared with the glass transition (T_gs) of 168 °C for the kurome lacquer LPU 100 film in this condition, all the T_gs of these blended films were lower, that is, 160 °C for LPU 90, 144 °C for LPU 80, 135 °C for LPU 70, 122 °C for LPU 60, 114 °C for LPU 50, and 110 °C for polyurethane. It seems that there is a relationship between the ratios of the LPU series and their corresponding glass transition temperature.

The typical structures and concentrations of urushiol used are shown in Scheme 10.7. The lacquer film can be regarded as a segmented block of copolymers having hard segments (HS) and soft segments (SS), like some polyurethanes with HS and SS through polyether or polyester chains linkages [32–34]. Although the structure of the lacquer film is very complex and unclear [35,2], as the polyurethane in the blends increases, the blended films become softer and their glass transition temperature decreases as a result.

The T_g results from DSC are a little different from RPT T_gs. In Figure 10.25, the highest T_g is 168 °C for LPU 50, and the T_g of LPU 100 is only 165 °C. Other values are 161 °C and 162 °C for LPU 90 and LPU 60, respectively; both LPU 80 and LPU

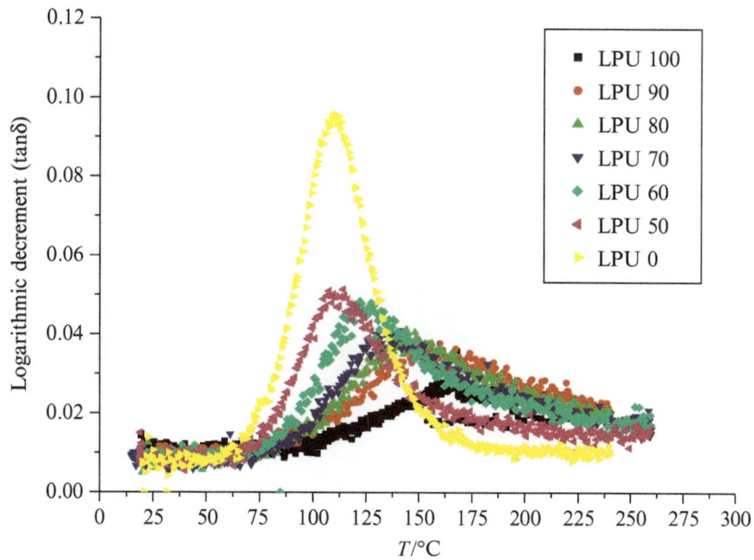

FIGURE 10.24

T_gs in RPT measurement of LPU series.

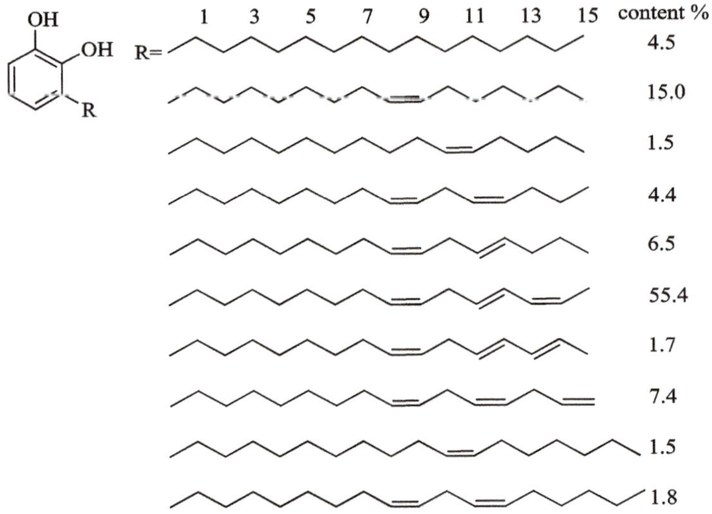

SCHEME 10.7

Typical structures and components of urushiol.

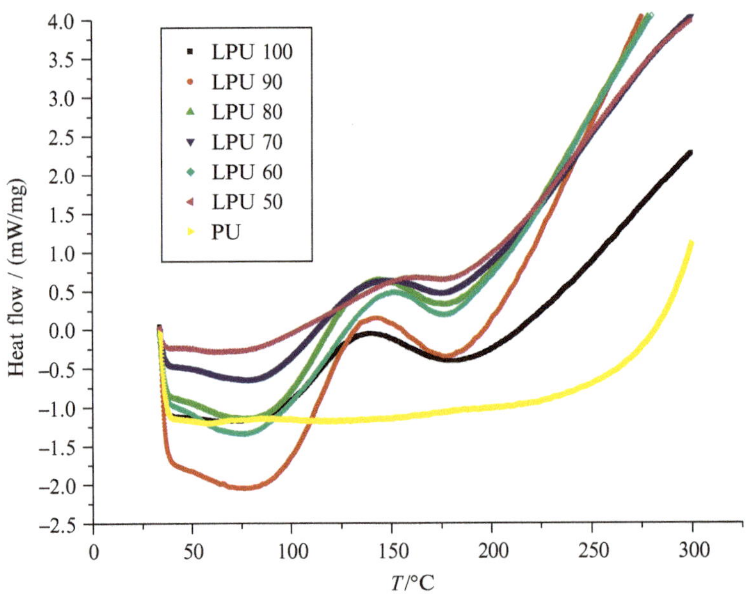

FIGURE 10.25

DSC curve of blended films of LPU series stored for 6 months at room temperature, oven rate rise: 10 °C/min.

70 are 159 °C and 114 °C for polyurethane. The difference between RPT and DSC results may be related to different methods and instruments as well as a difference between the surfaces of the films and the bulk films; this phenomenon seems widespread and has been reported in both practice and theory [16,15]. In fact, however, the differences between the DSC results of the LPU series are not apparent except for polyurethane, which can be softer than lacquer films according to the results of color, gloss, pencil hardness, RPT, and DSC examinations. When the ratio of lacquer to polyurethane is more than 50%, the quality of blended films is very similar by and large. The blended films from LPU 100 to LPU 50 have two endothermic peaks. The lower peaks are assigned to the onset of microdomain mixing of the noncrystalline hard and soft segments, and the higher peaks (around 175 °C) are the hard segments with a glass transition temperature higher than polyurethane and lacquer. This phenomenon was also found on SPU materials [36].

Thermogravimetric analysis of the LPU series (LPU 50-100) revealed a dependence of the degradation behavior on the chemical structure, although it is still difficult to predict the total structure, and it appeared to depend on the different ratios of polyurethane to lacquer (Table 10.14 and Figure 10.26). Table 10.14 shows that the temperatures at which 5%, 10%, 15%, 20%, and 50% weight loss occurred with the increasing ratio of lacquer. Temperatures at which 50% weight loss of LPU 100 to LPU 50 occurs are about 100 degrees higher than for polyurethane only.

As the ratio of lacquer increases, the temperatures for 50% weight loss increase. PU shows obvious two-step degradation, indicating that the material volatilizes with soft and hard segments, respectively, at this condition. The blended films LPU 100 to LPU 50 showed degradation in two and/or even three steps. These steps are predicted to involve soft segments and/or glycoprotein, and hard segments of the blended films are due to polyurethane and lacquer, respectively. It is interesting that the decomposition line of LPU 90 astonishingly overlaps that of LPU 100, indicating that the thermal properties of lacquer are not affected greatly when the concentration of polyurethane in the blended film is less than 10%.

It should be pointed out that all the properties, color, gloss, surface and bulk T_g, and decomposition temperatures, change with the drying time as the gel ratio shows (Table 10.13), because both polyurethane and lacquer films have some unreacted

Table 10.14 Temperature of LPU series at different weight loss measured by TGA

Weight loss (%)	Temperature (°C)						
	PU	LPU 50	LPU 60	LPU 70	LPU 80	LPU 90	LPU 100
5	276	229	242	236	215	238	226
10	298	275	283	283	276	292	289
15	311	302	314	317	314	332	327
20	321	331	351	356	352	373	366
50	367	458	464	467	465	463	466

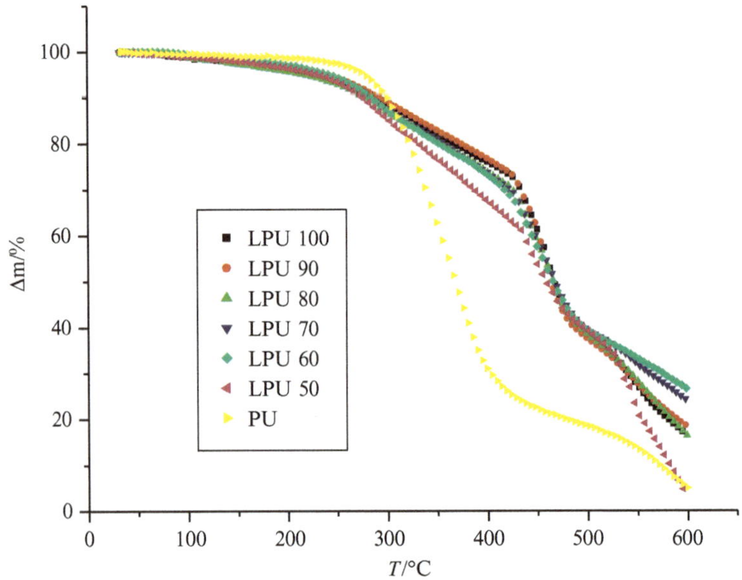

FIGURE 10.26

TGA of LPU series films. Specimens were first dried for 7 days at RH 60% and 20 °C, then stored for 6 months in air at room temperature, oven rate rise: 10 °C/min.

function groups and will continue autoxidizing in the drying process as time proceeds. This can be seen clearly from their IR and NMR spectra.

Infrared spectra (IR) were taken on a JASCO FTIR 460$^+$ spectrometer under dry air at 25 °C. Because it is possible that moisture in the lacquer sap would melt a KBr or NaCl disk, polyvinylidene chloride (PVC) film (Saran Wrap, Asahi Kasei, Japan) was used to support the IR sample. The lacquer was thinly coated on the PVC film with a spatula, and the same kind of PVC film was used as a blank. Spectra of samples were acquired at 2 cm^{-1} resolution ranged from 400 cm^{-1} to 4000 cm^{-1} under the same scanning and temperature conditions.

^{1}H NMR and ^{13}C NMR were recorded on a 500 MHz JEOL JNM α-500 spectrometer. Samples were dissolved in DMSO-d_6 and acetone-d_6 with chemical shifts referenced from tetramethylsilane (TMS).

Figure 10.27 shows the IR spectra of LPU 50 (A, 72 h), PU (B, 0 h), and lacquer/polyurethane (LPU 100, 0 h) films

As shown in Figure 10.27, after blending, the peak at 2275 cm^{-1} for the NCO group [23] of LPU 50 (other films have the same phenomenon) has disappeared completely after 72 h. The peak at 3426 cm^{-1} of refined lacquer (LPU 100) due to OH group of urushiol moves down to a frequency of 3348 cm^{-1} of LPU 50, but compared with polyurethane, it is clear that the intensity of this peak of the blended films increased because of the intermolecular interaction between urushiol and urethane. The weak peak around 3008 cm^{-1} is due to the double bonds on the side chains (R chains in Scheme 10.7)

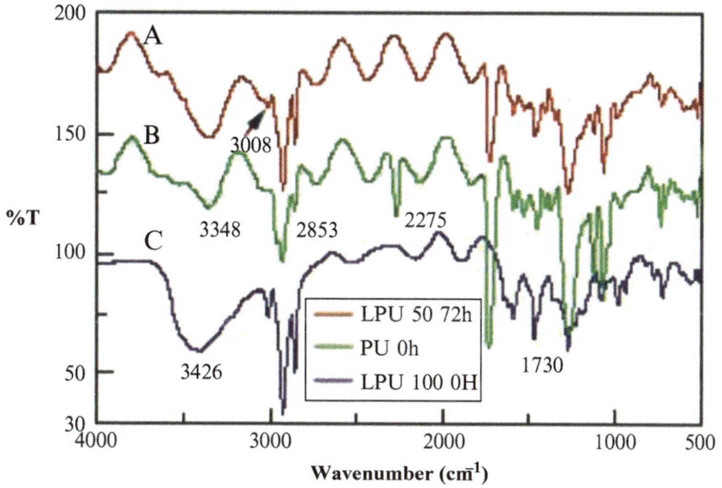

FIGURE 10.27

FT-IR spectra of LPU 50 (A, 72 h), PU (B, 0 h), and lacquer/polyurethane (LPU 100, 0 h) films.

of urushiol in LPU 100, and the reaction is not complete after 72 h. The frequency at 1730 cm^{-1} is the free C=O stretching vibration of urethane [28], and its intensity is gradually reduced due to the chain extension. Because of the reaction between isocyanate and urethane groups (Formulas 1 and 2), there are allophanate linkages in the blended films. However, due to the overlapping of the allophanate groups with the CH_2 and CH_3 at 1450-1460 cm^{-1} [28,37], it is difficult to distinguish it on FT-IR spectra.

If allophanate NH groups are formed in polyurethane and/or LPU, a characteristic signal at about 10.85 ppm will appear in ^1H NMR spectra [26]. However, it was not seen in this experiment (Figures 10.28 and 10.29). In fact, in Figures 10.28 and 10.29, the region from 8.7 ppm to 9.0 ppm at peak (1) can be assigned to allophanate NH groups adjacent to phenyl groups. The intensity in this region increases with the increasing reaction time, but the number of these linkages is relatively small. The region arising from carbamate NH groups, as shown at peak (2), also increases with the increased time.

In polyurethane film, the solvent peaks of xylene at regions (4) and (9), of ethylbenzene at regions (3), (8), and (10), and of toluene at regions (3) and (9) overlap and are difficult to distinguish. From the strong peaks at regions (7), (9), (10), (11), and (12), the solvents butyl acetate, ethyl acetate, and propylene glycol monomethylether acetate can be included in the polyurethane film, although the stronger intensity at these regions can be attributed to the CH_2 and/or CH groups adjacent to ester groups that are newly formed in isocyanate reactions.

The intensity of the region at peak (3) in Figures 10.28 and 10.29 is quite the opposite, increasing and decreasing or disappearing, respectively. The phenomenon shown in Figure 10.28 may just be the result of the increasing number of nitrogen-bound proton groups, such as carbamate, allophatate, urea, and/or the linkages of

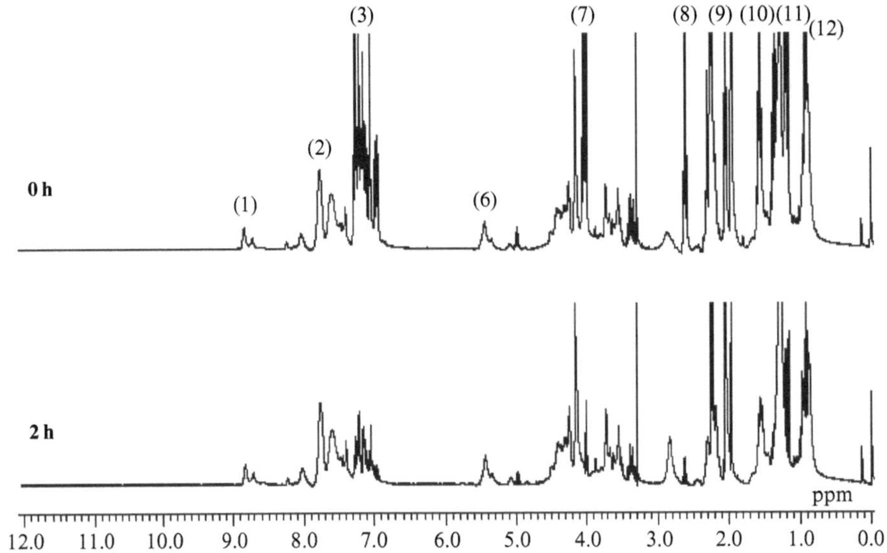

FIGURE 10.28

500 MHz ^1H NMR spectra of commercial polyurethane prepared at room temperature with different reaction times (h). Solvent: acetone-d_6.

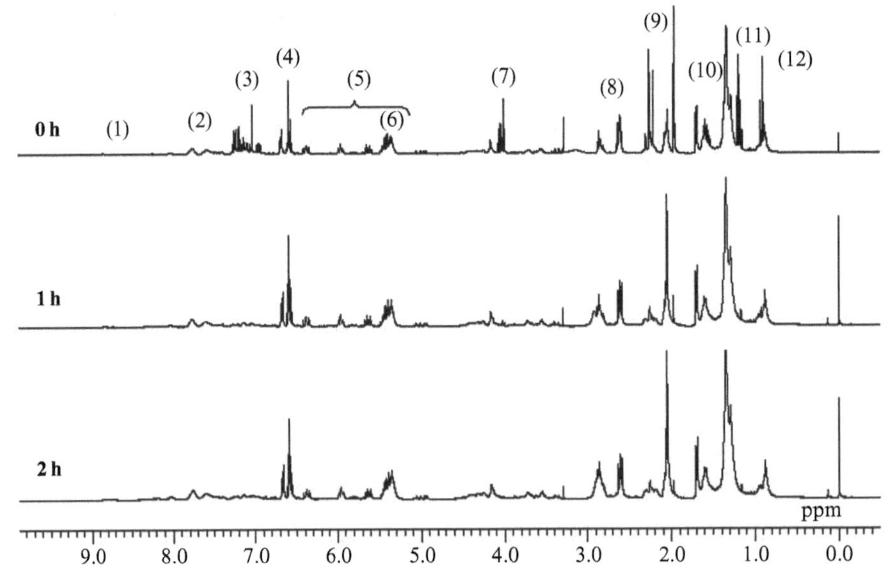

FIGURE 10.29

500 MHz ^1H NMR spectra of LPU 50 prepared at room temperature for different reaction times (h). Solvent: acetone-d_6.

no-proton nitrogen-containing groups, such as uretedinedione and isocyanurates, because these linkages will be helpful to the intensity of the aromatic proton region [26]. The decreasing/disappearing of peak (3) in Figure 10.29 was a little unusual. It may be that the newly formed polymer film LPU 50 is not dissolved in acetone. The region at (6) of both Figures 10.28 and 10.29 can be assigned to methyne protons of the main chains close to carbamate groups [26,28].

Region (4) in Figure 10.29 is the resonance of phenyl ring protons. The peaks at region (5) may be the olefin groups of urushiol [38], and it certifies that there are no reactions between isocyanate and olefin groups. The reaction may be between isocyanate and phenolic hydroxyl groups of urushiol [13], as shown in Scheme 10.8.

From Figure 10.29, it seems that the organic solvents are volatized very rapidly in the blending process. All the hazardous organic solvents contained in polyurethane resin A and its hardener B disappeared from the blended film LPU 50, which is important for use in indoor upholstery. In contrast to the increase in the number of esters, such as butyl acetate, ethyl acetate, and propylene glycol monomethylether acetate in the drying process of polyurethane film, as shown in Figure 10.29, the decreasing intensity at (11) is due to the release of these organic solvents in the blending process.

The gel ratio was detected in blended films that were stored for different numbers of months and dipped in acetone for 10 min at room temperature; after dipping, the films were dried in a Taiyou muffle furnace (Isuzu, Tokyo, Japan) for 1 h at 50 °C, cooled, and examined at room temperature. All data reported in this study are the mean values of five separate experiments except when noted, and the results are summarized in Table 10.15.

The soluble fractions of the LPU series were determined in acetone solvent. Over time, the gel ratio was improved, indicating that these two components of polyurethane and natural lacquer are time-dependent function materials. Hardening additives will change the gel ratio in the drying process, but in small amounts (±0.1 g), the difference can be ignored. After 6 months, most of the LPU series had gel ratios of more than 99%, regardless of the hardener; Table 10.15 shows the typical results from LPU 70.

As described above, a commercial polyurethane preparation was blended with natural lacquer, and a good result was obtained. The drying times of the LPU series, especially for dust-free dry, decreased about 1 h. The addition of polyurethane greatly

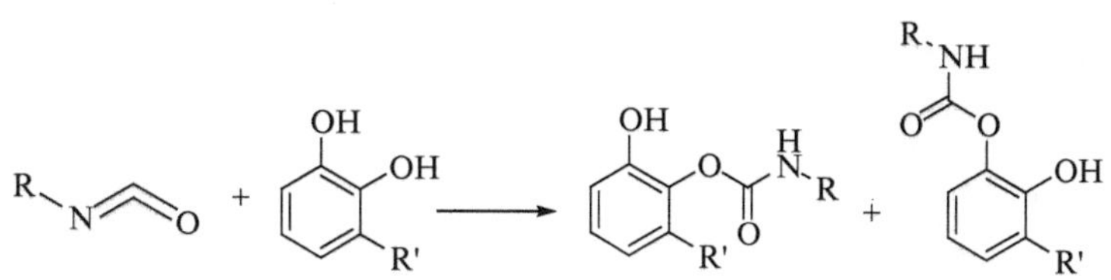

SCHEME 10.8

Reaction between isocyanate and urushiol.

Table 10.15 Gel ratios of LPU 70 with different drying times in air at room temperature

		Gel ratio (%)				
Names	Hardener (g)	2weeks	1month	2months	3months	8months
Lacquer	-	92.03	94.95	97.82	98.93	99.36
LPU 70	0.90	87.15	92.45	96.13	97.61	98.52
LPU 70	0.95	89.03	92.18	95.07	94.24	98.79
LPU 70	1.00	88.63	93.76	96.73	97.91	99.07
LPU 70	1.05	87.26	94.86	98.87	97.73	99.22
LPU 70	1.10	88.91	93.72	95.63	94.38	99.31

increased the color and gloss of the LPU series and also the water and weather resistances, especially against ultraviolet light. Based on the results of pencil lead hardness, RPT, DSC, and TGA analyses, the hardness of the blended films was also improved compared with polyurethane, as well as retaining the advantages of natural lacquer.

The structure of the blended films is very complex, and investigation of the reaction mechanism between urethane and urushiol is still under way. The results of these experiments indicate, however, that the new blended films are promising as a natural green coating in the future, especially LPU 70 and LPU 80.

REFERENCES

[1] K. Nagase, Y. Kamiya, K. Hodumi, T. Miyakoshi, The relationship between the change in urushiol by repeatitive *Kurome* of the urushi liquid and natural drying property occurring in a low humidity environment, Nippon Kagaku Kaishi 3 (2002) 377 (in Japanese).

[2] R. Lu, S. Harigaya, T. Ishimura, K. Nagase, T. Miyakoshi, Development of a fast drying lacquer based on raw lacquer sap, Prog. Org. Coat. 51 (2004) 238.

[3] K. Nagase, R. Lu, T. Miyakoshi, Studies on the fast drying hybrid urushi in low humidity environment, Chem. Lett. 33 (2004) 91.

[4] R. Lu, T. Ishimura, K. Tsutida, T. Honda, T. Miyakoshi, Development of a fast drying hybrid lacquer in a low relative-humidity environment based on *Kurome* lacquer sap, J. Appl. Polym. Sci. 98 (2005) 1055.

[5] R. Lu, T. Yoshida, T. Miyakoshi, Oriental lacquer: A natural polymer, Polym. Rev. 53 (2013) 153.

[6] F. Fleury, Recherches sur la laccase, Masson, Paris, 1924.

[7] R. Lu, T. Honda, T. Ishimura, T. Miyakoshi, Study of a naturally drying lacquer hybridized with organic silane, Polym. J. 37 (2005) 309.

[8] T. Ishimura, R. Lu, T. Miyakoshi, Studies on the reaction mechanism between urushiol and organic silane, Prog. Org. Coat. 55 (2006) 66.

[9] T. Honda, R. Lu, T. Miyakoshi, Chrome-free corrosion protective coating based on lacquer hybridized with silicate, Prog. Org. Coat. 56 (2006) 279.

[10] R. Lu, T. Moteki, T. Miyakoshi, Development and characterization of micro-dispersed lacquer, J. Chin. Lac. 29 (2010) 1.

[11] T. Terada, K. Oda, H. Oyabu, T. Asami, Lacquer—the Science and Practice, Rikou Publisher, Tokyo, Japan, 1999, p. 200–211.
[12] Y. Nakao, Dyeing with ultra fine noble metal particle I, Sen'i gakkaishi 47 (1991) 38.
[13] R. Lu, M. Ono, S. Suzuki, T. Miyakoshi, Studies on a newly designed natural lacquer, Mater. Chem. Phys. 100 (2006) 158.
[14] R. Lu, Y. Wan, T. Honda, T. Ishimura, Y. Kamiya, T. Miyakoshi, Design and characterization of modified urethane lacquer coating, Prog. Org. Coat. 57 (2006) 215.
[15] A.M. Mayes, Glass transition of amorphous polymer surfaces, Macromolecules 27 (1994) 3114.
[16] N. Satomi, K. Tanaka, A. Takahara, T. Kajiyama, Surface molecular motion of monodisperse α, ω-diamino-terminated and α,ω-dicarboxy-terminated polystyrenes, Macromolecules 34 (2001) 8761.
[17] S. Harigaya, T. Honda, R. Lu, T. Miyakoshi, C. Chen, Enzymatic dehydrogenative polymerization of urushiols in fresh exudates from the lacquer tree, Rhus vernicifera DC, J. Agric. Food Chem. 55 (2007) 2201.
[18] B.M. Sioeberg, M. Sahlin, Handbook of Proteins: Structure, Function and Methods, in: M.M. Cox, G.N. Philips (Eds.), vol. 1, Wiley, Chichester, UK, 2007, p. 403.
[19] Y. Nakao, Dyeing with ultra-fine noble metal particle II, Sen'i gakkaishi 47 (1991) 392.
[20] R. Takei, R. Lu, T. Miyakoshi, Dimer structures and laccase catalyzed polymerization mechanism of laccol in fresh *Rhus succedanea* lacquer sap, Int. J. Polym. Anal. Charact. 18 (2013) 199–210.
[21] T. Honda, X.-M. Ma, R. Lu, D. Kanamori, T. Miyakoshi, Preparation and characterization of a new lacquer based on blending urushiol with thitsiol, J. Appl. Polym. Sci. 121 (2011) 2734–2742.
[22] P.J. O'Malley, The reaction profile for hydrogen atom transfer from phenol to peroxyl free radicals, Chem. Phys. Lett. 364 (318–322) (2002) 2002.
[23] B.K. Kim, B.U. Ahn, M.H. Lee, S.K. Lee, Design and properties of UV cured polyurethane dispersions, Prog. Org. Coat. 55 (2006) 194.
[24] M. Tielemans, P. Roose, P.D. Groote, J.-C. Vanovervelt, Colloidal stability of surfactant-free radiation curable polyurethane dispersions, Prog. Org. Coat. 55 (2006) 128.
[25] V. Acharya, C.R. Prabha, C. Narayanamurthy, Synthesis of metal incorporated low molecular weight polyurethanes from novel aromatic diols, their characterization and bactericidal properties, Biomaterials 25 (2004) 4555–4562.
[26] A.M. Heintz, D.J. Duffy, S.L. Hsu, S. Wu, W. Chu, C.W. Paul, Effects of reaction temperature on the formation of polyurethane prepolymer structures, Macromolecules 36 (2003) 2695–2704.
[27] B.H. Lee, H.J. Kim, Influence of isocyanate type of acrylated urethane oligomer and of additives on weathering of UV-cured films, Polym. Degrad. Stab. 91 (2006) 1025–1035.
[28] A.L. Daniel da Silva, J.M. Martin-Martinez, J.C. Moura Bordado, Influence of the free isocyanate content in the adhesive properties of reactive trifunctional polyether urethane quasi-prepolymers, Int. J. Adhesion & Adhesives 26 (2006) 355–362.
[29] M. Melchiors, M. Sonntag, C. Kobusch, E. Jurgens, Recent developments in aqueous two-component polyurethane (2 K-PUR) coatings, Prog. Org. Coat. 40 (2000) 99–109.
[30] X. Zhang, S. Tasaka, N. Inagaki, Studies on surface molecular motion of oligomeric polystyrene by differential scanning calorimetry, Polym. Adv. Tech. 11 (2000) 40–47.
[31] R. Lu, K. Ishimura, K. Tsutida, T. Honda, T. Miyakoshi, Development of a fast drying hybrid lacquer in a low relative-humidity environment based on *Kurome* lacquer sap, J. App. Polym. Sci. 98 (2005) 1055–1061.

[32] Y. Kamiya, T. Miyakoshi, Synthesis of urushiol components and analysis of urushi sap from *Rhus vernicifera*, J. Oleo Sci. 50 (2001) 865–873.

[33] N. Niimura, T. Miyakoshi, Characterization of natural resin films and identification of ancient coating, J. Mass Spectr. Soc. Jpn 51 (2003) 439–457.

[34] A. Frick, A. Rochman, Characterization of TPU-elastomers by thermal analysis (DSC), Polym. Test. 23 (2004) 413–417.

[35] N. Niimura, Y. Iijima, T. Miyakoshi, Hardening process and surface structure of lacquer films studied by X-ray photoelectron spectroscopy, Surf. Inter. Anal. 24 (1996) 237.

[36] R. Ogawa, Y. Iwasaki, K. Ishihara, Thermal property and processability of elastomeric polymer alloy composed of segmented polyurethane and phospholipid polymer, J. Biomed. Mater. Res. Part A 62 (2002) 214–221.

[37] M.A. Semsarzadeh, A.H. Navarchian, Effects of NCO/OH ratio and catalyst concentration on structure, thermal stability, and crosslink density of poly(urethane-isocyanurate), J. Appl. Polym. Sci. 90 (2003) 963–972.

[38] T. Miyakoshi, K. Nagase, T. Yoshida, Progress of Lacquer Chemistry, first ed., IPC Publishing, Tokyo, Japan, 2000.

CHAPTER

Lacquerware

11

11.1 LACQUER FILM

Lacquer is a natural coating material polymerized by laccase or thermal curing. The transparency, color, gloss, and hardness of a lacquer film vary deeply depending on the drying conditions. In general, lacquer film is a durable film that has high resistance to acid, water, and organic solvents.

Lacquer sap is water-in-oil (W/O) emulsion. After drying to form a film, the polysaccharides and glycoproteins are submerged in an urushiol ocean: the water has evaporated and left small cavities in the film as shown in Figures 11.1 and 11.2.

In general, once a lacquer film begins to deteriorate, the gloss is reduced or disappears. The main reason is because a reaction between urushiol and oxygen in air has occurred and formed peroxides and radicals, which cause lacquer film degradation. Raw lacquer film has a very low gloss value of about 20-40. In order to improve the film gloss, raw lacquer is usually refined or a drying oil such as tung oil and/or linseed oil is added. Another way to increase the gloss of lacquer film is polishing the surface, and some results of these various methods are summarized in Table 11.1.

Front-face SEM photos of lacquer films are shown in Figure 11.3. Many pits and holes can be observed in the film surface; they are left when the water evaporates from the lacquer emulsion. It also can be seen that the front face has a low gloss and the back face has a high gloss (Table 11.1). In the drying process, the front faces shrinks and is deformed by the enzymatic polymerization. After it is polished, the front face becomes smooth and the gloss is increased. In addition, modifying the lacquer by adding a drying oil causes a reaction between urushiol and the oil to occur, which then forms an oil mask that increases the gloss.

Although lacquer is a natural resin, it is often treated as a special coating material based on its drying properties and uses. In Japan, in the Japanese Industrial Standards (JIS) definition of lacquer, it is JIS-K-5900 and JIS-K-5400, and in China it is defined as Guo Biao (GB/T-14703, Chinese national standard), as described in the Appendix.

It is very difficult to compare lacquer and synthetic resin paints by the characteristics of the film, and the properties also differ due to different drying conditions such as natural or heat drying. Therefore, JIS-K-5400 is defined by ten aspects of coating film as follows.

(1) Gloss
(2) Adhesion

Lacquer Chemistry and Applications. http://dx.doi.org/10.1016/B978-0-12-803589-4.00011-0

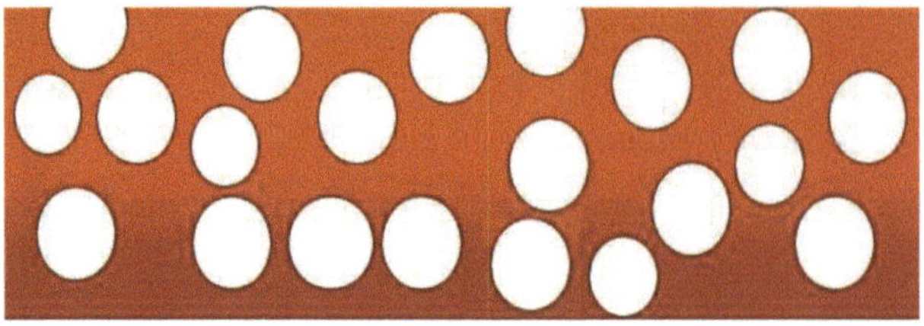

FIGURE 11.1

Structure of lacquer film.

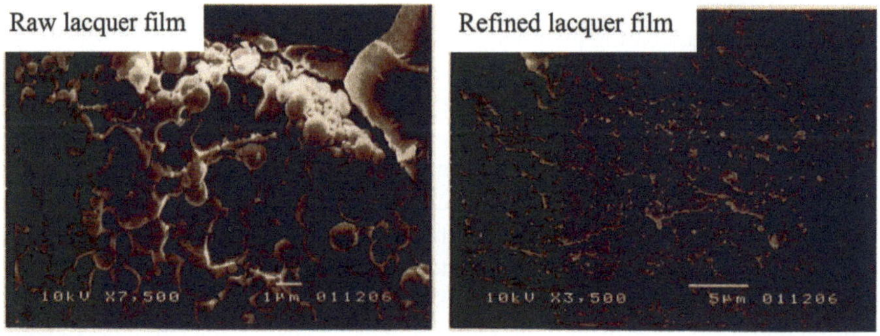

FIGURE 11.2

SEM photos of raw (left) and refined (right) lacquer films.

Table 11.1 Gloss values of raw, refined, refined (polished), and modified lacquers

Lacquer Name	Front (air interface)	Back (glass contact face)
Raw	23.6%	90.8%
Refined	70.2%	92.5%
Refined (be polished)	91.6%	92.5%
Modified	97.6%	98.0%

(3) Water resistance
(4) Alkali resistance
(5) Acid resistance
(6) Impact resistance
(7) Solvent resistance
(8) Crosscut adhesion test
(9) Pencil hardness
(10) Buckling resistance

We have measured the ten aspects of some Asian lacquer films and synthetic paint films, and the results are summarized in Table 11.2.

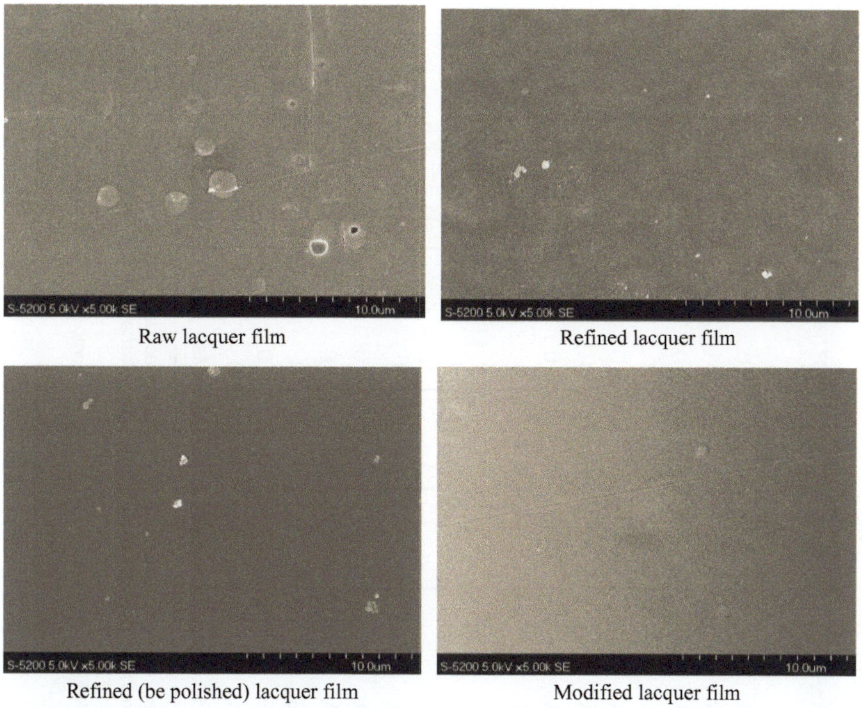

FIGURE 11.3

SEM photos of raw, refined, refined (polished), and modified lacquer films.

The dynamic viscoelasticity of lacquer has a low-temperature dependence, which is not found in synthetic resin coatings. This specificity is evaluated by the viscoelastic storage modulus (E', dyne/cm^2) and the loss modulus (E'', dyne/cm^2).

The elastic storage modulus of lacquer film indicates a flexibility of about $E' = 10^9$-10^{11} dyne/cm^2, and shows low-temperature dependence in a wide range (20-200 °C), as shown in Figure 11.4. This is typical glass elasticity with excellent heat resistance.

The elastic loss modulus indicates the hardness is about $E'' = 10^9$ dyne/cm^2, and has subtransitions at 40, 80, and 100 °C, as shown in Figure 11.5. The subtransition at the low temperature is due to the reaction of the urushiol side chain, and the subtransition at the high temperature is due to polymerization of the catechol ring. It can be speculated that lacquer film obtains its physical properties of high hardness and good flexibility from the small subtransition at the high temperature. These properties are not found in commercial synthetic resin coatings, even in cashew nut shell liquid (CNSL), which is the closest to the lacquer coating, has a high-temperature dependence of the storage modulus. Meanwhile, the temperature dependence of raw lacquer is higher than that of refined lacquer. This phenomenon can be considered due to the different degrees of polymerization. Refined lacquer

Table 11.2 Properties comparison of various lacquer and synthetic resin films

Type	DC	Gl	NA	WR	AlR	AcR	IR	SR	CT(P)	PH	BR(mm)
Jap-raw	rt	20	A	B	D	A	×	A	4	2B	2
	Heat	20	A	A	E	A	O	A	10	7H	6
Jap-R	rt	88	A	A	A	A	×	A	0	H	2
	Heat	91	A	A	E	A	×	A	10	6H	8
Ch-raw	rt	19	A	C	D	A	×	A	8	3B	2
	Heat	88	A	A	D	A	×	A	10	6H	6
Ch-R	rt	88	A	A	A	A	×	A	0	2B	2
	Heat	90	A	A	E	A	×	A	10	6H	6
T-raw	rt	26	A	D	C	A	×	A	4	6B	2
	Heat	75	A	A	E	A	×	A	8	5H	3
T-R	rt	90	A	B	A	A	O	A	0	HB	2
	Heat	82	A	A	E	A	×	A	10	5H	2
Ch-M	rt	94	A	A	B	A	O	A	2	3B	2
	Heat	94	A	A	B	A	×	A	10	5H	2
Ure	rt	100	A	D	B	A	O	A	10	2H	2
Epo	rt	100	A	A	A	B	O	A	0	H	2
Alk	rt	100	A	E	E	E	O	A	10	F	2
Cash	rt	98	A	E	C	A	×	A	6	H	2
CM	rt	Gla	Gla	Gla	Gla	Gla	Msp	Msp	Msp	Msp	Tin plate

Jap-raw: Japanese raw lacquer; Jap-R: Japanese refined lacquer; Ch-raw: Chinese raw lacquer; Ch-R: Chinese refined lacquer; T-raw: Taiwan raw lacquer; T-R: Taiwan refined lacquer; Ch-M: Chinese modified lacquer added in oil; Ure: urethane; Epo: epoxy; Alk: alkyd; Cash: Cashew nut shell liquid; CM: Coating material.

DC: Drying condition; Gl: Gloss; NA: No adhesion; WR: Water resistance; AlR: Alkali resistance; AcR: Acid resistance; IR: Impact resistance; SR: Solvent resistance; CT (P): Crosscut adhesion test (point); PH: Pencil hardness; BR (mm): Buckling resistance.

Rt: room temperature drying; Heat: heating drying.

Gla: glass plate; Msp: mild steel plate.

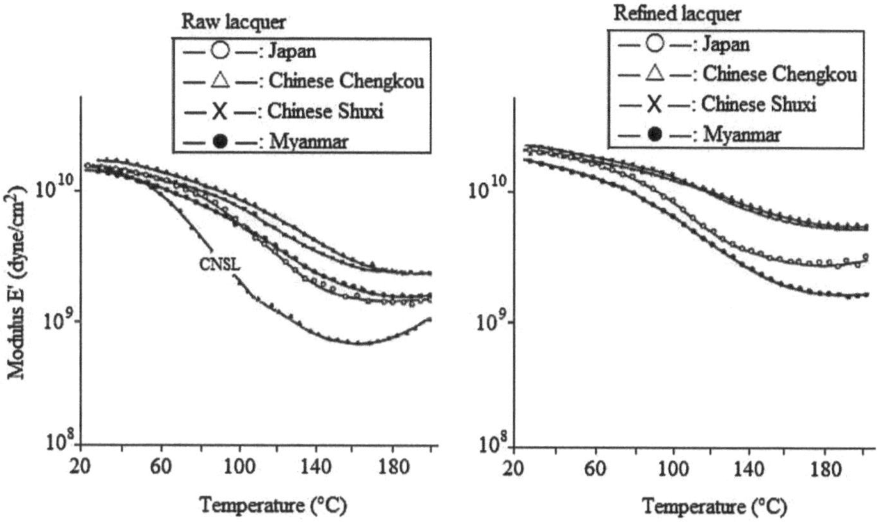

FIGURE 11.4

Storage modulus (E') of lacquer films.

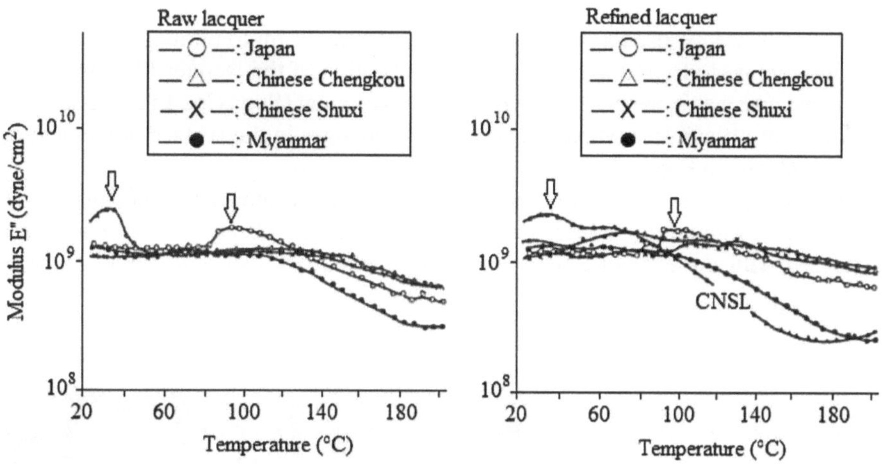

FIGURE 11.5

Loss modulus (E'') of lacquer films.

has a higher degree of polymerization than raw lacquer. These results suggest that the traditional lacquer refining methods, such as "nayashi" and "kurome" in Japan, and "Jiaoban Tuoshui" in China, not only disperse the lacquer emulsion homogeneously and accelerate prepolymerization but also can improve the durability of the lacquer film.

11.2 ANALYSIS OF LACQUER FILMS

11.2.1 X-RAY

Polymerization of lacquer sap will form a durable film that is resistant to water, organic solvents, dilute acids, and alkalis. Cultural treasures coated with lacquer have maintained their beautiful surfaces without loss of their original beauty for more than a thousand years. However, because lacquer sap is a water-in-oil emulsion and dries by oxidation of urushiol catalyzed by laccase, a thin film of urushiol polymer coats the polysaccharide granules, and each granule is impregnated by the glycoprotein, as shown in Figure 11.6. This kind of a structure makes the lacquer film very susceptible to damage by ultraviolet irradiation [1–4]. Exposure to ultraviolet rays will cause the lacquer coating film to deteriorate, for example, the film will flake off, lose its gloss, and crack [5,6]. Due to these characteristics, protection and restoration of lacquer-coated cultural treasures has become an important problem for museums, and identification of the lacquer types in ancient coatings has also become very important for conservation and restoration studies. In order to reveal the mechanism of deterioration of lacquer film, as well as develop methods of conservation and restoration of lacquerware, many analyses of lacquer film have been reported [7–10]. Further, experimental studies on a single-layer lacquer film on a glass sheet were also reported [11,12], but the deterioration mechanism of the lacquer film has not been clarified.

Recently, we exposed a polished four-layer lacquer film to ultraviolet irradiation for different periods, and quantified the deterioration of the lacquer films [13]. The UV resistance was measured with a FAL-AU instrument (Suga Co., Ltd., Japan). The durations of ultraviolet ray exposure and microchanges in the surface of the lacquer film are shown in Figure 11.7.

The different UV exposure times produced different interface states in the lacquer film. A minute ruggedness and toroidal shapes are observed at 48 hours, and increased with the exposure time. Not only cracks, but also with round pinholes

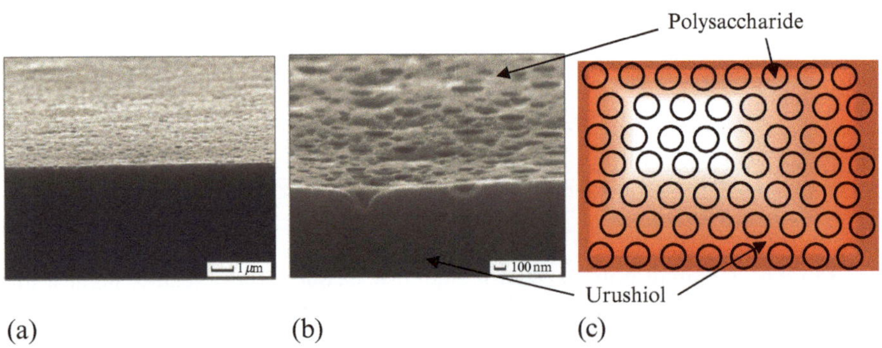

FIGURE 11.6

Photos of raw lacquer film (A×10,000, B×50,000) and a structural model (C).

FIGURE 11.7

Micrograph of lacquer coating film irradiated by UV for different periods.

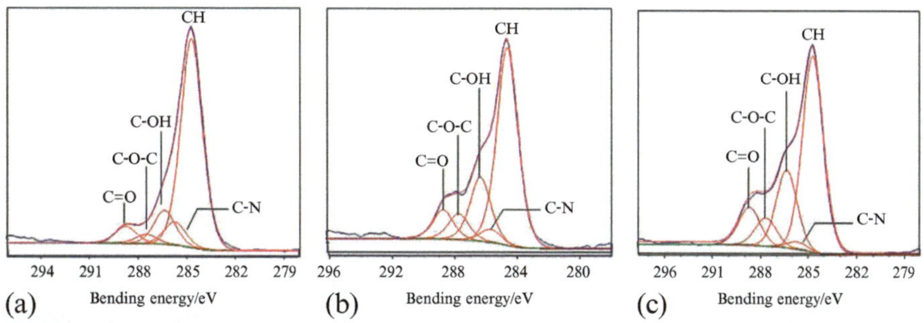

FIGURE 11.8

Curve-fitting of C ls spectra of the black lacquer film, (A) 0 h, (B) 192 h, and (C) 384 h (θ=90°).

appeared in the toroidal white parts after irradiation for 384 hours. That is to say, as the ultraviolet ray exposure time increase, the particulate material and the toroidal-type material increased on the surface of the lacquer films, decreasing the smoothness of the surface and reducing the gloss. The X-ray photoelectron spectroscopy (XPS) is shown in Figure 11.8. It can be seen that the ratios of the functional groups C=O, C–O–C, C–OH, and C–N in the initial lacquer film (Figure 11.8A, UV irradiation time 0) were lower than those after UV irradiation times of 192 hours (Figure 11.8B) and 384 hours (Figure 11.8C), and the ratio was observed to be highest at the UV irradiation time of 384 hours. In other words, these functional groups on the surface of lacquer films increased with an increase in UV irradiation time. In contrast, the functional group of C–N decreased as the UV irradiation time increased.

The concentration ratios of each group and the C 1 s and O 1 s are summarized in Table 11.3. As the UV irradiation time increased, C 1 s decreased and O 1 s increased. The main component of lacquer sap, urushiol, is polymerized by a laccase-catalyzed nucleus-side chain coupling to C–O as well as C–C, and UV irradiation damaged the C–O and C–C coupling. The deterioration of the surface of the lacquer film due to UV irradiation was considered to occur as follows: first, the projection of a particulate material appeared on the surface of the lacquer film. This particulate material has

Table 11.3 Proportion of functional groups (%) and elemental (at %) analysis of black lacquer film for each ultraviolet ray exposure time

UV exposure time (h)	C=O	C–O–C	C–OH	C–N	C–H	O 1s	C 1s
0	5.828	3.191	11.693	8.035	71.250	20.384	79.615
192	8.937	8.012	19.410	4.023	59.579	29.395	70.604
384	10.996	8.224	22.387	2.105	56.285	32.838	67.161

been demonstrated to have C=O, C–O–C, C–OH, and C–N groups. Subsequently, the toroidal white material increased. This white material can be considered to be due to the polysaccharides in the lacquer film, following the appearance of cracks and pinholes. The C–H and C–C groups decreased, and the C=O, C–O–C, C–OH, and C–N groups increased as the UV irradiation time increased, suggesting that the lacquer film was oxidized and the C–H and C–C bonds were cleaved, and these results also were confirmed by IR and pyrolysis analysis. It is important to understand not only that lacquer deteriorates due to light exposure, but also how the deterioration happens on the surface of the lacquer film.

As described above, lacquer film is a cross-linked polymer polymerized by laccase that is insoluble in most solvents. In order to analyze the structure of lacquer film and the source of lacquer sap, only a few analytical techniques, such as solid NMR, FT-IR, and XPS, are suitable. However, most of these conventional techniques are time consuming, demand large amounts of sample, and frequently require several pretreatments [14]. Further, they do not always provide clear results [15]. During our research work, we have found that pyrolysis-gas chromatography/mass spectrometry is an effective method for analyzing lacquer film because this method can discriminate between the pyrolysis product of a lacquer sample and that of other natural resins [8,16]. It is sometimes extremely important to determine whether objects that are claimed to be lacquerware are actually created from lacquer sap or some other resins. As a part of the lacquer research project in our laboratory, we have synthesized urushiol, laccol, and thitsiol analogs and successfully analyzed their films by pyrolysis-gas chromatography/mass spectrometry [17,18]. We also have applied this method for the analysis of Ryukyu and other ancient lacquer films, and the detail analyzed results will be shown in Section 11.3.

11.2.2 PYROLYSIS GC/MS

Lacquer film is a cross-linked polymer that is first polymerized by laccase and subsequently by autoxidation of unsaturated side chains. Because lacquer films are insoluble in most solvents, only a few analytical techniques, including solid NMR spectroscopy, FT-IR spectroscopy, and X-ray photoelectron spectroscopy are appropriate. However, most of these conventional techniques are time consuming, demand large amounts of sample, and frequently require several pretreatments; further, they do not always obtain clear results. Pyrolysis-gas chromatography/mass spectrometry (Py-GC/MS) is effective for analyzing lacquer film because this method can discriminate between the product of pyrolysis of a lacquer sample and that of other natural resins. In this section, we describe analysis of lacquer films of typical Oriental lacquer saps, including ancient lacquerwares, by Py-GC/MS. Urushiol, laccol, and thitsiol, which are the main constituents of the Oriental lacquers, produce, respectively, characteristic pyrolysis products.

Py-GC/MS is a comparatively old analytical method of the techniques developed to analyze polymers. A schematic diagram and configuration are showed in Figure 11.9. In the primary stage, it was not used very much because it has problems with reproducibility and long analysis times, though pyrolysis has begun to be used to analyze natural and synthetic fibers. The technique for combining this pyrolysis method with gas chromatography and mass spectrometry was reported by Davison in 1954, and it has been appreciated in recent years as a significant technique to analyze three-dimensional network structures. This technique involves high-temperature treatment of the polymer film in order to isolate each element of the polymer using the solvent for gas chromatography and to perform mass spectrometry of the gas chromatography peak at the same time.

The advantage of this technique is the convenience of the device and pretreatment of the analysis sample. In particular, a polymer can be analyzed using a very small amount of sample (0.01-1 mg). The best feature of Py-GC/MS is pyrolysis. The thermal cracking unit can be heated quickly to a high temperature, and this is directly related to the reproducibility of the analysis performance. A microfurnace is usually used in pyrolysis, as shown in Figure 11.10.

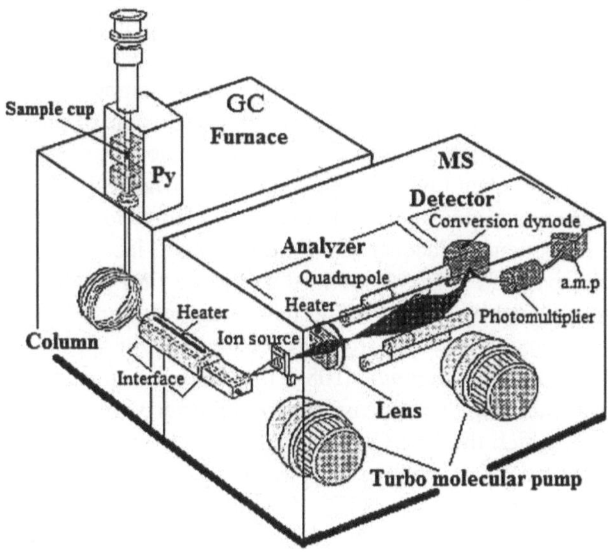

FIGURE 11.9

Configuration of Py-GC/MS system.

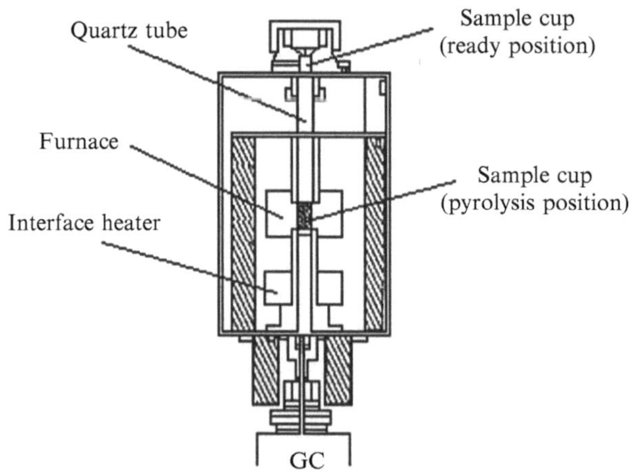

FIGURE 11.10

Internal structure of pyrolyzer.

The method involves putting the sample into the inactivated sample holder in the microfurnace, dropping it into the reactor core from the state set in the device, which is filled with helium as the carrier gas using a switch, and then is pyrolyzed. A feature of this technique is that the measurement results are steady because the small capacity keeps the temperature dispersion comparatively low [19].

There are mainly three kinds of Py-GC/MS: single-shot, a direct method, evolved gas analysis (EGA), and double-shot method. Single-shot is the easiest method in which only one heating performs the thermal decomposition. The range of temperatures is usually set at 50-1000 °C. After the sample is thermally decomposed, it is gasified and introduced into a gas chromatograph. The gas separated by GC is measured by MS, as shown in Figure 11.11. The decomposition temperature of lacquer film is 500 °C. The GC column is a type of dimethyl polysiloxane or phenyl methylpolysiloxane because of the high polarity of the lacquer catechol configuration. Setting the temperature of the oven high enough and using a heat-resistant column are also important because urushiol is undetectable at temperatures lower than 280 °C.

The main component of lacquer is a lipid (urushiol, laccol, and thitsiol) that is a catechol derivative, and it is polymerized to form a film by the laccase. When a lacquer film is analyzed by Py-GC/MS, a mass fragment $m/z = 108$ from alkyl phenol and a mass fragment $m/z = 123$ from alkyl catechol will appear as characteristic fragment ions. The fragment ion $m/z = 108$ is the hydrogen attached to the 3-position carbon of the catechol ring that appears due to McLafferty transference. $m/z = 123$ is a fragment ion in which the first CH_2 from a catechol ring enters another catechol ring, and constitutes a seven-member ring, as shown in Figure 11.12.

The total ion chromatography (TIC) of standard lacquer tapped from a *T. vernicifluum* lacquer tree is shown in Figure 11.13. Many components are detected in the TIC, and this is due to the complex structure of the lacquer film and various components contained in a lacquer sap. The ion chromatographs of (A) $m/z = 108$ and (B) $m/z = 123$ are shown in Figure 11.14. The highest peak is the component that has

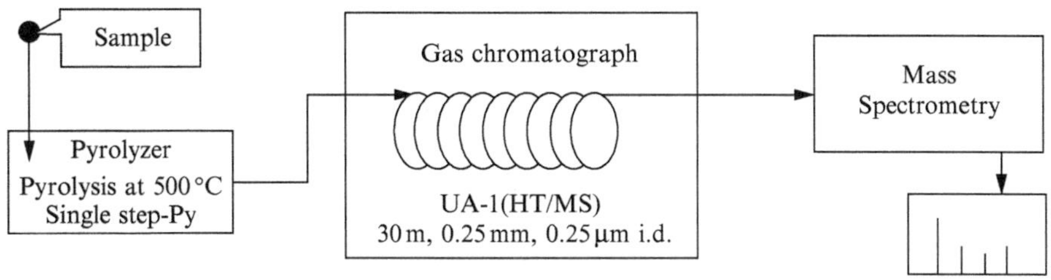

FIGURE 11.11

Configuration of Py-GC/MS.

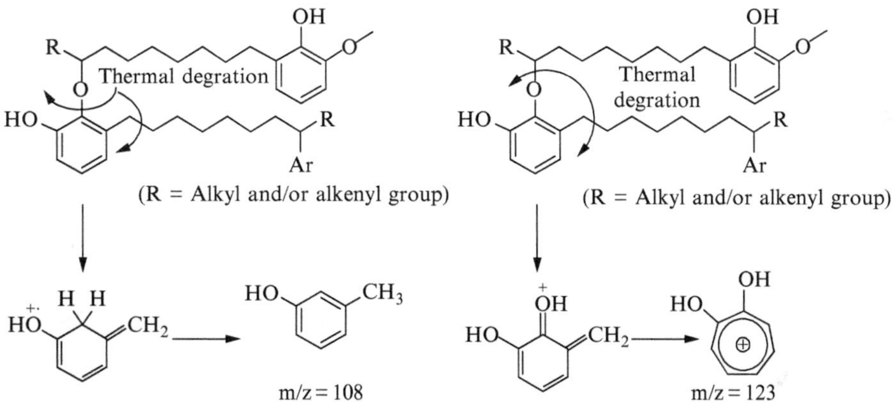

FIGURE 11.12

Formation mechanism of fragment *m/z* = 108 and *m/z* = 123.

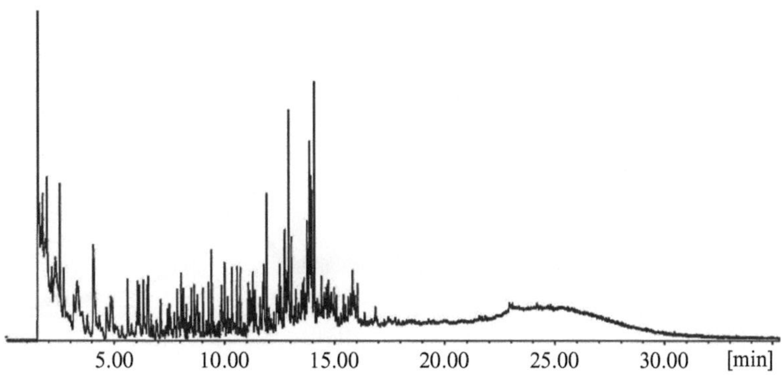

FIGURE 11.13

TIC of *T. vernicifluum* lacquer film.

carbon number 7 in the side chain. In the case of urushiol and thitsiol (Figure 11.15), carbon number 7 is the most frequent component, but since the position of the double bond changes in laccol, the alkyl phenol of carbon number 9 is detected as the greatest peak (Figure 11.16) [20,21].

The principal constituent of lacquer lipid is a catechol ring. Catechol has two hydroxyl groups of a phenol nature, and due to its high polarity, it tends to adhere to the column, decreasing analytical sensitivity. In order to solve this problem, derivatization is often employed. There are reports of the use of tetramethylammonium hydroxide (TMAH) in the analysis of a lignin, which generates a component containing the hydroxyl group during thermal decomposition and is an example of analysis of a polysaccharide [22–27]. TMAH can also be used effectively at the time of the thermal decomposition of lacquer. The reaction can be considered as shown in Scheme 11.1.

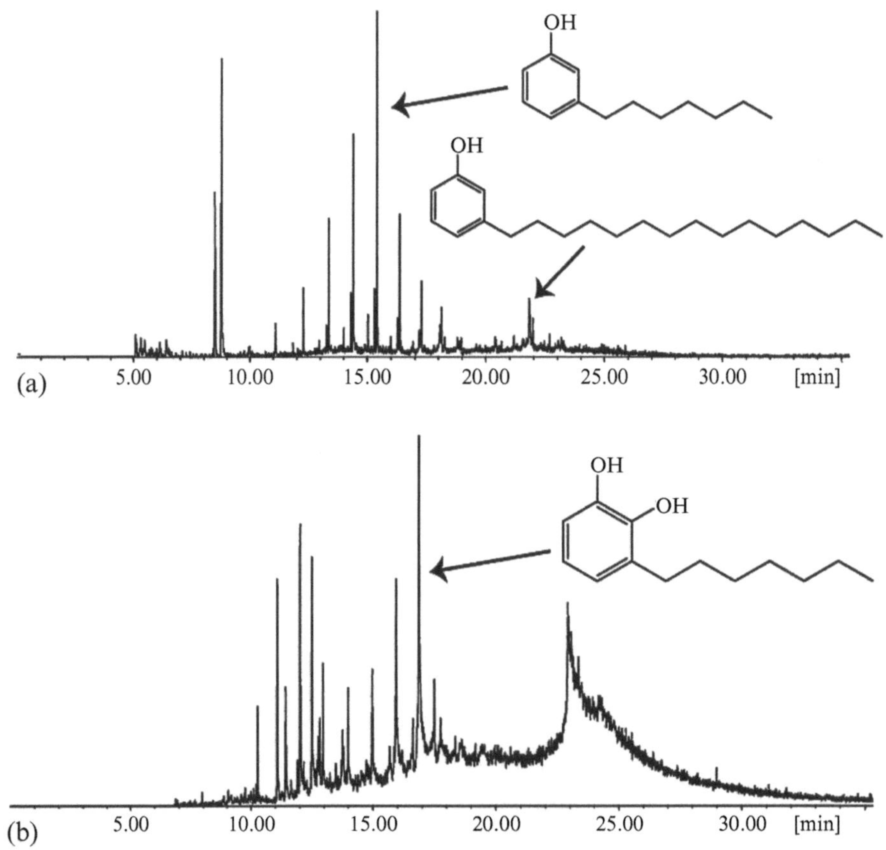

FIGURE 11.14

Ion chromatography of *T. vernicifluum* lacquer film: (A) $m/z=108$, and (B) $m/z=123$.

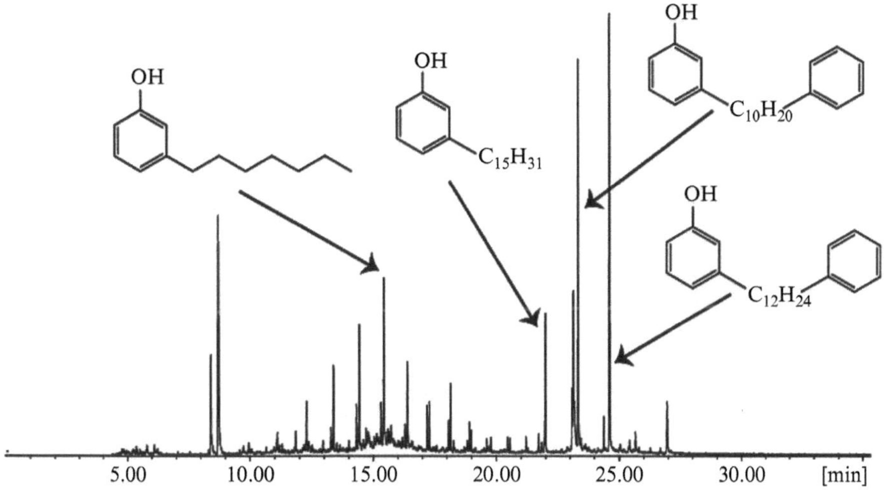

FIGURE 11.15

Ion chromatography of *G. usitata* lacquer film.

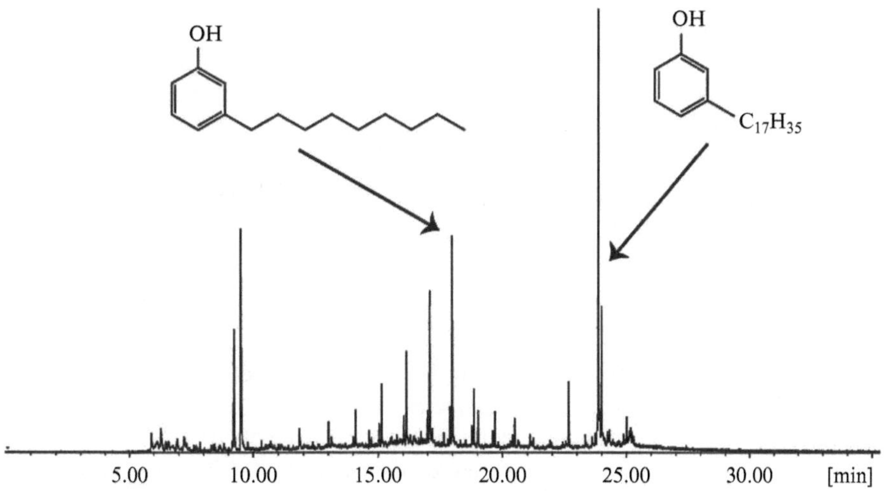

FIGURE 11.16

Ion chromatography of *T. succedanea* lacquer film.

Figure 11.17 shows the pyrolysis results of *T. vernicifluum* lacquer film derivatization by TMAH. It can be seen that 1,2-dimethoxy-3-pentadecylbenzene was detected due to a methylation reaction. This method is useful for the analysis of cultural properties that can provide only small samples.

Another important pyrolysis method is EGA, a method of analyzing the gas emitted while heat is applied to a sample [28–31]. The mechanical configuration

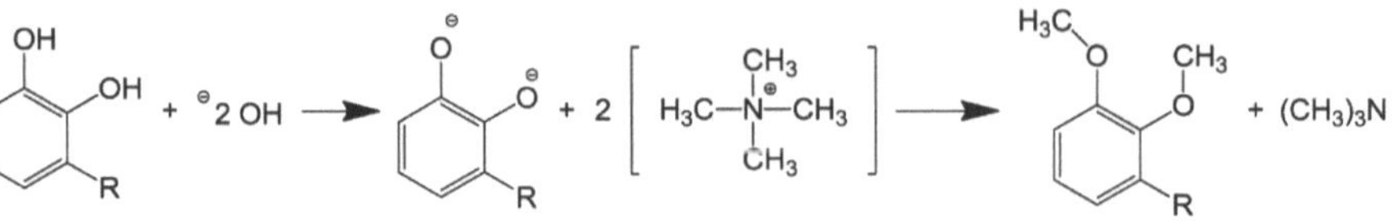

SCHEME 11.1

Methylation reaction of hydroxyl group by TMAH.

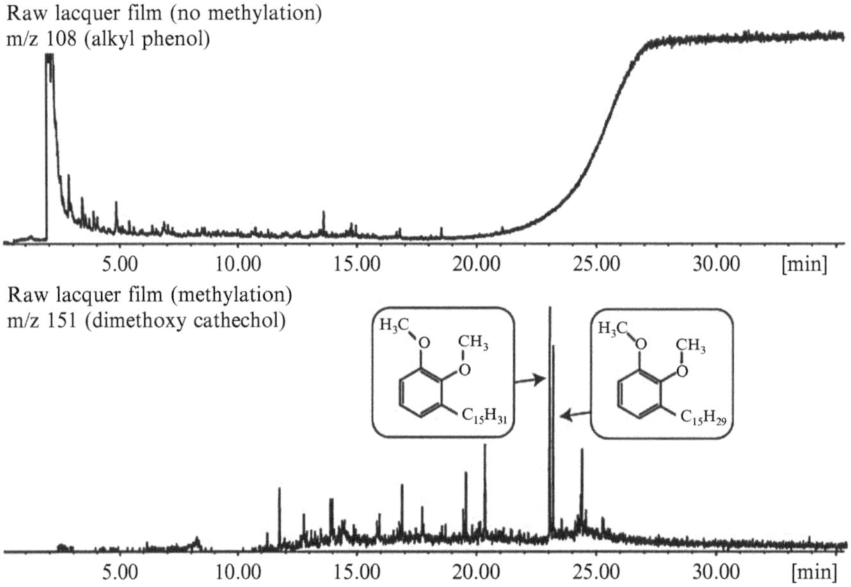

FIGURE 11.17

Comparison of sensitivity of methylation.

characteristically uses a very short column. The column is about 1 m in length and does not have a fixed zone. Since many organic macromolecules are decomposed at less than 1000 °C, a set of unstring temperatures in the range of 50-1000 °C is often used. In the case of a lacquer film, the temperature range is set at 50-650 °C (Figure 11.18).

EGA analysis decomposes a lacquer film in three steps (Figure 11.19). When the heat decomposition points were divided into three steps, two were the original peaks of $m/z = 60$ and $m/z = 108$, which is a thermal decomposition peak of urushiol, and one was $m/z = 123$. The first peak of $m/z = 60$ at a lower temperature was a short carboxylic acid with various carbon chains. The following peak was considered to be due to the decomposition of sugars. The acetone powder (AP) taken from lacquer sap also was analyzed by EGA, and the peak position of acetone powder was mostly in agreement with the second EGA peak of *T. vernicifluum* lacquer film (Figure 11.20).

Double-shot pyrolysis is a method that can be performed using the results of EGA. It applies heat to the same sample gradually and analyzes only the gas generated in a certain temperature range (Figure 11.21).

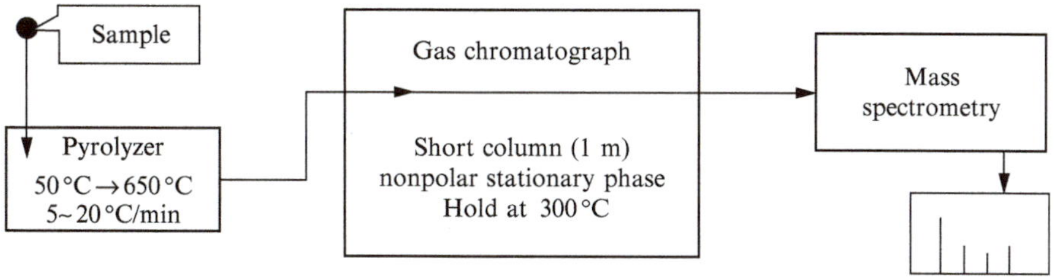

FIGURE 11.18

Configuration of EGA system.

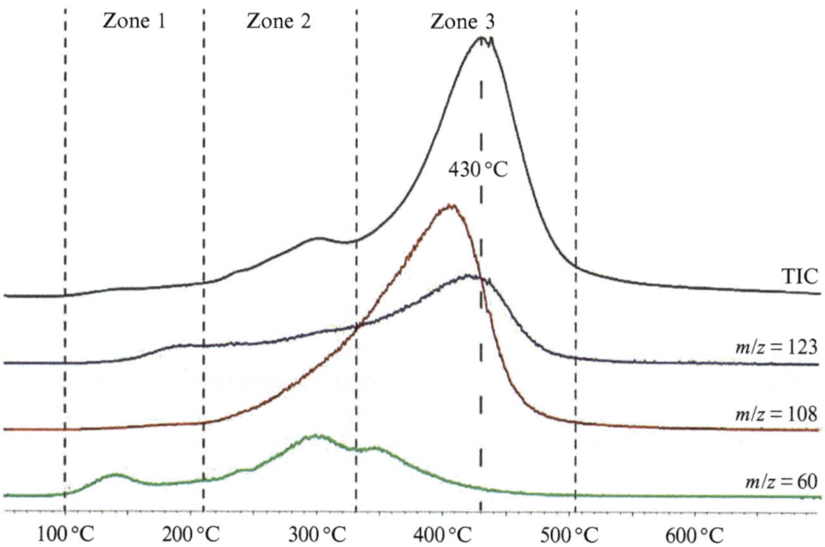

FIGURE 11.19

EGA of *T. vernicifluum* lacquer film.

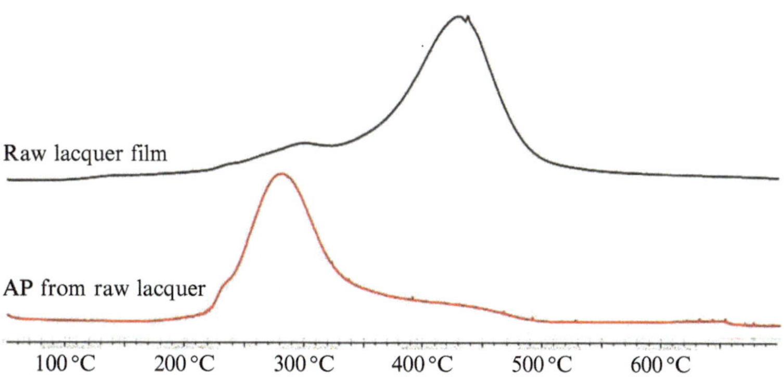

FIGURE 11.20

EGA of *T. vernicifluum* lacquer film and AP.

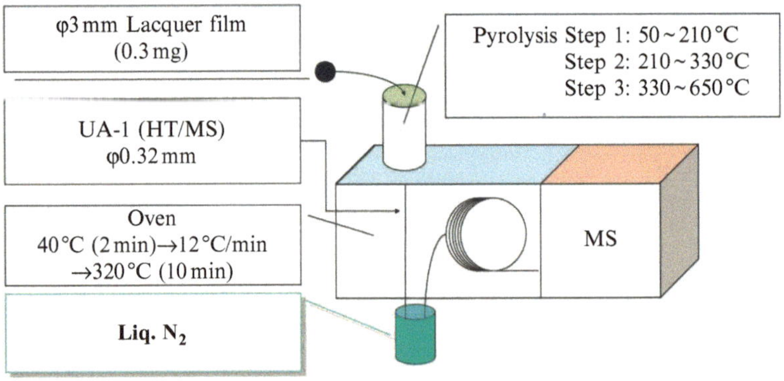

FIGURE 11.21

Configuration of double-shot pyrolysis.

A sample is placed in the pyrolyzer and the heating program is started. The first part of the column is dipped in liquid nitrogen, and the gas chromatograph is not moved. If thermal decomposition occurs at a predetermined temperature, the pyrolyzed sample is removed from the heating block immediately. After raising the sample from the oven, the GC column is removed from the liquid nitrogen, and then gas chromatography is started. Follow this procedure by the number of daylights of a thermal decomposition. In this way, the Py-GC/MS will store a record of each EGA peak that is attained [32–34].

This technique is convenient to remove a plasticizer previously added to a sample or to remove a solvent that remains in very small quantity. However, this method cannot be used when a sample reacts with the energy of heating. In the case of a lacquer, various additive admixtures such as pine drying oil, tung oil, linseed oil, tar, etc., may be present. Thus, this technique is very effective when components with clearly different heat decomposition points are present.

Three kinds of lacquers from *T. vernicifera*, *T. succedanea*, and *G. usitata* were coated on 70 cm × 40 cm glass plates. They were allowed to polymerize in a humidity-controlled chamber at a relative humidity of 70-90% at 20 °C for 12 h, and then removed from the chamber and stored in the open air.

Pyrolysis GC/MS measurements were carried out using a vertical microfurnace-type pyrolyzer PY-2010D (Frontier Lab, Japan), an HP 6890 gas chromatograph (Hewlett-Packard), and an HPG 5972A mass spectrometer (Hewlett-Packard). A stainless steel capillary column (0.25 mm i.d. × 30 m) coated with 0.25 μm of Ultra Alloy PY-1 (100% methylsilicone) was used for the separation. The lacquer sample (~1.0 mg) was placed in the platinum sample cup, and the cup was placed on top of the pyrolyzer at near ambient temperature. The sample cup was introduced into the furnace at 500 °C, and then the temperature program of the gas chromatograph oven was started. The gas chromatograph oven was programmed to provide a constant temperature increase of 20 °C per min from 40 °C to 280 °C, then held for 10 min at 280 °C. The flow rate of the helium gas was 18 ml/min. All the pyrolysis products were identified by mass spectrometry. The mass spectrometry ionization energy was 70 eV (EI-mode). The results of three kinds of lacquer detail are described in detail below.

Pyrolysis of Urushiol Polymer

The lacquer tapped from *T. vernicifluum* lacquer tree was pyrolyzed at 500 °C. The total ion chromatogram (TIC) and mass chromatogram (m/z = 320) are shown in Figure 11.22. Although a complex TIC was obtained, the mass chromatogram and mass spectrum revealed that urushiol (MW = 320) was the main component of the lacquer (peak 1).

The pyrolysis products obtained in mass chromatograms and mass spectra of m/z = 123 and m/z = 108 of urushiol lacquer film are shown in Figure 11.23. 3-Heptylcatechol (C7) and 3-heptylphenol (C7) with the highest relative intensity

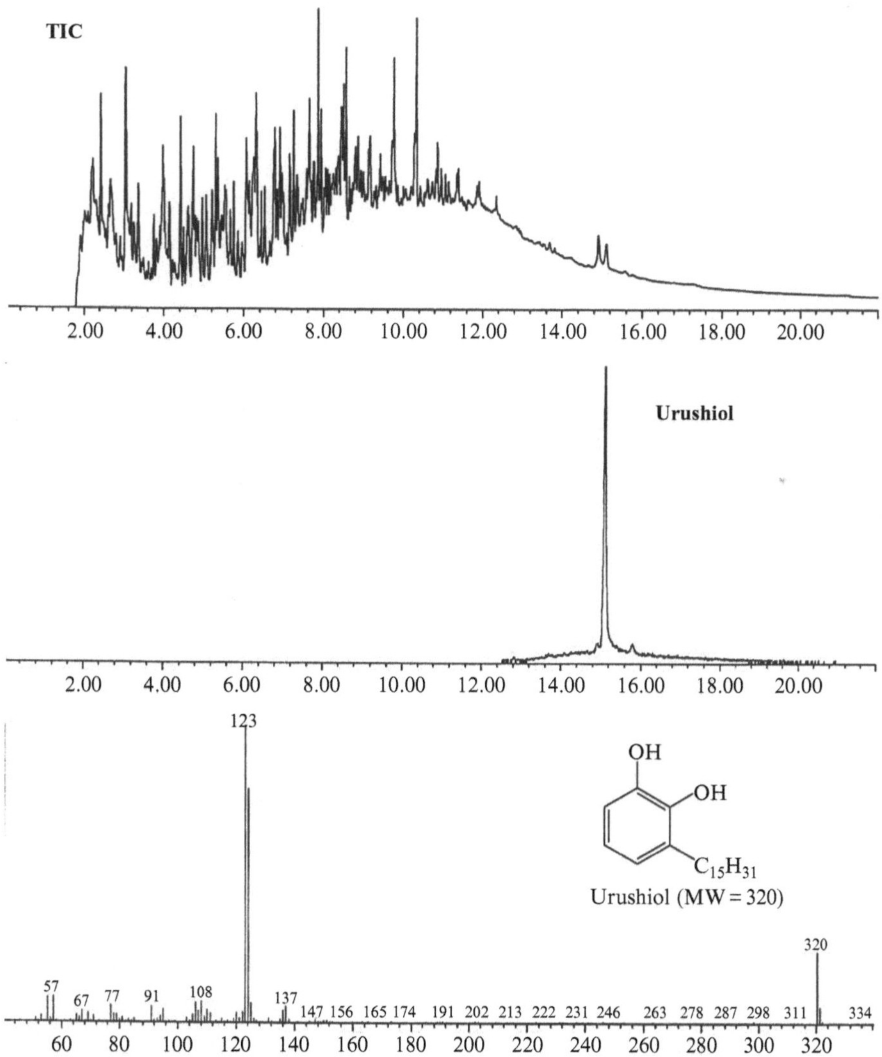

FIGURE 11.22

TIC, mass chromatogram (m/z=320), and mass spectrum of urushiol lacquer film.

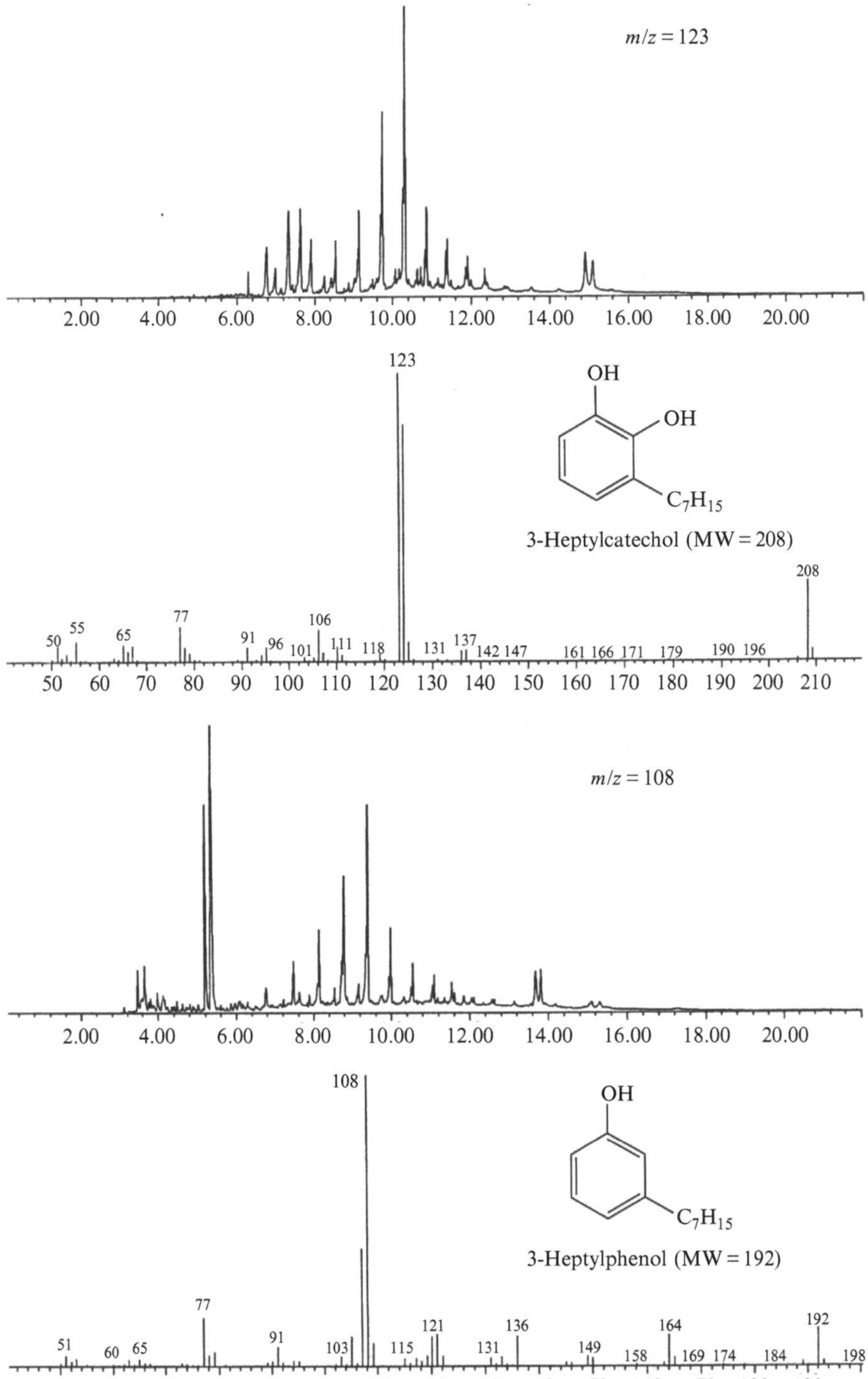

FIGURE 11.23

Mass chromatograms (*m*/*z* = 123 and 108) and mass spectra of urushiol lacquer film.

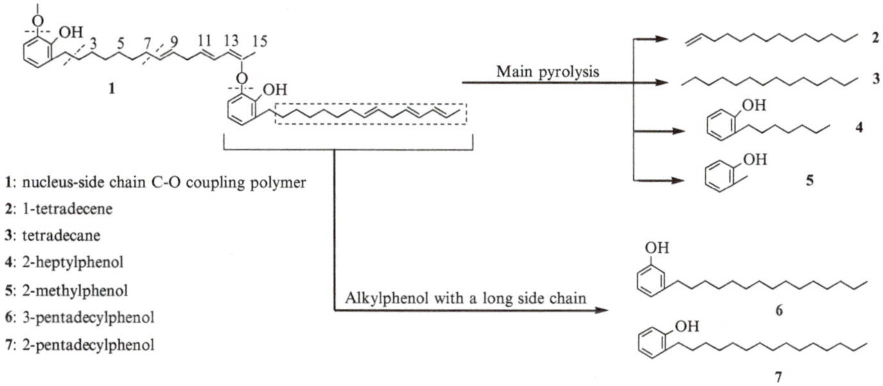

SCHEME 11.2

Pyrolysis mechanism of urushiol polymer.

were detected in the mass chromatogram at $m/z=123$ and $m/z=108$, respectively. It has been reported that the double bonds of olefins at α and β positions are most susceptible to thermal cleavage [35,36]. Therefore, as shown in Scheme 11.2, the highest yield of 3-heptylcatechol (C7) can be attributed to the preferential cleavage at the α-position of the double bonds of the nucleus-14th chain C–O couplings of the urushiol polymers.

Pyrolysis of Laccol Polymer

Lacquer tapped from *T. succedanea* lacquer tree was also pyrolyzed at 500 °C like the urushiol lacquer film. The total ion chromatogram and mass chromatogram ($m/z=320$) are shown in Figure 11.24. Although a complex TIC was obtained, the mass chromatogram and mass spectrum confirmed that laccol (MW = 348) was the main component of the lacquer film (peak 1).

The pyrolysis products obtained in mass chromatograms and mass spectra of $m/z=123$ and $m/z=108$ of laccol lacquer film are shown in Figure 11.25. 3-Nonylcatechol (C9) and 3-nonylphenol (C9) with the highest relative intensity were detected in the mass chromatogram at $m/z=123$ and $m/z=108$, respectively.

Pyrolysis of Thitsiol Polymer

G. usitata is a lacquer tree that grows in Myanmar, Thailand, and Cambodia, and the main component is thitsiol, which contains 3- and 4-heptadecadienylcatechols as well as a series of α- and β-position.

The thitsiol lacquer film was pyrolyzed at 500 °C. Figure 11.26 shows the TIC and individual mass chromatograms at $m/z=346$, 348, 310, 326, 338, and 354. Peaks 1, 2, 3, 4, 5, and 6 of the mass chromatograms were identified as 4-hepatadecenylcatechol (MW = 346), 4-hepatadecenylcatechol (MW = 348), 3-(10-phenyldecyl)phenol (MW = 310), 3-(10-phenyldecyl)catechol (MW = 326), 4-(10-phenyldodecyl)phenol (MW = 338), and 4-(12-phenyldodecyl)catechol (MW = 354), respectively. Because

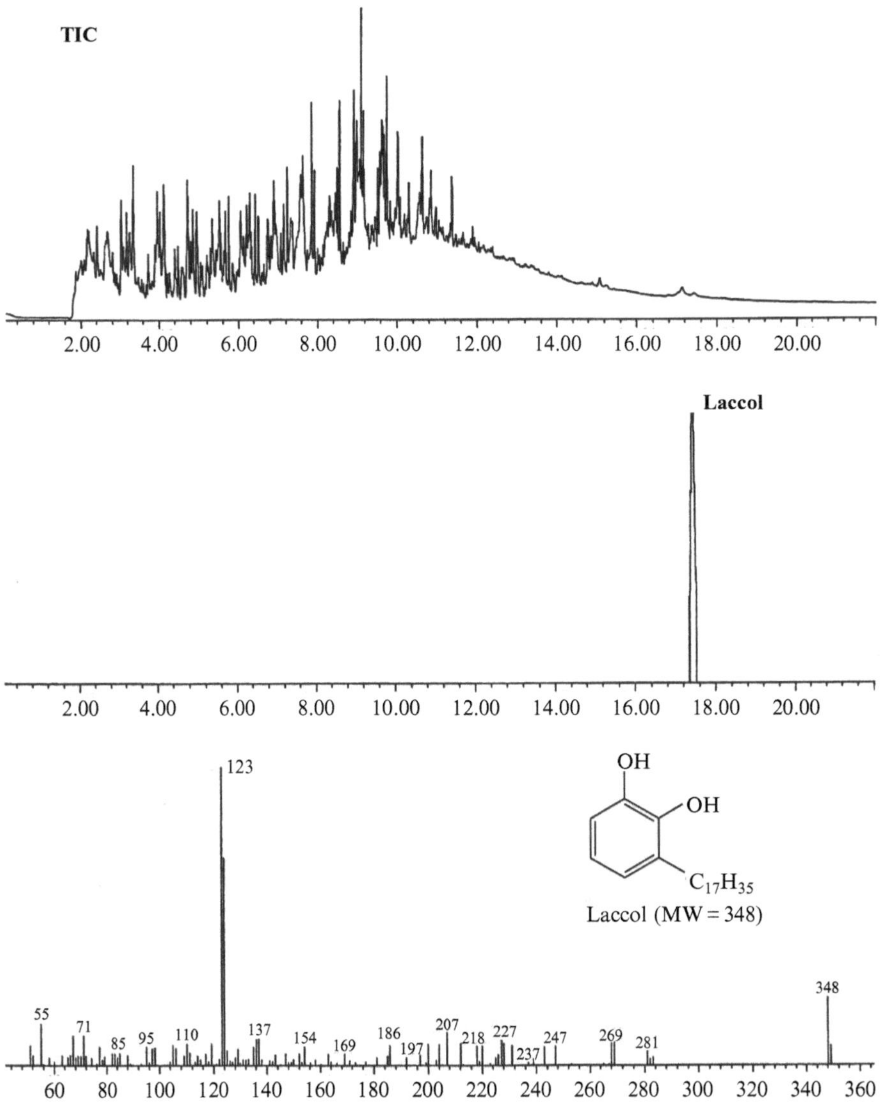

FIGURE 11.24

TIC, mass chromatogram (*m*/*z*=348), and mass spectrum of laccol lacquer film.

the C–O coupling polymers are terminated with 4-hepatadecenylcatechol, these compounds can be formed from such terminal groups.

In the *m*/*z*=123 mass chromatograms of the thitsiol lacquer film, a pair of peaks of the 3- and 4-alkylcatechols were detected (Figure 11.27). The relative intensities of the 3- and 4-heptylcatechols (C7) were the highest in the pyrolysis products of the thitsiol lacquer. The highest yields of the 3- and 4-heptylcatechols were

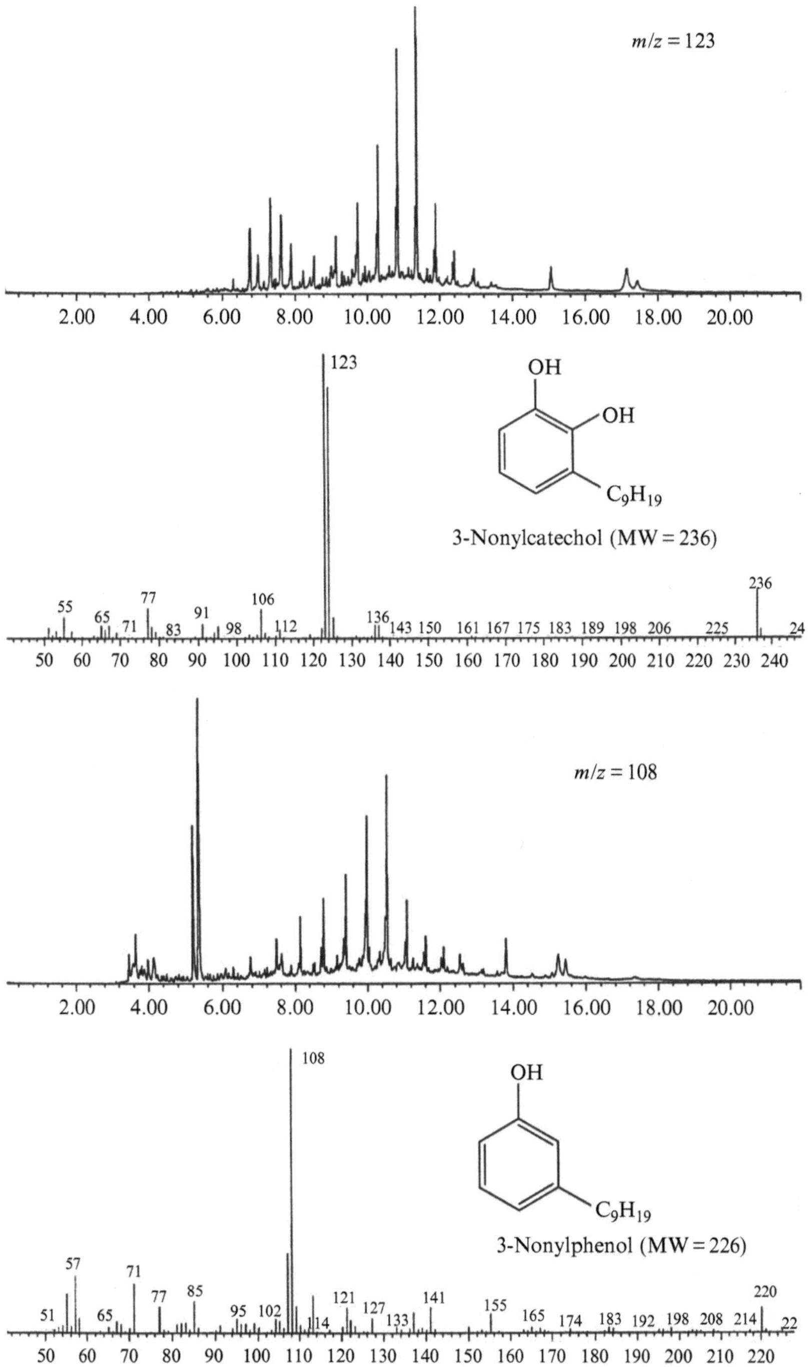

FIGURE 11.25

Mass chromatograms (*m/z* = 123 and 108) and mass spectra of laccol lacquer film.

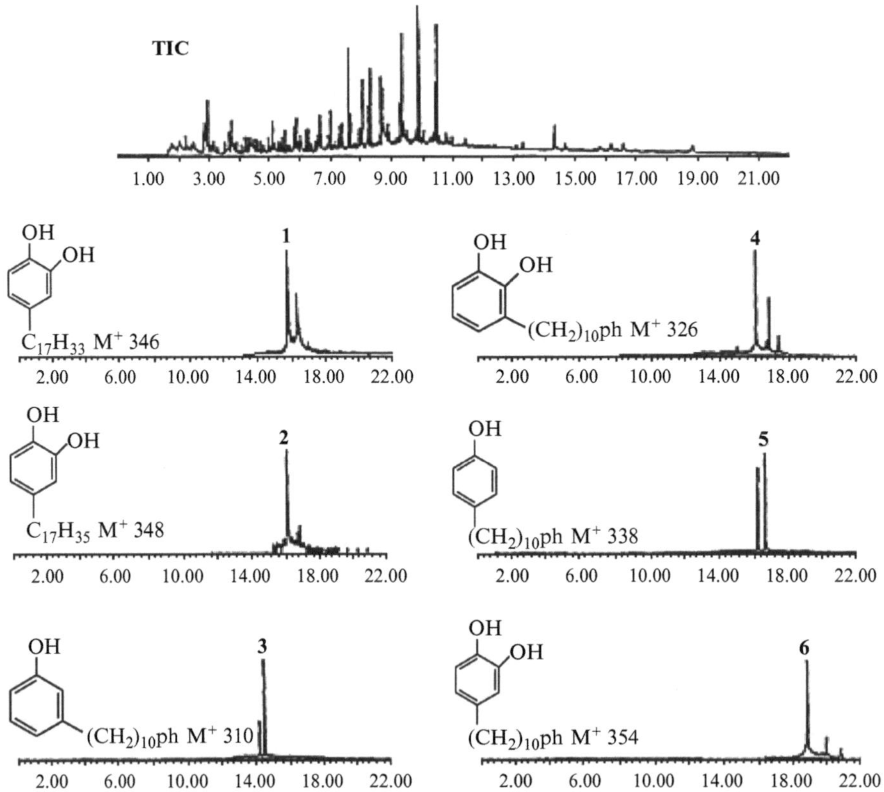

FIGURE 11.26

TIC and mass chromatograms (m/z=346, 348, 310, 326, 338, 354) of thitsiol lacquer film.

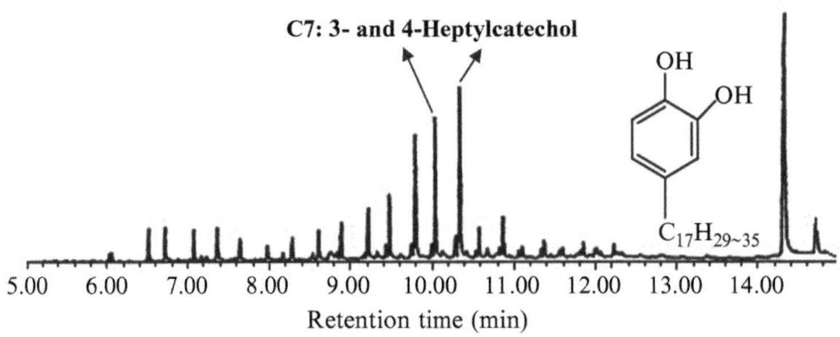

FIGURE 11.27

m/z=123 mass chromatograms of thitsiol lacquer film.

Table 11.4 Characteristic pyrolysis products of lacquer films

	Features of pyrolysis products		
Lacquer species	**Monomer**	**Alkylcatechol**	**Alkylphenol**
T. vernicifluum	Urushiol	3-Heptylcatechol	3-Heptylphenol
T. succedanea	Laccol	3-Nonylcatechol	3-Nonylphenol
G. usitata	Thitsiol	3-Heptylcatechol, 4-Heptylcatechol	3-Heptylphenol

considered to be mainly due to cleavage at the α-position of the double bonds of the nucleus—8th and 12th chain C=O couplings for the thitsiol polymers. The alkylphenols detected are likely to be the products of pyrolysis of the nucleus-side chain coupling of the thitsiol polymers. Dimerization of the lacquer monomers is considered to proceed through the laccase-catalyzed nucleus-side chain coupling as well as the C–C coupling. The yields of 3-heptylphenol (C7) were the highest, as shown in Figure 11.26. The α and β positions of the double bonds of the olefins are susceptible to thermal cleavage so that these highest yields are thought to be produced primarily by cleavage at the α position of the double bonds of thitsiol, such as the 3- and 4-(8,11-heptadecadienyl)catechols%%.

The 3-pentadecylcatechol (MW=320) (urushiol), 3-heptadecylcatechol (MW=348) (laccol), and 4-heptadecylcatechol (MW=348) (thitsiol) are characteristic pyrolysis products of the three kinds of lacquer films, as summarized in Table 11.4. Furthermore, the data acquired for the alkylcatechols and alkylphenols in the pyrolysis products also can help to determine the lacquer species.

11.3 APPLICATION OF Py-GC/MS IN LACQUER SCIENCE

11.3.1 IDENTIFICATION OF LACQUER SPECIES

As described above, the main pyrolysis products of urushiol are 3-heptylcatechol and 3-heptylphenol, of laccol are 3-nonylcatechol and 3-nonylphenol, and of thitsiol are 3-heptylcatechol, 3-heptylphenol, and 4-heptylcatechol, respectively. In order to confirm these results, we synthesized urushiol [37], laccol [38], and thitsiol [39] lacquer films and analyzed them by Py-GC/MS measurement.

Analysis of Natural and Synthesized Urushiol Films by Py-GC/MS

Compounds **1**, **2**, and **3**, the major components of urushiol, shown in Figure 11.28, were synthesized in good yield via the Witting reaction from a ylide derived from (3E,5Z)-3,5-heptadienyltriphenylphosphonium iodide with 3-(8-oxo-1-octyl)catechol diacetate, using a stepwise procedure based on repeated protection and deprotection of the hydroxyl group of catechol, using the technique we reported previously [17]. The trienyl urushiol compounds **2** and **3** were synthesized in a similar method using ylides derived from (3E,5E)-3,5-heptadienyltriphenylphosphonium iodide and 3E-3,5-heptadienyltriphenylphosphonium iodide, respectively.

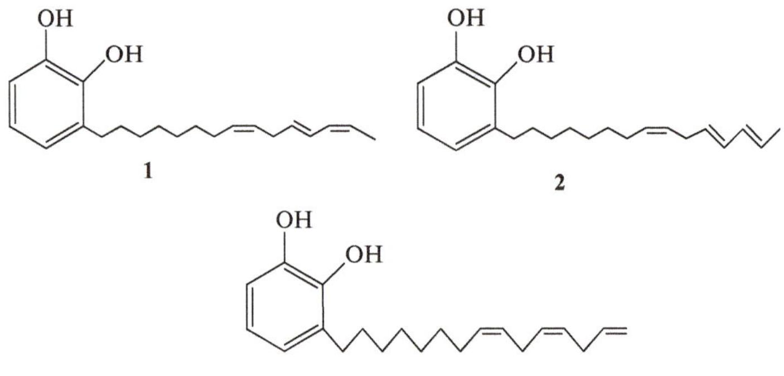

FIGURE 11.28

Stereo-structure of trienylurushiols: (1) 3-[(8Z,11E,13Z)-8,11,13-pentadecatrienyl] catechol, (2) 3-[(8Z,11E,13E)-8,11,13-pentadecatrienyl]catechol, (3) 3-[(8Z,11E)-8,11,14-pentadecatrienyl]catechol.

Synthesized lacquer films were prepared as follows: 50 mg of compound 1 was added to 10 mg water-isopropyl alcohol (1:1, v/v) containing acetone powder (10 mg), which had been separated as an acetone-insoluble material from *T. vernicifluum* lacquer sap. The resulting mixture was stirred for about 15 min. The reaction mixture had a viscosity suitable for coating and was coated onto a glass plate and dried in a humidity-controlled chamber with a relative humidity (RH) of 80% at 25-30 °C for 10 h. The film was then removed from the chamber and stored in air for 6 months. Films of a single component were prepared. Compounds **2** and **3** were treated in the same manner to obtain the corresponding lacquer films. The films made from the synthesized trienyl urushiol components **1**, **2**, and **3** are called synthesized lacquer films A, B, and C, respectively.

The preparation of natural lacquer was as follows. The sap exuded from a *T. vernicifluum* lacquer tree grown in Japan, and composed of the following lipid components: 5% saturated urushiol, 18% monoenyl urushiol, 12% dienyl urushiol, and 65% trienyl urushiol. This lacquer sap was coated onto a glass plate and dried in a humidity-controlled chamber under an RH of 80% at 25-30 °C for 10 h. The film was then removed from the chamber and stored in air for 6 months like the synthesized lacquer films.

The three synthesized lacquer films and the natural lacquer film were pyrolyzed at 500 °C, and then the resulting pyrolysis products were characterized by GC-MS analysis. The specific ions at *m/z* 123 and 108 are fragment ions of the alkylcatechols and alkylphenols that were produced during the electron ionization process in the mass spectrometer, and the mechanism of thermal decomposition shown is as before (Scheme 11.2).

The TIC of the natural and synthesized lacquer films are shown in Figure 11.29. The major components in the TIC peaks were defined as alkanes, alkenes, and alkylbenzenes, respectively. The mass chromatogram of selective scanning of $m/z = 123$ is shown in Figure 11.30. Peaks 1, 2, and 3 were identified as 3-pentylcatechol ($M^+ = 180$),

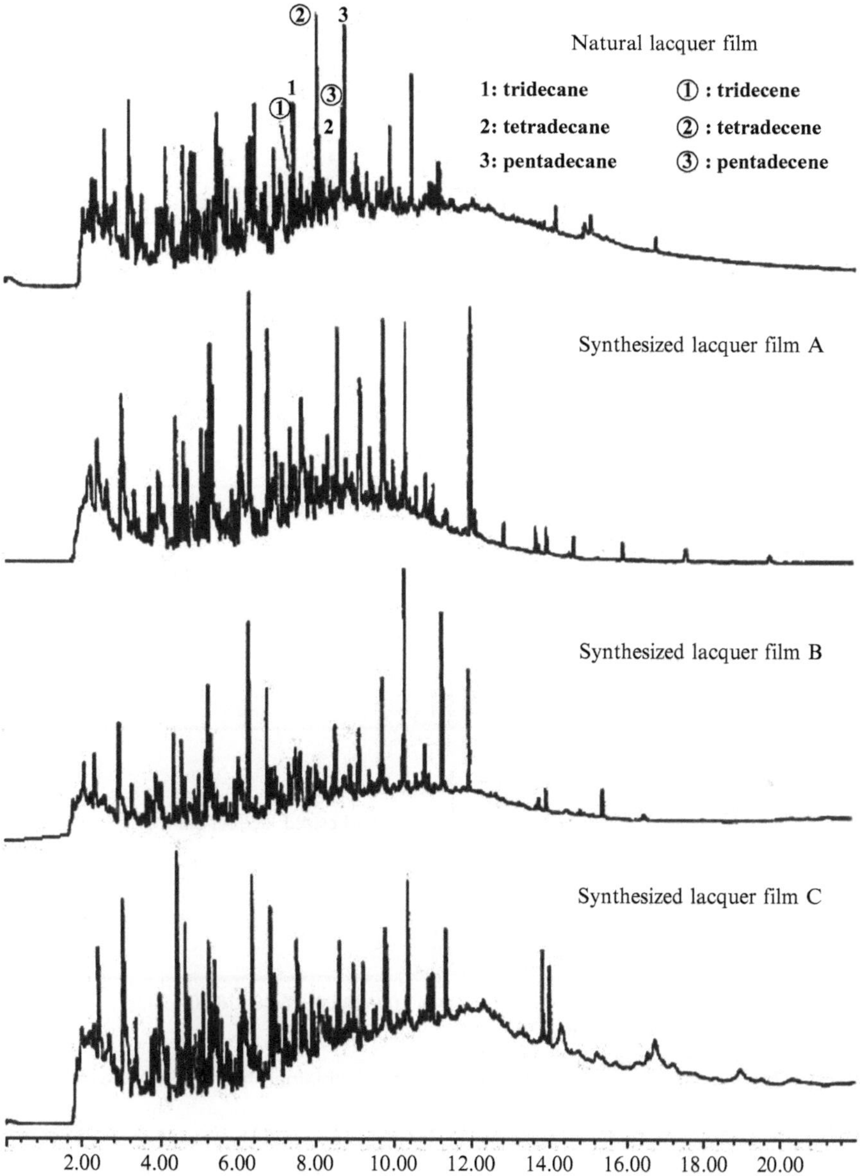

FIGURE 11.29

TIC of natural and synthesized lacquer films due to pyrolysis at 500°C.

3-hexylcatechol ($M^+=194$), and 3-heptylcatechol ($M^+=208$), respectively. Peaks 4 and 5 were observed in only the natural lacquer film, and were identified as the urushiol monomer components 3-pentadecenylcatechol ($M^+=318$) and 3-pentadecylcatechol ($M^+=320$), based on analysis of their mass spectra, shown in Figure 11.31.

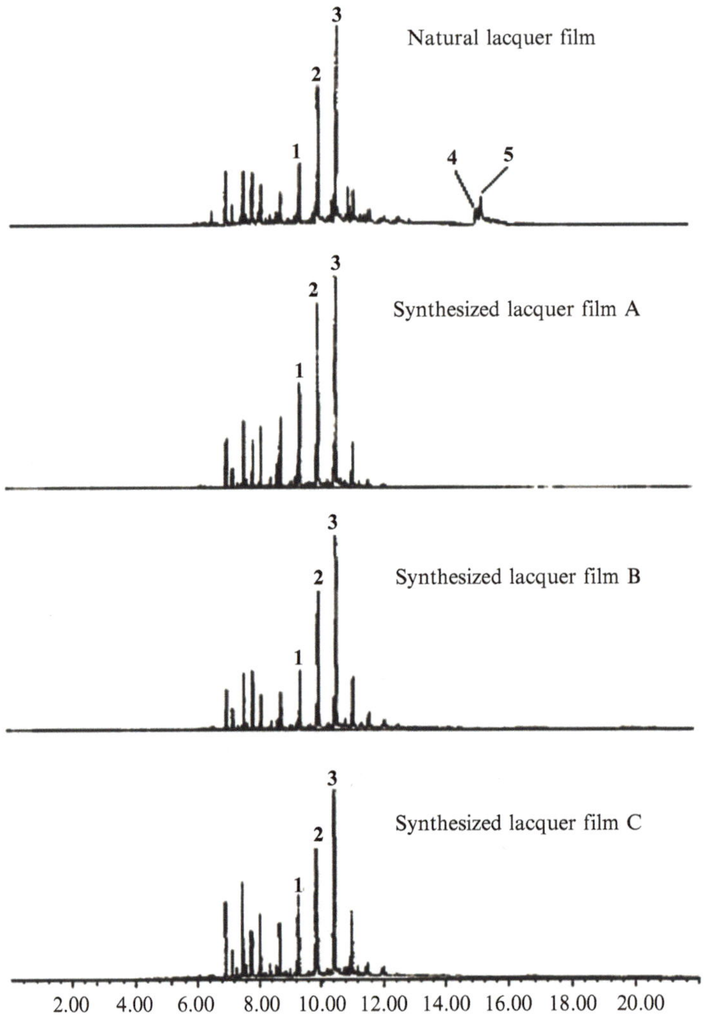

FIGURE 11.30

Selective plotting of ion species ($m/z = 123$) in spectra obtained from TIC of one natural and three synthesized lacquer films.

Figure 11.31 shows the mass spectra of compounds giving rise to peaks 1-5 in the selective plotting of $m/z = 123$ ion species in the spectra from the pyrolysis of synthesized and natural lacquer films shown in Figure 11.30. It is evident from Figure 11.30 that saturated urushiol (peak 5) and monoenyl urushiol (peak 4) were detected by pyrolysis of the natural lacquer film, but not of the synthesized lacquer films. These urushiols likely produced thermally decomposed fragments from the terminal alkylcatechol and alkenylcatechol side chains of the natural lacquer film.

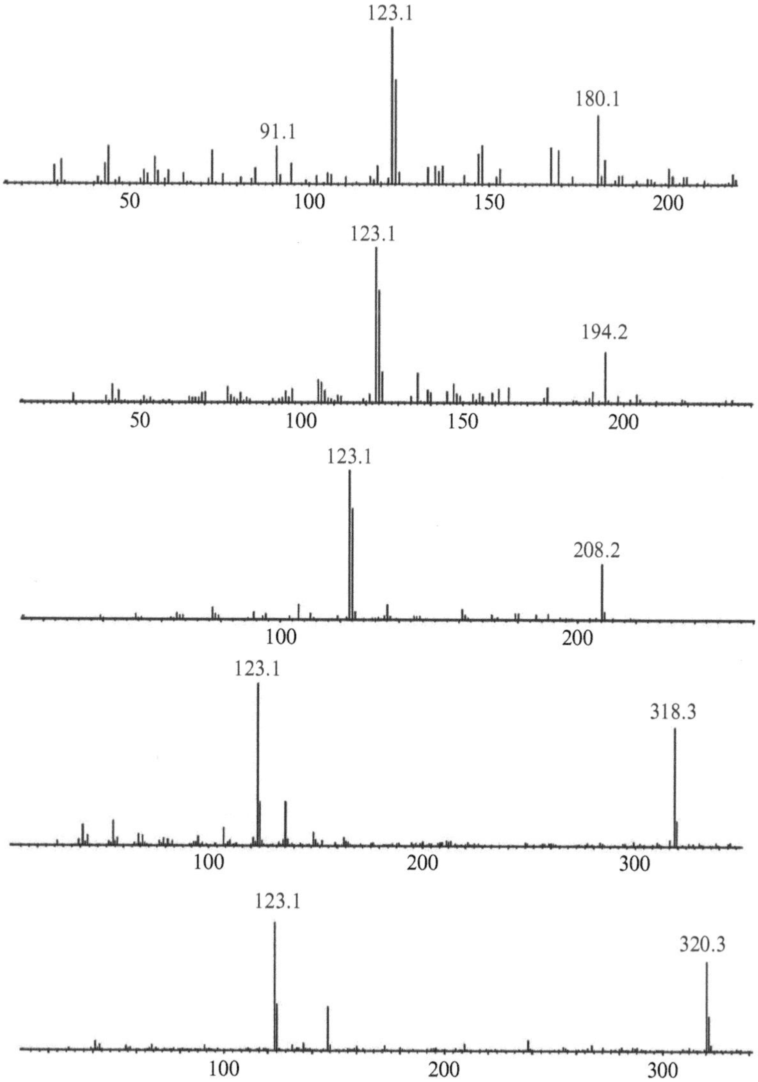

FIGURE 11.31

Mass spectra of alkylcatechols from synthesized lacquer film A (peaks 1-3) and natural lacquer film (peaks 4-5) in mass chromatogram (*m*/*z*=123).

Therefore, it was confirmed that the natural lacquer film has urushiol components with both saturated and monoenyl side chains, unlike the synthesized lacquer films.

The highest peak intensity in each mass chromatogram was peak 3, because most urushiol components have a double bond at position 8 of their side chain, and the bond between C7 and C8 is easily thermally decomposed after polymerization, as shown in Figure 11.32. The highest peak intensity of *T. vernicifluum* was 3-heptylcatehol (M^+=208).

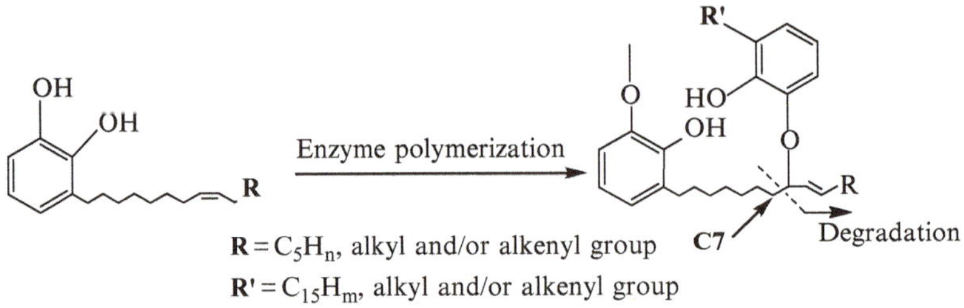

FIGURE 11.32

Postulated location of lacquer film in thermal decomposition due to pyrolysis.

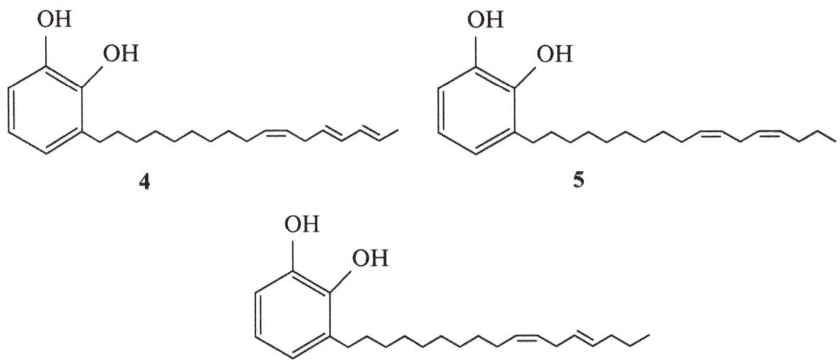

FIGURE 11.33

Stereo-structures of trienyl and dienyl laccols: (4) 3-[(10Z,13E,15E)-10,13,15-heptadecatrienyl]catechol, (5) 3-[(10Z,13Z)-10,13-heptadecadienyl]catechol, and (6) 3-[(10Z,13E)-10,13-heptadecadienyl]catechol.

Analysis of Natural and Synthesized Laccol Films by Py-GC/MS

The trienyl and dienyl laccols, compounds **4**, **5**, and **6**, shown in Figure 11.33 were synthesized by a method similar to the synthesis of urushiol described in Section 4.1.1. Compound **4**, the major component of laccol, was synthesized via the Witting reaction from an ylide derived from (3E,5E)-3,5-heptadienyltriphenylphosphonium iodide with 3-(8-oxo-1-decyl)catechol diacetate. Compounds **5** and **6** were synthesized from 3Z-heptadienyltriphenylphosphonium iodide and 3E-heptadienyltriphenylphosphonium iodide, respectively.

Synthesized and natural laccol lacquer films also were prepared by methods similar to those of urushiol lacquer film described above. However, because the laccase activity of *T. succedanea* is lower than that of *T. vernicifluum*, the laccol lacquer films were dried in a humidity-controlled chamber with RH of 90% at 30 °C for 10 h. Then they were removed from the chamber and stored in air for 6 months. The films from the synthesized laccol components **4**, **5**, and **6** are called synthesized lacquer films D, E, and F, respectively.

The three synthesized laccol lacquer films and one natural *T. succedanea* lacquer film were pyrolyzed at 500 °C, and the resulting pyrolysis products were characterized by GC-MS analysis. The specific ions at $m/z = 123$ and 108 are fragment ions of alkylcatechols and alkylphenols that were produced during the electron ionization process in the mass spectrometer, like those of the urushiol lacquer film. The TIC of natural and synthesized laccol lacquer films are shown in Figure 11.34.

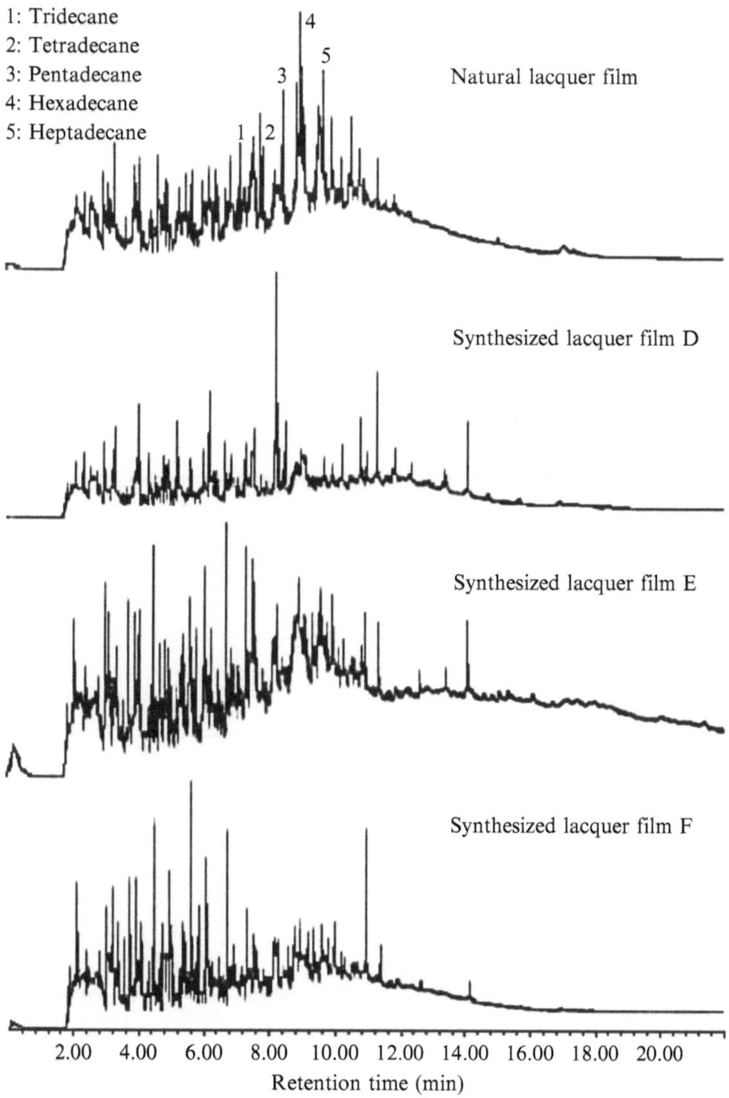

FIGURE 11.34

TIC of natural and synthesized laccol lacquer films due to pyrolysis at 500 °C.

The major components in the TIC peaks were defined as alkanes, alkenes, and alkylbenzenes, respectively.

The mass chromatogram of the selective scanning of $m/z=123$ is shown in Figure 11.35. Peaks 1, 2, and 3 were 3-heptylcatechol ($M^+=208$), 3-octylcatechol

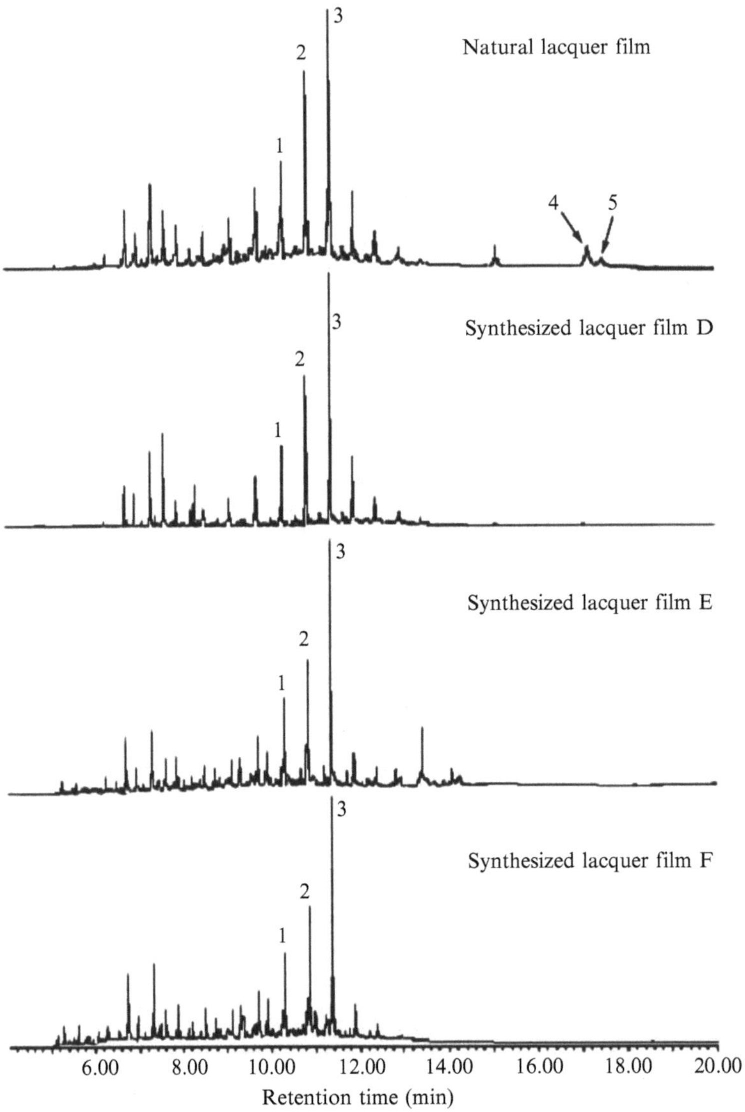

FIGURE 11.35

Selective plotting of ion species ($m/z=123$) in spectra obtained from TIC of one natural and three synthesized lacquer films.

(M^+=222), and 3-nonylcatechol (M^+=236), respectively. Peaks 4 and 5 were observed in only the natural lacquer film and were identified as laccol monomer components, 3-heptadecenylcatechol (M^+=346) and 3-heptadecylcatechol (M^+=348), based on their mass spectra (Figure 11.36).

These laccols were likely produced as thermally decomposed fragments from the terminal alkyl and alkenylcatechol of the natural *T. succedanea* lacquer film. Therefore, the natural lacquer sap has laccol components with both saturated and monoenyl side chains, but the synthesized laccols do not. This was the same difference as that of the natural lacquer film from *T. vernicifluum*. The highest peak intensity was peak 3 in each mass chromatogram because most laccol components have a double bond at position 10 of their side chain, and the bond between C9 and C10 is easily thermally decomposed after polymerization. This result was different from that of the lacquer film from *T. vernicifluum*, which has a double bond at position 8 of its side chain. The highest peak intensity of *T. vernicifluum* was 3-heptylcatechol (M^+=208) of *T. succedanea* was 3-nonylcatechol (M^+=236).

Analysis of Natural and Synthesized Thitsiol Films by Py-GC/MS

The main component of thitsiol, 3-(10-phenyldecyl)catechol, was synthesized by the reaction of dimethyl ether with catechol and 1-phenyl-10-iodododecane with n-BuLi, followed by deprotection of the hydroxyl groups of catechol, as shown in Figure 11.37. The detailed experimental process can be found in the literature (3 and 4).

The preparation of synthesized thitsiol lacquer films is as follows: 50 mg 3-(10-phenyldecyl)catechol was added to 50 mg of a water-isopropyl alcohol mixture (1:1, v/v) containing 20 mg acetone powder obtained from a natural *G. usitata* lacquer sap. The resulting mixture was stirred for about 15 min, and then applied to a glass sheet and dried in a humidity-controlled chamber with a relative humidity of 90% at 30 °C for 10 h. The film was then removed from the chamber and stored in air for 6 months. The natural lacquer film also was prepared by the same method.

The synthesized thitsiol lacquer films and a natural *G. usitata* lacquer film were pyrolyzed at 500 °C, and then the resulting pyrolysis products were characterized by GC-MS analysis. Figure 11.38 shows the TIC of the natural and synthesized lacquer films. The main pyrolysis products of the natural lacquer film were identified by the mass spectra as alkanes, alkenes, alkylbenzenes, and alkenylbenzenes. On the other hand, alkanes and alkenes were not detected in the TIC of the synthesized lacquer film because the alkanes and alkenes were derived from 3- or 4-alkenylcatechols contained in the natural lacquer sap. Each peak in TIC is a pair of peaks of saturated and unsaturated components having one double bond in the termination. It was considered that the alkanes and alkenes were formed by decomposition of the terminal of the lipid components in the polymer.

Figure 11.39 are the mass chromatograms (m/z=123) of natural and synthesized thitsiol lacquer films. Alkylcatechol was detected in the mass chromatogram of the natural lacquer film, and detected as a pair of peaks of 3-alkylcatechol and 4-alkylcatechol. Due to the analysis of each mass spectrum, the detailed assignment

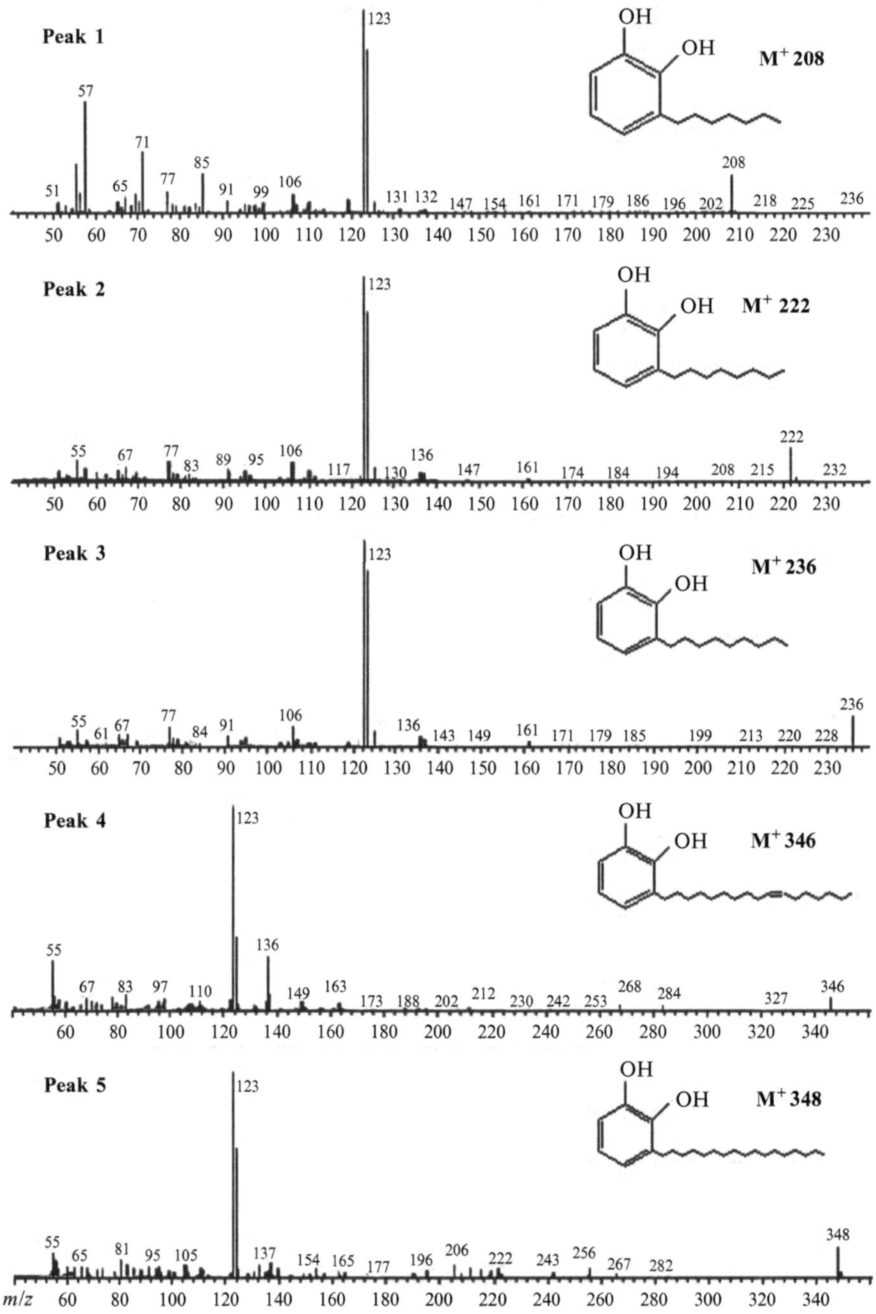

FIGURE 11.36

Mass spectra of peaks 1-5 in the mass chromatogram (m/z=123).

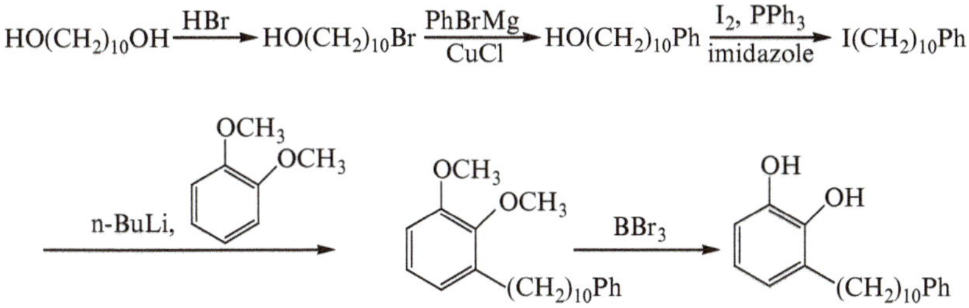

FIGURE 11.37

Synthesis of 3-(10-phenyldecyl)catechol.

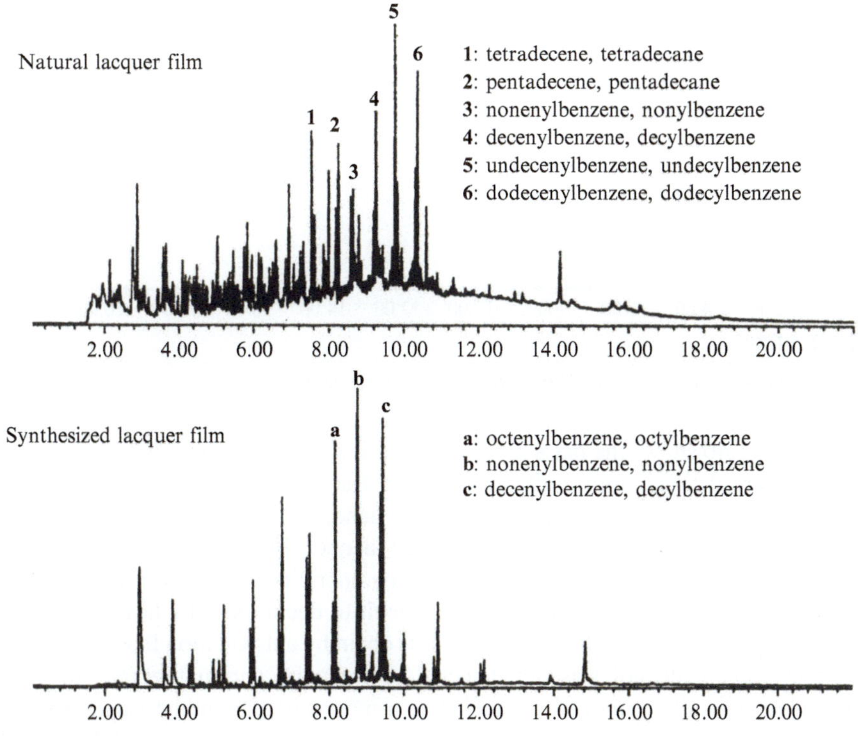

FIGURE 11.38

TIC of natural and synthesized thitsiol lacquer films.

was as follows: ① 3-methylcatechol, 1: 4-methylcatechol; ② 3-ethylcatechol, 2: 4-ethylcatechol; ③ 3-propylcatechol, 3: 4-propylcatechol; ④ 3-butylcatechol, 4: 4-butylcatechol; ⑤ 3-pentylcatechol, 5: 4-pentylcatechol; ⑥ 3-hexylcatechol, 6: 4-hexylcatechol; ⑦ 3-heptylcatechol, 7: 4-heptylcatechol; and ⑧ 3-octylcatechol, 8: 4-octylcatechol; 9: 3-pentadecylcatechol; 10: 3-(10-phenyldecyl)catechol.

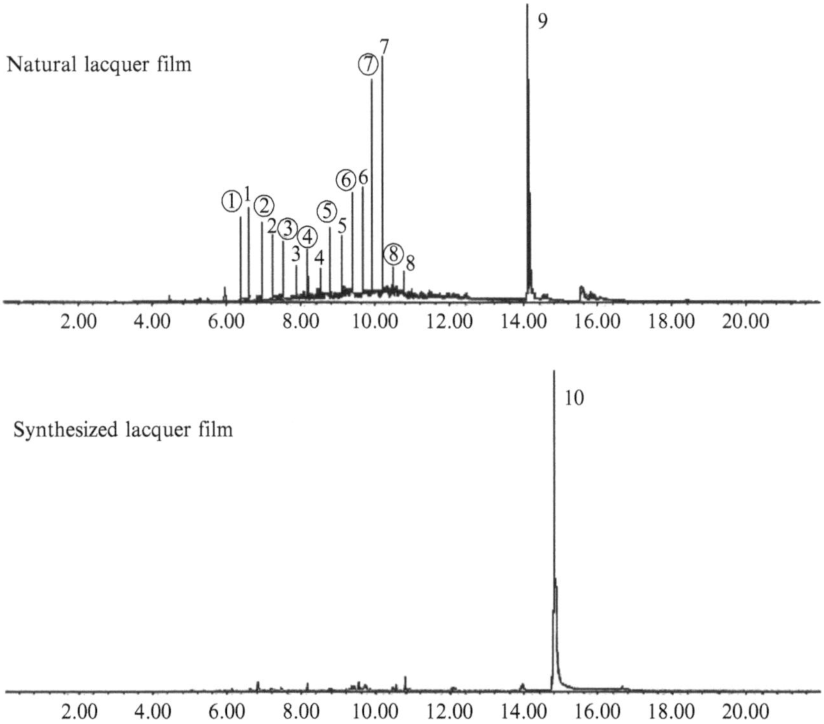

FIGURE 11.39

Mass chromatogram ($m/z = 123$) of natural and synthesized films.

Figure 11.40 shows the mass spectra of a pair of peaks ⑥-6 and ⑦-7. These components were derived from 3-alkylcatechol or 4-alkylcatechol contained as monomeric components in *G. usitata*. On the other hand, alkylcatechol was not detected in the mass chromatogram of the synthesized lacquer film because the synthesized lacquer film included no alkylcatechols as monomeric components. The side chain of the polymerized ω-alkylcatechol was difficult to decompose by heating because ω-alkylcatechol is not polymerized at the side chain by autoxidation like a fatty acid. Therefore, alkylcatechol was not detected in the mass chromatogram. However, 3-(10-phenyldecyl)catechol was detected as a monomer. The highest pair-peak intensity of *G. usitata* was 3-heptylcatechol and 4-heptylcatechol ($M^+ = 208$, peaks ⑦ and 7 in Figure 11.39).

The synthesized lacquer films and corresponding natural lacquer films were characterized using Py-GC/MS. The saturated and monoenyl components were detected in the natural *T. vernicifluum* and *T. succedanea* lacquer films, respectively, but not in the synthesized lacquer films. However, alkylphenols, alkenylphenols, alkanes, and alkenes having longer carbon chains than the side chains of the synthesized lacquers were detected by the Py-GC/MS analysis. Comparing the peak intensity of the mass chromatograms with that of *T. succedanea*, *T. vernicifluum*, and *G. usitata*

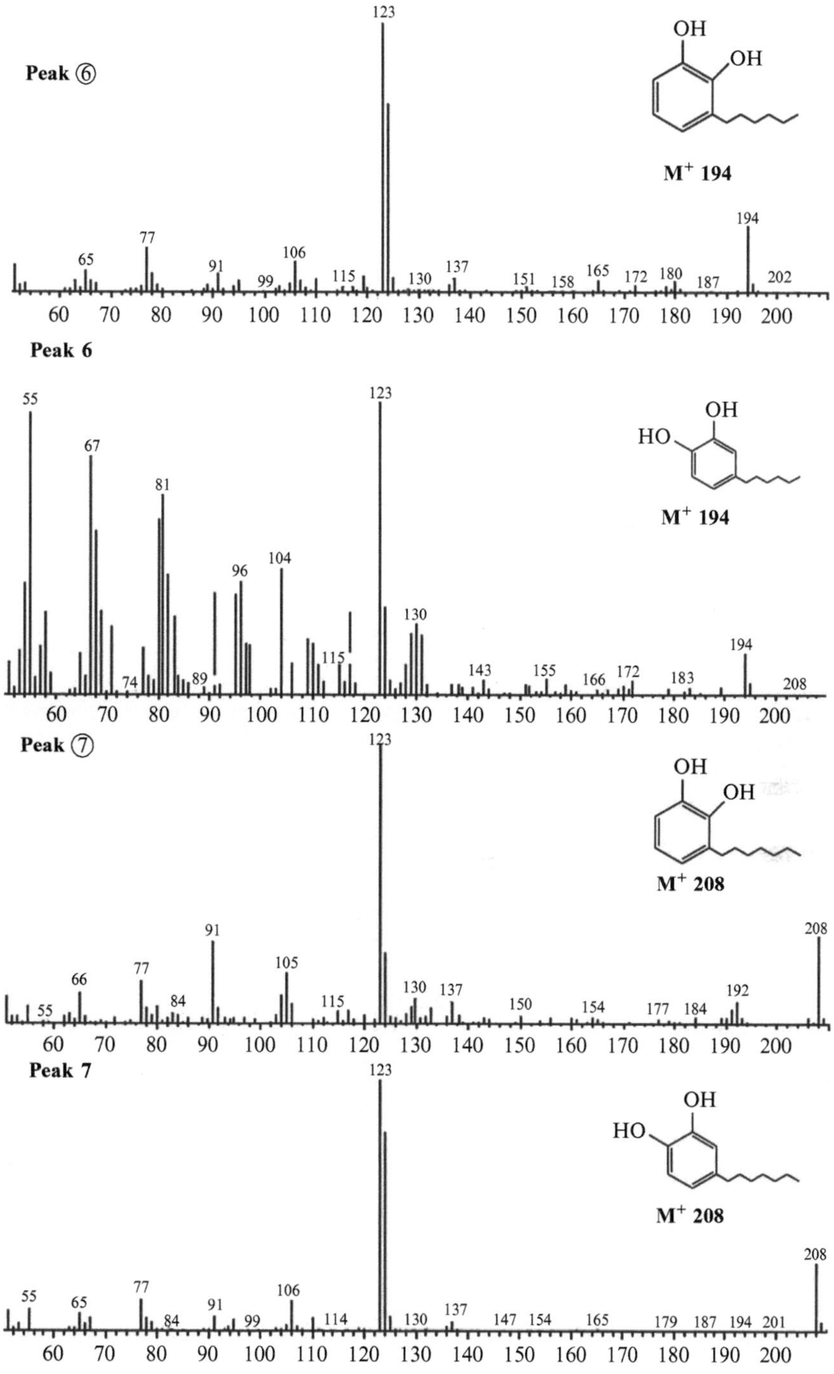

FIGURE 11.40

Mass spectra of pair of peaks of ⑥-6 and ⑦-7.

lacquer films showed that the highest peak of *T. vernicifluum* represented the C7 components (heptylcatechol and heptylphenol), and the highest peak of *T. succedanea* represented the C9 components (nonylcatechol and nonylphenol) due to a double bond in position 10 of its side chain of laccol. On the other hand, a pair of peaks of 3-alkylcatechol and 4-alkylcatechol was detected in the *G. usitata* lacquer film. The highest peaks were 3-heptylcatechol and 4-heptylcatechol due to the side chain attached in position 3 and/or position 4 of catechol ring of thitsiol.

11.3.2 IDENTIFICATION OF ANCIENT LACQUERWARE

Py-GC/MS is a powerful and versatile technique to identify Oriental lacquers as described above. In this section, we describe analysis of several ancient lacquerwares by Py-GC/MS, and compare them with the results of natural lacquer films to determine the kind of lacquer.

Ryukyu Lacquerware

We have analyzed six kinds of Ryukyu lacquerware by Py-GC/MS to determine the lacquer source [40]. Six pieces of lacquer belonging to the Urasoe Art Museum are summarized in Table 11.5, and photos of each lacquerware object are shown in Photo 11.1. Each sample was removed from the ancient lacquerware objects during restoration. All pieces of Ryukyu lacquerware in Table 11.5 were analyzed by pyrolysis-gas chromatography/mass spectrometry at 500 °C.

Figure 11.41 shows the TIC and mass chromatogram ($m/z=320$, $m/z=123$, $m/z=108$) of Sample 2. The peak at the retention time of 14.326 min ($m/z=320$) is urushiol, at 10.038 min ($m/z=123$) is 3-heptylcatechol, and at 9.139 min ($m/z=108$) is 3-heptylphenol, respectively. The mass spectra are shown in Figure 11.42. Samples 1, 4, and 5 showed TIC, mass chromatograms, and mass spectra similar to that of Sample 2, suggested that these four lacquerware objects were made from *T. vernicifluum*.

The TIC and mass chromatogram of Sample 3 are shown in Figure 11.41, and the mass spectrum is shown in Figure 11.44. The results showed that the peak at the retention time of 16.072 min ($m/z=348$) is laccol, at 11.105 min ($m/z=123$) is 3-nonylcatechol, and at 10.282 min ($m/z=108$) is 3-nonylphenol. Sample 6 produced the same pyrolysis results as Sample 3, suggesting that these two lacquerware objects were made from *T. succedanea*. The pyrolysis products of the six samples are summarized in Table 11.6.

Table 11.5 Samples used for pyrolysis

No.	Collection method	Art object (Museum I.D. number)
1	Naturally flaked off	Black lacquer wood box with *Raden* (43)
2	Naturally flaked off	Black lacquer table box with *Raden* (66)
3	Naturally flaked off	Black lacquer box with *Hakue* (220)
4	Naturally flaked off	Black lacquer box with *Hakue* (416)
5	Collection during restored	Black lacquer box with *Raden* (26)
6	Collection during restored	Red lacquer fabric bowl with *Hakue* (100)

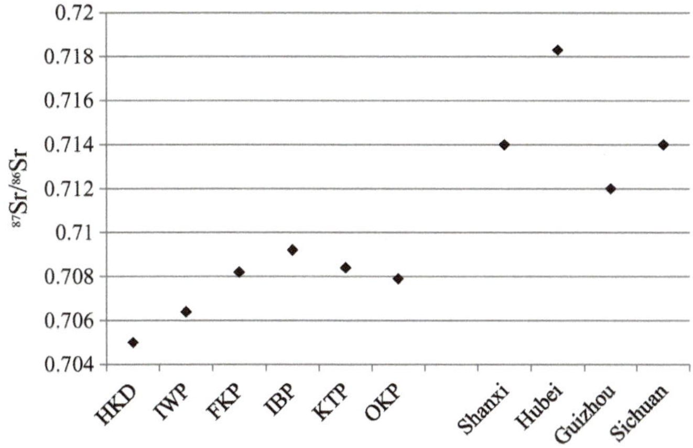

PHOTO 11.1

Six lacquer objects belonging to Urasoe Art Museum.

FIGURE 11.41

TIC, mass chromatograms (m/z=320, 123, and 108) of sample 2.

The Py-GC/MS results of Ryukyu ancient lacquerware confirmed that four lacquerware objects belonging to the Urasoe Art Museum were made from *T. vernicifluum* lacquer sap and the other two were made from *T. succedanea* lacquer sap. Because Py-GC/MS analysis revealed what kind of lacquer sap was used to produce Ryukyu lacquerware, the procurement source and production system of the materials of Ryukyu lacquerware were clarified, and will be very useful in the conservation and restoration of other valuable ancient lacquerware.

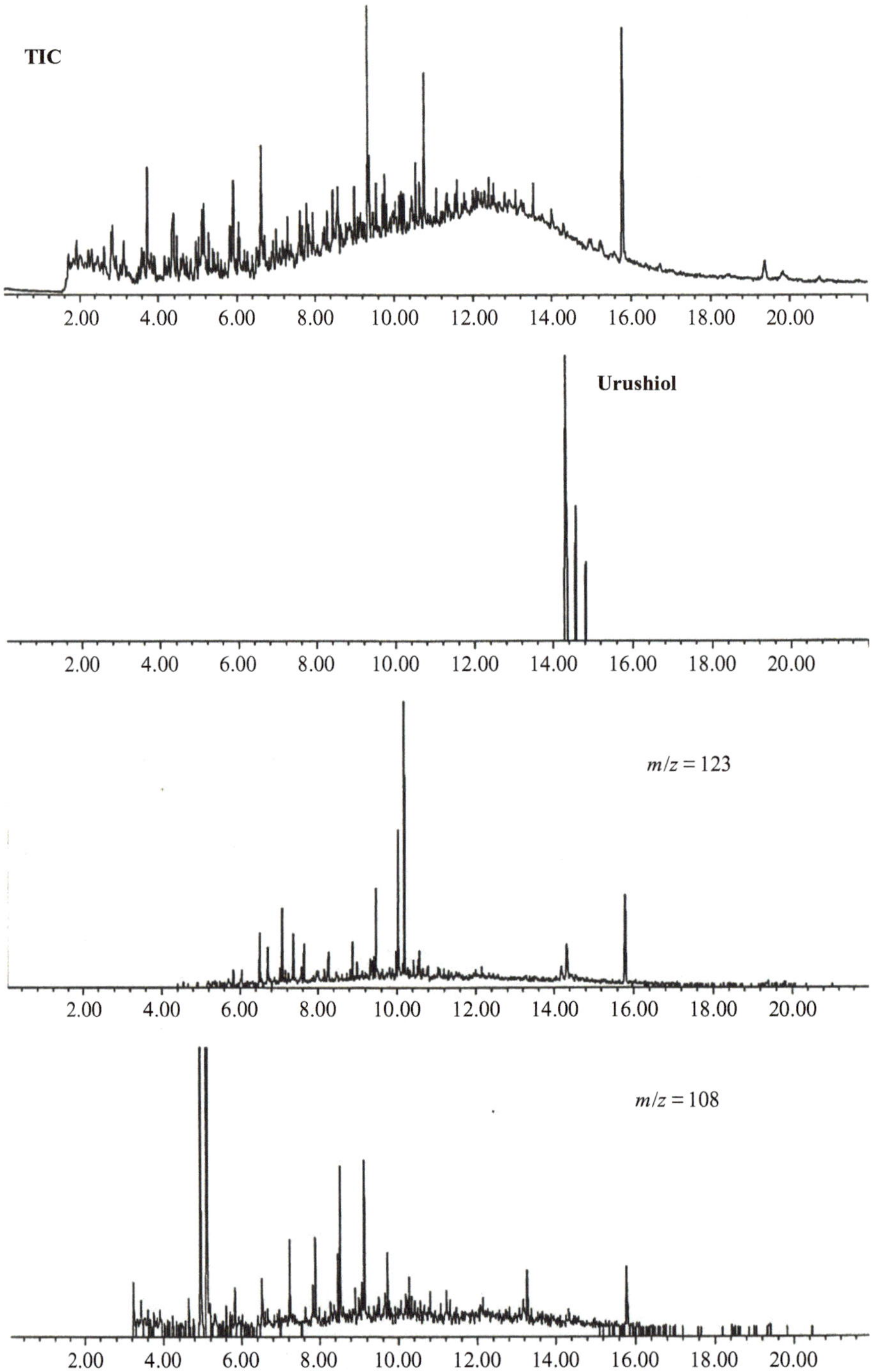

FIGURE 11.42

Mass spectra of sample 2.

Table 11.6 Pyrolysis products of Ryukyu lacquer films and their species

		Pyrolysis products		
No.	**Lacquer species**	**Monomer**	**Alkylcatechol**	**Alkylphenol**
1	*T. vernicifluum*	Urushiol	3-Heptylcatechol	3-Heptylphenol
2	*T. vernicifluum*	Urushiol	3-Heptylcatechol	3-Heptylphenol
3	*T. succedanea*	Laccol	3-Nonylcatechol	3-Nonylphenol
4	*T. vernicifluum*	Urushiol	3-Heptylcatechol	3-Heptylphenol
5	*T. vernicifluum*	Urushiol	3-Heptylcatechol	3-Heptylphenol
6	*T. succedanea*	Laccol	3-Nonylcatechol	3-Nonylphenol

Other Ancient Lacquerware

We also used this analytical pyrolysis method on the characterization of five ancient lacquerwares objects that were obtained from the surface of a wooden dish excavated from a site that dates back to the seventeenth to eighteenth century A.D. at Kinenkanmae Iseki (the ruins in the front of the museum) on the Meiji University campus in Tokyo, Japan (Sample 1); a Nanban lacquer film from the seventeenth century A.D. (Sample 2); an old lacquer film obtained from the surface of wooden crafts imported from an Asian country during the seventeenth to eighteenth century A.D. (Sample 3); a piece of lacquer taken from a four-eared jar that is a Japanese National Important Cultural Property found in sixteenth to seventeenth century Kyoto ruins (Sample 4), and Baroque and Rococo lacquer films obtained from the wood surfaces of the Rococo church St. Alto in Altomunster, Munich, Germany (Sample 5). It is sometimes extremely important to determine whether objects that are claimed to be lacquerware are actually created from lacquer sap or other resins. Whether an object is lacquerware can be precisely determined by the presence of urushiol, laccol, or thitsiol with alkylcatechols and alkylphenols using Py-GC/MS.

The TIC and mass chromatograms of $m/z=320$ of Sample 1 are shown in Figure 11.43. Urushiol, 3-pentadecylcatechol (MW=320), was identified as the monomer of the lacquer film based on the mass spectrum and retention time. This result was compared to those of the three types of Oriental lacquers. 3-Pentadecylcatechol with MW=320 is the saturated urushiol component, which is the monomer of the *T. vernicifluum* lacquer. The monomers of the *T. succedanea* lacquer are laccol components such as 3-hepatadecylcatechol with MW=348. This was not detected in Sample 1 except for the 3-pentadecylcatechol with MW=320. The monomer of *Melanorrhoea usitata* lacquer is thitsiol, which has saturated and monoenyl side chains, such as 4-hepatadecylcatechol (MW=348) and 3- and 4-(ω-phenylalkyl) phenols and catechols, and these components, except for the 3-pentadecylcatechol of MW=320, were not detected in Sample 1. It was concluded that Sample 1 lacquer was made from *T. vernicifluum* lacquer sap.

The TIC and mass chromatograms of $m/z=60$, $m/z=123$, and $m/z=108$ from the pyrolysis products of Sample 2 are shown in Figure 11.44. The mass chromatograms of $m/z=60$ indicated that the Sample 2 lacquerware included a drying oil, which was

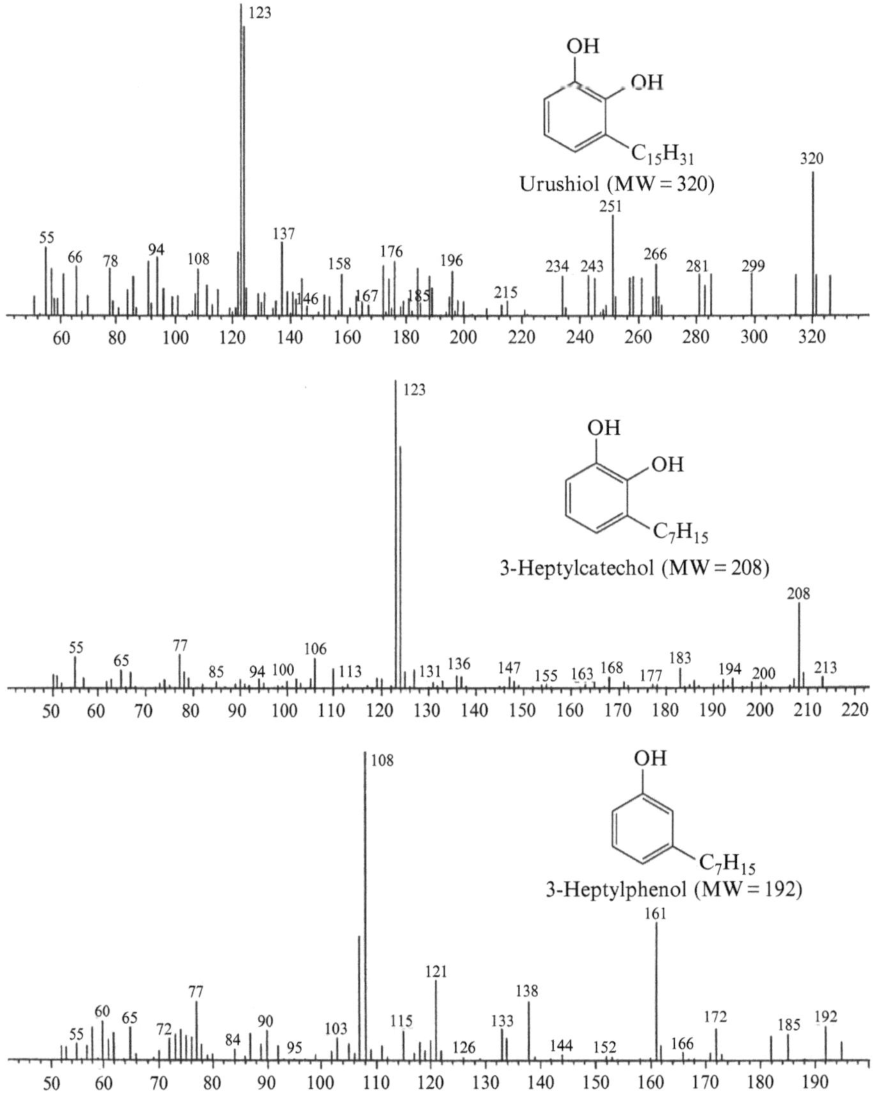

FIGURE 11.43

Pyrolysis data of Sample 1 excavated at Meiji University, Tokyo, Japan.

added to retard the rate of hardening and affected the physical properties of the film. The mass chromatograms of $m/z = 108$ and $m/z = 123$ of the pyrolysis products of the lacquerware are also shown in Figure 11.44. The greatest number of peak was from 3-heptylphenol (C7) and 3-heptylcatechol (C7), as revealed by the mass spectra. It was concluded that the Sample 2 lacquer was made from *T. vernicifluum* lacquer sap. Urushiol was not detected because the surface of the Sample 2 lacquerware was

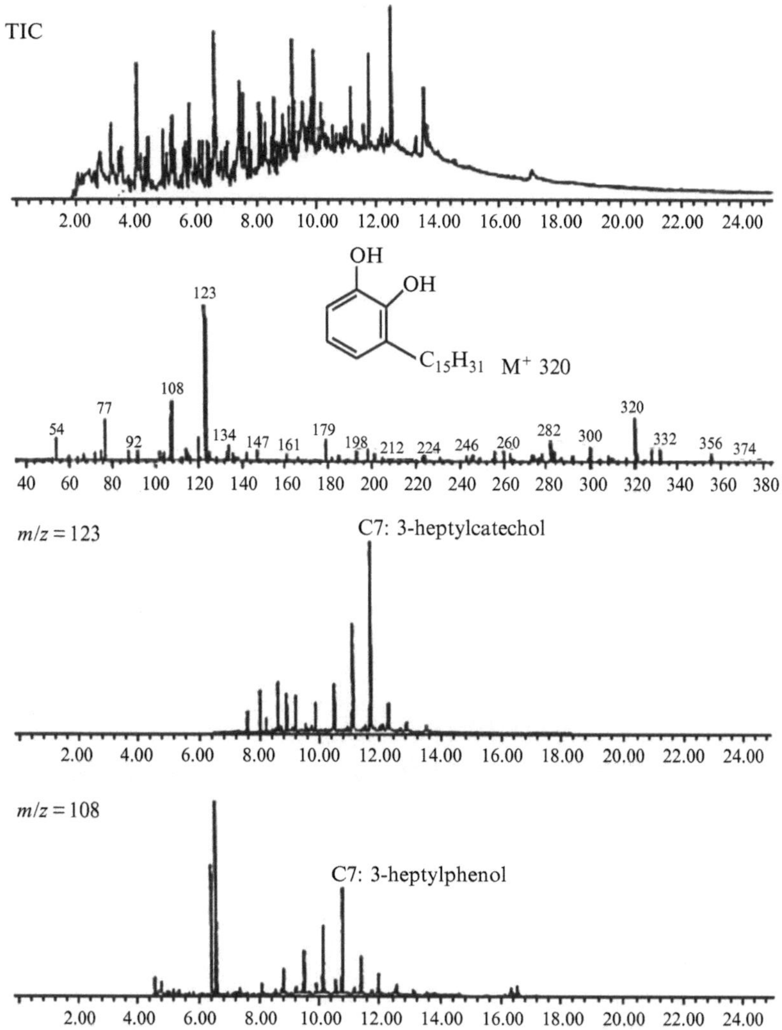

FIGURE 11.44

Pyrolysis data of Sample 2 lacquer film.

oxidized by oxygen and light. A type of wax was detected in the mass spectrum of the TIC from the pyrolysis products. It is considered that the wax was used to polish the surface of the lacquerware.

The TIC and mass chromatograms of $m/z = 123$ and $m/z = 108$ for the pyrolysis products of Sample 3 are shown in Figure 11.45. Alkylcatechols and alkylphenols were detected in the mass chromatograms. The greatest number of peaks was due to the 3- and 4-nonylcatechols (C9) and 3-heptylphenol (C7), as determined from the mass spectra. It was concluded that the Sample 3 lacquerware was produced using

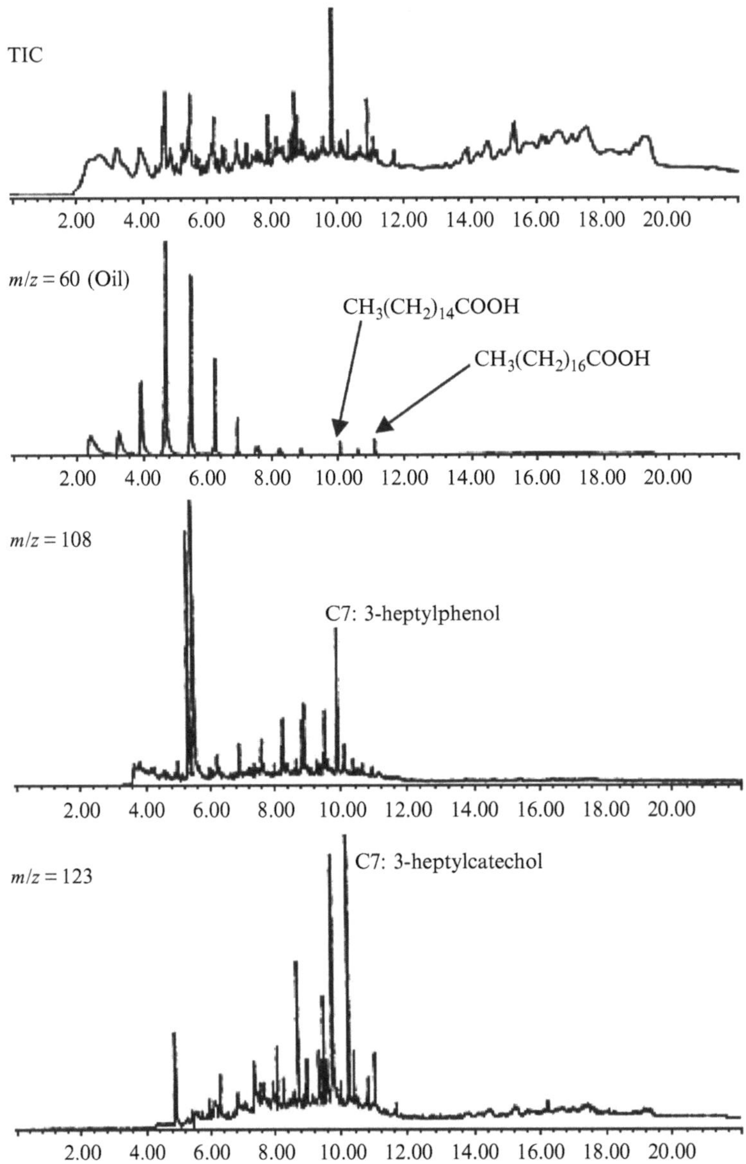

FIGURE 11.45

Pyrolysis data of Sample 3.

G. usitata lacquer sap. Thitsiol was not detected in the pyrolysis data because the surface of the lacquerware was oxidized by oxygen and exposure to light. In the TIC and mass spectra of the pyrolysis products, a type of wax was detected. The wax was revealed to be a type of beeswax by the mass spectra. The beeswax was used to protect and polish the surface of the lacquerware.

The TIC and mass chromatograms of $m/z = 123$ and $m/z = 108$ for the pyrolysis products of Sample 4 are shown in Figure 11.46. The peak at the retention time of 17.3 min ($m/z = 123$, peak 1) is 3-(10-phenyldecyl)catechol and at 20.4 min ($m/z = 123$, peak 2) is 3-(10-phenyldodecyl)catechol, according to the analysis of the mass chromatogram. It was concluded that the Sample 4 lacquer was made from *G. usitata* lacquer sap.

Figure 11.47 shows the Py-GC/MS products of the Sample 5 (Baroque and Rococo) lacquer films. Lacquer components were not detected in the TIC on the mass chromatograms of the alkylcatechols of $m/z = 123$ and alkylphenols of $m/z = 108$

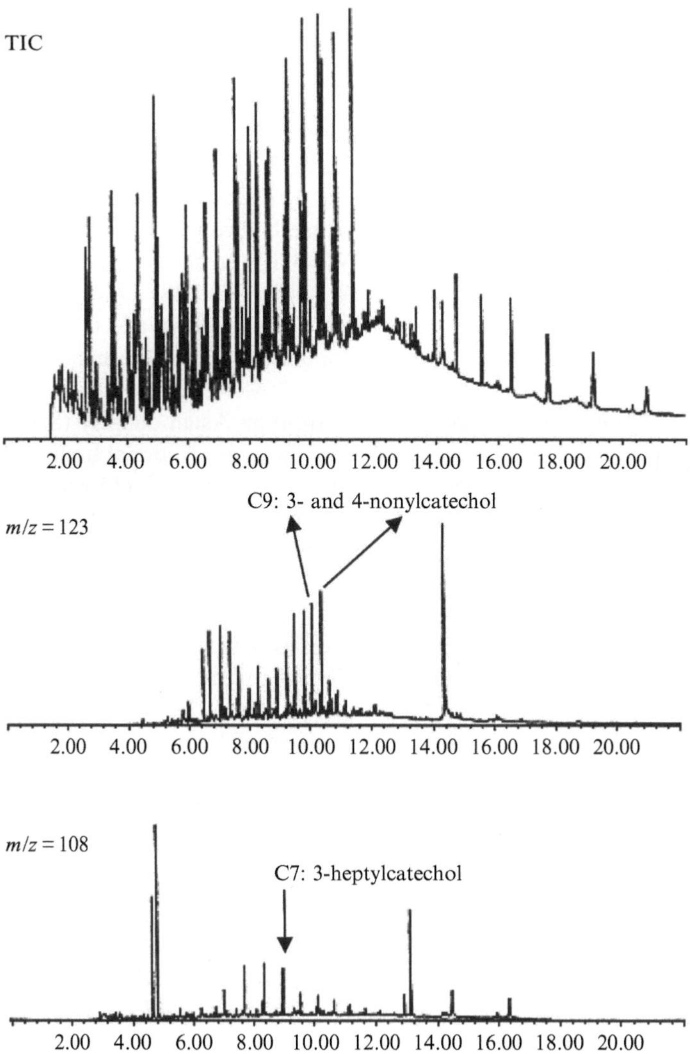

FIGURE 11.46

Mass chromatograms of Sample 4.

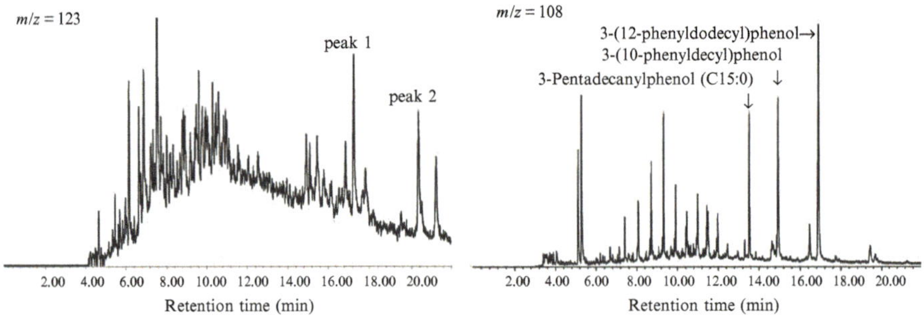

FIGURE 11.47

Pyrolysis data of Baroque and Rococo lacquer films.

of the pyrolysis products. Monoterpene and sesquiterpene components were detected in the pyrolysis products of both lacquerwares. It was concluded that the lacquerwares were made from natural resins, but the pyrolysis products of the Baroque and Rococo lacquer and natural resins using this method were not clear.

In summary, 3-pentadecylcatechol (MW=320) (urushiol), 3-heptadecylcatechol (MW=348) (laccol), and 4-heptadecylcatechol (MW=348) (thitsiol) are the main products of the pyrolysis of *T. vernicifluum*, *T. succedanea*, and *G. usitata*, respectively. Compared with the results of the natural lacquer film, the ancient lacquer film (Sample 1) and Nanban lacquer film (Sample 2) were assigned to *T. vernicifluum*. Both the old lacquerware objects imported from an Asian country (Sample 3) and the lacquer taken from a four-eared jar that is a Japanese National Important Cultural Property obtained from sixteenth to seventeenth century Kyoto ruins (Sample 4) were assigned to *G. usitata*. However, although they were also called "lacquer," the Baroque and Rococo (Sample 5) lacquer films were identified as being made from a natural resin. The pyrolysis products clearly showed a good correspondence to the components of the lacquer sap.

This chapter describes the chemical properties of lacquer saps and films that were analyzed by pyrolysis-gas chromatography/mass spectrometry measurement. It was revealed that the advanced Py-GC/MS analytical method is useful for identifying and evaluating the lacquer components and the origin of lacquer species. Py-GC/MS is a well-known method applied to various areas. Due to its ease of control, speed of analysis, and good reproducibility, the Py-GC/MS method not only can be applied to lacquer films, organic coatings, and other materials that cannot be dissolved in solvents, but also to discriminate between lacquer and other resins for the conservation or restoration of lacquerware.

11.4 IDENTIFICATION OF LACQUER PROVENANCE BY ^{87}SR/^{86}SR ISOTOPE RATIO

As described above, lacquer trees grown in the Chinese mainland and Japan belong to the same species, *T. vernicifluum*, and the main component of the sap is urushiol. Although the Py-GC/MS method can identify lacquer trees and/or natural resins by their different pyrolysis products, it is not able to discriminate Chinese and Japanese lacquers that come from the same tree species, have almost the same components, and each component has almost the same chemical structure. Ancient lacquerwares excavated from China was dated to 7000 years ago, and from Japan also can date to 5000 years ago. If the provenance of lacquer sap coating ancient Chinese and Japanese lacquerwares can be determined, the botanical distribution of *T. vernicifluum* lacquer trees in the past over the Chinese mainland and Japanese islands and the tradition and spread of lacquer crafts from China to Japan should be revealed. In addition, Ryukyu lacquerwares were actively traded in the fifteenth to sixteenth centuries, and the process was used to produce the lacquerwares exported to Western European countries, where it was called "japaning." It is important to identify the provenance of lacquer trees to clarify the trade routes and movement of these lacquerwares.

In the identification of the lacquer provenance, analysis of the structure of the main lipid components by Py-GC/MS is the principal method, as described above. However, this method cannot distinguish Chinese and Japanese lacquer. Therefore, we focused on the strontium isotopes in the lacquer, and attempt to compare their concentration in the Chinese and Japanese lacquer sap and film.

Strontium has four stable isotopes in nature, ^{84}Sr, ^{86}Sr, ^{87}Sr, and ^{88}Sr, with natural relative percentages of about 0.56%, 9.86%, 7.00%, and 82.58%, respectively.[8] The ^{87}Sr concentration increases over time due to the β-radioactive decay of ^{87}rubidium (^{87}Rb), while the ^{86}Sr concentration does not change, so the $^{87}Sr/^{86}Sr$ ratio is used as an indicator of the strontium isotope composition. In general, material with a higher and longer Rb/Sr ratio has a faster Sr isotope evolution and larger Sr isotope ratio. The ancient continental crust generated by partial melting of mantle rocks has a higher Rb/Sr ratio than the mantle composed of the Earth's rocks; this is because Rb is incorporated into the mantle of melting magma at a lower temperature than Sr, so the evolution of the Sr isotope ratio is faster than in mantle rocks, as shown in Figure 11.48.

The continental crust from the old mantle has a higher $^{87}Sr/^{86}Sr$ isotope ratio than the one differentiated from the young mantle. The Japan islands comprise rocks of various origins and ages, but on average they are younger than on the Chinese mainland, so the $^{87}Sr/^{86}Sr$ isotope ratio is lower compared with Chinese mainland.

The lacquer samples were first pretreated in a subclean box, then in a clean laboratory, and the detail treatment process as follows.

Step 1: Removal of organic materials. Lacquer samples (100 mg, lacquer sap, film, bark, and branches) were treated with 6 ml of 14 M nitric acid at 120 °C for 11 h in a Teflon beaker with a lid to dissolve the organic matter, and the lid was opened in dry heat to remove the organic matter. Then, 5 ml of 14 M nitric acid, 0.5 ml perchloric acid, and 0.5 ml hydrogen peroxide were added to the beaker and continuously heated at 120 °C for 9 h covered, and then the lid was opened in dry heat to remove the organic matter. This sample treatment was repeated at least for four times in order to completely remove the organic matter. Then, 1 ml of 7 M nitric acid was added to the samples, and shaken to dissolve in nitric acid; these were named sample A. Soil

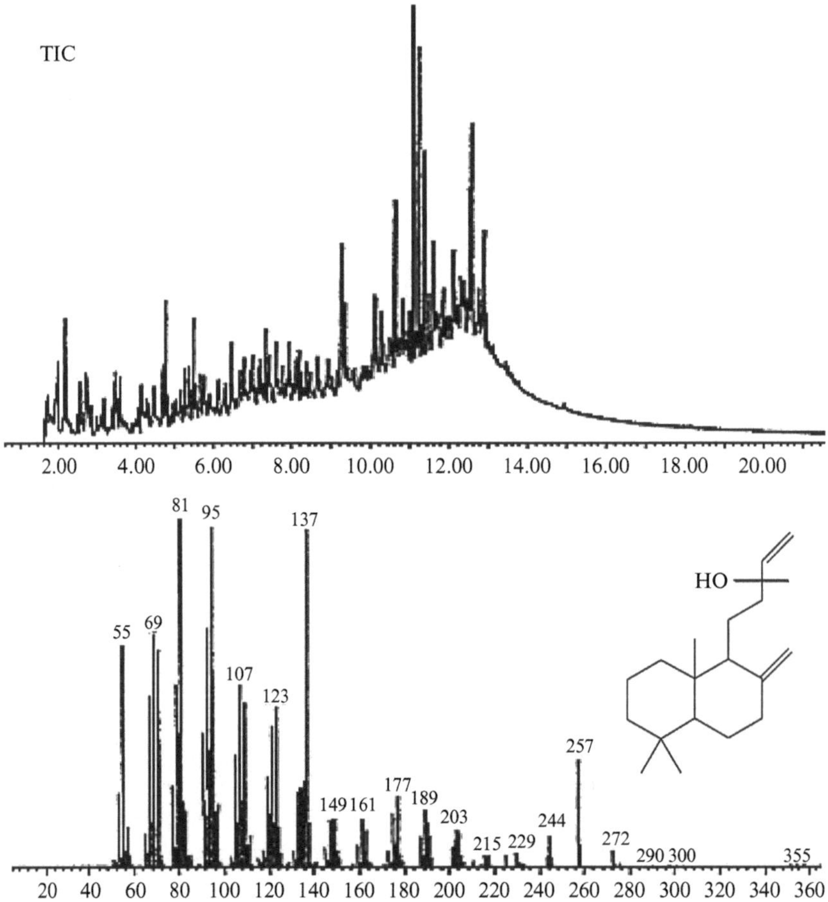

FIGURE 11.48

Sr isotope ratios in continental crust.

in which lacquer trees were growing was first cleaned ultrasonically with 10 ml of 1 M ammonium acetate for 1 h at room temperature, and then centrifuged at 9200 rpm for 15 min. The supernatant was treated in the same manner as a lacquer sample.

Step 2: Column separation of Sr. Sr resin (0.5 ml, particle size: 50-100 μm, Eichrom Co., Ltd., USA) was added to a prepackaged column, and washed with 15 ml 2% nitric acid and 4 ml 3 M nitric acid in order to eliminate Sr residues. Sample A (1 ml) was added to this column, and eluted with nitric acid in the order of (1) 3 M, 10 ml, (2) 7 M, 6 ml, and (3) 3 M, 2 ml to eliminate other elements, and finally eluted with 7.5 ml, 2% nitric acid to collect Sr; these were named sample B.

Step 3: Inductively coupled plasma mass spectrometry. Sample B was analyzed by isoprobe multicollector inductively coupled plasma mass spectrometry (ICP-MS, Micromass Ltd., UK) at the Earthquake Research Institute, University of Tokyo, Japan. Mass numbers ^{83}Kr, ^{84}Sr, ^{85}Rb, ^{86}Sr, ^{87}Sr, and ^{88}Sr were measured. The signal strength of ^{84}Kr and ^{86}Kr were estimated from the ^{83}Kr signal strength, and the signal strengths of ^{84}Sr and ^{86}Sr were corrected based on ^{84}Kr and ^{86}Kr signal strengths. Similarly, the signal strength of ^{86}Rb was estimated from the ^{85}Rb signal strength, and the signal strength of ^{87}Sr was corrected based on the ^{86}Rb signal strength. The internal standard control was 50 ppb NIST SRM 987 (strontium carbonate, $^{87}Sr/^{86}Sr = 0.71025$).

The concentration of strontium isotopes contained in agricultural products is related to the soil in which they are grown. Because the Sr isotope ratios of the ancient Chinese mainland and relatively young Japanese islands are different, the ratio has been used in Japan to identify the provenance of green onions [41]. Based on this idea, we have measured the strontium isotope ratio of lacquer sap (lacquer film), lacquer tree bark, and lacquer tree branches grown in the Daigo area of Ibaraki Prefecture, Japan. Although the results showed a very small difference in the $^{87}Sr/^{86}Sr$ ratio, the value was within the range of error; that is, the Sr isotope concentrations in bark, branches, and lacquer of the lacquer tree were almost identical, as shown in Figure 11.49.

The Daigo soil in which lacquer trees were grown was also analyzed. The soil was treated with acetic acid ammonium, nitric acid, and hydrofluoric acid, and the extracts were named EAA (extracted by acetic acid ammonium), ENA (extracted by nitric acid), and EFA (extracted by hydrofluoric acid), respectively. The results were shown in Figure 11.50.

The strontium isotope concentrations in lacquer sap and EAA showed the same values. Because the EAA component is easily leached from soil by groundwater, this result indicated that the strontium dissolved in soil water is absorbed by lacquer trees and supplied to lacquer sap. The higher ENA value may be due to the addition of some inorganic components dissolved together in nitric acid. Because hydrofluoric acid is able to dissolve silica salt rocks, EFA showed the highest $^{87}Sr/^{86}Sr$ ratio.

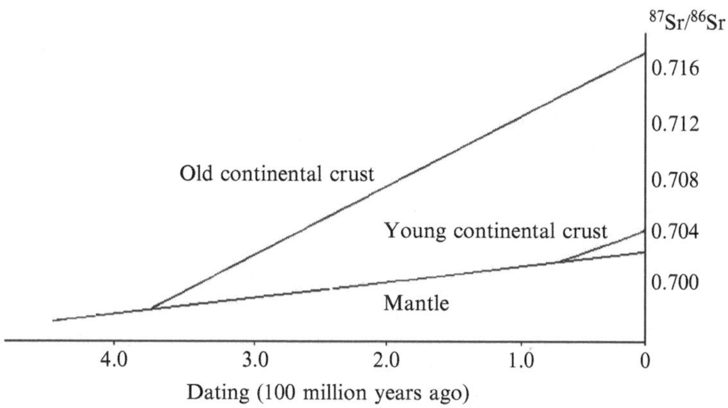

FIGURE 11.49

Strontium isotope ratios in lacquer film, bark, and branches.

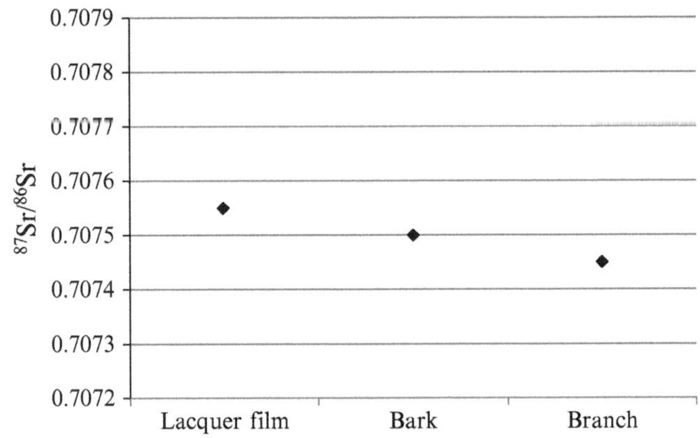

FIGURE 11.50

Strontium isotope ratio of lacquer sap and extracts from soil.

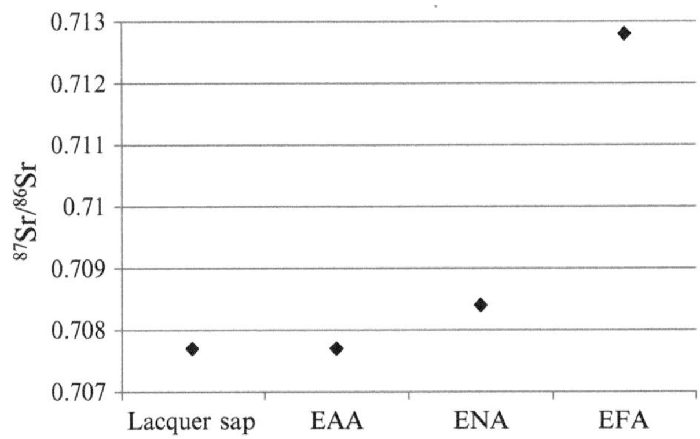

FIGURE 11.51

Strontium isotope ratios of Japanese and Chinese lacquer.

After revealing the relationship between the $^{87}Sr/^{86}Sr$ concentrations of a lacquer tree and the soil in which it is grown, we investigated the $^{87}Sr/^{86}Sr$ isotope ratios of lacquer sap tapped from lacquer trees grown in Hokkaido (HKD), Iwate Prefecture (IWP), Fukushima Prefecture (FKP), Ibaraki Prefecture (IBP), Kyoto Prefecture (KTP), and Okayama Prefecture (OKP) of Japan; and the Shanxi, Hubei, Guizhou, and Sichuan Provinces of China. The results are shown in Figure 11.51.

It can be seen that the $^{87}Sr/^{86}Sr$ isotope ratios of Japanese lacquer are below 0.710, but those of the continental Chinese mainland are above 0.712. Thus, $^{87}Sr/^{86}Sr = 0.710$ can be used to distinguish lacquer produced in the Japanese islands from lacquers produced on the Chinese continent. However, although the $^{87}Sr/^{86}Sr$ isotope ratio is a powerful tool for identifying the provenance of a lacquer, it is necessary to combine it with the results of other methods of analysis.

REFERENCES

[1] K. Toishi, T. Kenjou, Deterioration of coated film of Japanese lacquer, Sikizai 40 (1967) 92.

[2] T. Ogawa, T. Oyabu, M. Sakamoto, Deterioration analysis of lacquer film surface by exposure to outdoor, Kobunkazai no kagaku 38 (1993) 37.

[3] T. Egasira, T. Ichikawa, M. Sakamoto, T. Ogawa, Improvement of drying for black urushi lacquer and of light-resistance for its film, Mater. Life 7 (1995) 78.

[4] T. Ogawa, H. Jinnai, S. Osawa, Stabilization in appearance of oriental lacquer film by bromination, Polymer Preprints, Japan 59 (2002) 127.

[5] M. Sasamori, T. Egashira, K. Kuwada, Research on improving the urushi project, Ishikawaken kogyoushikenjyou kenkyuhoukoku 39 (1991) 9.

[6] J. Kumanotani, Durable oriental lacquer (Urushi) -A self-surface renewable coating through effloresence in degradation of the core-shell type microgel densely packed lacquer, Sikizai 70 (1997) 808.

[7] N. Niimura, Y. Iijima, T. Miyakoshi, Hardening process and surface structure of lacquer films studied by X-ray photoelectron spectroscopy, Surf. Interface Anal. 24 (1996) 237.

[8] N. Niimura, T. Miyakoshi, J. Onodera, T. Higuchi, Structural studies of *Melanorrhoea usitate* lacquer film using two-stage pyrolysis/gas chromatography/mass spectrometry, Rapid Commun. Mass Spectr. 10 (1719) 1996.

[9] N. Niimura, T. Miyakoshi, Characterization of synthesized lacquer analogue films using X-ray photoelectron spectroscopy, Surf. Interface Anal. 29 (2000) 381.

[10] I. Shiihara, Studies on confirming the type of dried urushi film and its formation mechanism (urushi hardening study report II), Bull. Chem. Soc. Jpn. 75 (2002) 2061.

[11] K. Toyoshima, Study of deterioration of urushi film with ultraviolet radiation, Mater. Life 8 (1996) 28.

[12] Y. Oyabu, T. Asami, T. Ogawa, Deterioration process of urushi films by accelerated weathering test, Mater. Life 10 (1998) 43.

[13] Y. Kamiya, R. Lu, T. Kumamoto, T. Honda, T. Miyakoshi, Deterioration of surface structure of lacquer films due to ultraviolet irradiation, Surf. Interface Anal. 38 (2006) 1311.

[14] J. Lambert, J. Frye, G. Carriveau, The structure of oriental lacquer by solid state nuclear magnetic resonance spectroscopy, Archaeometry 33 (1991) 87.

[15] J. Kumanotani, Super-durable structure of oriental lacquer films, Polym. Mater. 59 (1988) 278.

[16] N. Niimura, T. Miyakoshi, J. Onodera, T. Higuchi, Characterization of *Rhus vernicifera* and *Rhus succedanea* lacquer films and their pyrolysis mechanisms studied using two-stage pyrolysis-gas chromatography/mass spectrometry, J. Anal. Appl. Pyrolysis 37 (1996) 199.

[17] R. Lu, Y. Kamiya, T. Miyakoshi, Characterization of lipid components of *Melanorrhoea usitata* lacquer sap, Talanta 71 (2007) 1536.

[18] R. Lu, Y. Kamiya, T. Miyakoshi, Applied analysis of lacquer films based on pyrolysis-gas chromatography/mass spectrometry, Talanta 70 (2006) 370.

[19] M. Herreraa, G. Matuschek, Fast identification of polymer additives by pyrolysis-gas chromatography/mass spectrometry, J. Anal. Appl. Pyrolysis 70 (2003) 35–42.

[20] T. Honda, R. Lu, N. Kitano, Y. Kamiya, T. Miyakoshi, Applied analysis and identification of ancient lacquer based on pyrolysis-gas chromatography/mass spectrometry, J. Appl. Polym. Sci. 118 (2010) 897–901.

[21] Y. Wan, R. Lu, Y. Du, T. Honda, T. Miyakoshi, Does Donglan lacquer tree belong to *Rhus vernicifera* species? Inter. J. Biol. Macromol. 41 (2007) 497–503.
[22] K. Urakami, A. Higashi, K. Umemoto, M. Godo, C. Watanabe, K. Hashimoto, Compositional analysis of copoly(DL-lactic/glycolic acid) (PLGA) by pyrolysis-gas chromatography/mass spectrometry combined with one-step thermally assisted hydrolysis and methylation in the presence of tetramethylammonium hydroxide, Chem. Pharm. Bull. 49 (2001) 203–205.
[23] M. Fukushima, M. Yamamoto, T. Komai, K. Yamamoto, Studies of structural alterations of humic acids from conifer bark residue during composting by pyrolysis-gas chromatography/mass spectrometry using tetramethylammonium hydroxide (TMAH-py-GC/MS), J. Anal. Appl. Pyrolysis 86 (2009) 200–206.
[24] X. Chai, T. Shimaoka, Q. Guo, Y. Zhao, Characterization of humic and fulvic acids extracted from landfill by elemental composition, 13C CP/MAS NMR and TMAH-Py-GC/MS, Waste Manage. 28 (2008) 896–903.
[25] R. Widyorini, T. Higashihara, J. Xu, T. Watanabe, S. Kawai, Self-bonding characteristics of binderless kenaf core composites, Wood Sci. Tech. 39 (2005) 651–662.
[26] R. Martine, G. Nicolas, C. Cecile, R. Christian, New analytical methodology for investigating both volatile and nonvolatile constituents from archaeological ceramic vessels, Abstracts of Papers, 225th ACS National Meeting, New Orleans, LA, United States, March 23–27, 2003, GEOC-107.
[27] H. Araki, N. Tatarazako, K. Kishi, K. Kuroda, Evaluation of bioaccumulation potential of 3,4,5-trichloroguaiacol in a zooplankton (Daphnia magna) by pyrolysis-GC/MS in the presence of tetramethylammonium hydroxide (TMAH), J. Anal. Appl. Pyrolysis 55 (2000) 69–80.
[28] T. Arii, Complex thermal analysis: Evolved gas analysis-mass spectrometry EGA-MS, Rigaku Janaru 39 (2008) 17–25.
[29] T. Arii, Evolved gas analysis-mass spectrometry (EGA-MS) using skimmer interface system equipped with pressure control function, J. Mass Spectro. Soc. Jpn. 53 (2005) 211–216.
[30] T. Tsugoshi, M. Furukawa, M. Ohashi, Y. Iida, Comparison of capillary and skimmer interfaces in evolved gas analysis-mass spectrometry (EGA-MS) with regard to impurities in ceramic raw materials, J. Therm. Anal. Calorimetry 64 (2001) 1127–1132.
[31] G. Matuschek, A. Kettrup, A. Prior, EGA/MS investigations on the thermal degradation of diammonium hexachloroplatinate, J. Therm. Anal. Calorimetry 56 (1999) 471–477.
[32] J. Lee, G. Chang, J. Kwag, M. Rhee, A. Buglass, G. Lee, Fast analysis of nicotine in tobacco using double-shot pyrolysis-gas chromatography-mass spectrometry, J. Agric. Food Chem. 55 (2007) 1097–1102.
[33] K. Quenea, S. Derenne, F. Gonzalez-Vila, J. Gonzalez-Perez, A. Mariotti, C. Largeau, Double-shot pyrolysis of the non-hydrolysable organic fraction isolated from a sandy forest soil (Landes de Gascogne, South-West France), J. Anal. Appl. Pyrolysis 76 (2006) 271–279.
[34] J. Lee, C. Lee, J. Kwag, A. Buglass, G. Lee, Determination of optimum conditions for the analysis of volatile components in pine needles by double-shot pyrolysis-gas chromatography-mass spectrometry, J. Chromatogr. A 1089 (2005) 227–234.
[35] N. Niimura, T. Miyakoshi, J. Onodera, T. Higuchi, Structural studies and polymerization mechanisms of synthesized lacquer films using two-stage pyrolysis-gas chromatography/mass spectrometry, Inter. J. Poly. Anal. Charact. 4 (1998) 309–322.

[36] N. Niimura, T. Miyakoshi, J. Onodera, T. Higuchi, Identification of ancient lacquer film using two-stage pyrolysis-gas chromatography/mass spectrometry, Archaeometry 41 (1999) 137–149.

[37] Y. Kamiya, Y. Niimura, T. Miyakoshi, Evaluation of synthesized lacquer films using pyrolysis-gas chromatography/mass spectrometry, Bull. Chem. Soc. Jpn. 73 (2000) 2621–2626.

[38] R. Lu, Y. Kamiya, Y. Wan, T. Honda, T. Miyakoshi, Synthesis of *Rhus succedanea* lacquer film and analysis by pyrolysis-gas chromatography/mass spectrometry, J. Anal. Appl. Pyrolysis 78 (2007) 117–124.

[39] R. Lu, Y. Kamiya, T. Miyakoshi, Preparation and characterization of *Melanorrhoea usitata* lacquer film based on pyrolysis-gas chromatography/mass spectrometry, J. Anal. Appl. Pyrolysis 78 (2007) 172–179.

[40] R. Lu, X. Ma, Y. Kamiya, T. Honda, Y. Kamiya, A. Okamoto, T. Miyakoshi, Identification of Ryukyu lacquerware by pyrolysis-gas chromatography/mass spectrometry, J. Anal. Appl. Pyrolysis 80 (2007) 101–110.

[41] Ministry of Agriculture, Forestry and Fisheries of Japan: Development of techniques for determination of the variety and the geographic provenance of agricultural products by using analysis of minor elements and DNA makers, in: Agriculture, Forestry and Fisheries Research Council Secretariat of Japan, 2005.

CHAPTER

Use of lacquer

12

12.1 LACQUER COATING

Lacquer sap had been used thousands of years ago by humans according to practices based on their experience. Lacquer was perhaps first used as an adhesive, and then used to paint dishes, production tools, ritual tools, and weapons such as bows and arrows. Hutermann reported the Chinese had begun using lacquer to produce artwork more than 6000 years ago [1], but in fact, the very famous ZhuQi MuWan (Vermilion lacquer wooden bowl) that was found in 1977 at the Hemudu site in Zhejiang Province, China, was manufactured 7000 years ago, during the Neolithic period [2]. In Japan, a red lacquer comb and several other lacquer artifacts found in Torihama Kaitsuka (ancient tomb) in Fukui Province were manufactured about 5000 years ago, in the late *Jōmon* period [3]. In the year 2000, some artifacts such as a woven headband, shoulder rest, bracelets, and foot ornament coated with lacquer excavated from the Kakinoshima-B site in Minamikayabe of Hokkaido, Japan, were found to be 9000 years old by radiocarbon dating [4]. However, all the samples were destroyed by a fire on December 28, 2002; thus, the carbon dating cannot be remeasured [5]. Therefore, in this book, we believed that the Hemudu lacquerware is the oldest one in the history and culture of lacquer.

Just as "china" is a common name for Chinese ceramics, "japan" is an old name for Japanese lacquerware. The technique of lacquer, which sometimes is called "japanning," involves applying several coats of varnish, each of which is dried and subsequently polished, to a substrate such as wood or bamboo. In this sense, the word "japan" implies the advanced lacquer culture in Japan, and in reality, this is true. Although China is the major producer of raw lacquer, the lacquer culture is not strong in modern Chinese life, and 98% of Chinese raw lacquer sap is exported to Japan. In Japan, lacquer culture has a strong mass base unlike China or other Asian lacquer-producing countries. The following material on the application of lacquer is mainly to introduce Japanese lacquer techniques.

12.2 LACQUER USED IN ART WORKS

Lacquerware refers to an object of wood, paper, or other carrier that is coated with lacquer sap. Lacquerware production can be divided roughly into three processes: (1) making the object, such as a bowl, (2) treating the surface of the object with a

Lacquer Chemistry and Applications. http://dx.doi.org/10.1016/B978-0-12-803589-4.00012-2

base coat, and (3) coating the object several times with lacquer. The lacquer coating process is a technique performed in a variety of localities including Japan [6,7], China [8], Thailand [9], and Myanmar [10]. Although the number of process steps, coating names, detail work, and materials may be slightly different, the coating processes can be considered to be essentially similar. A typical section of the Japanese lacquer coating procedure is shown in Figure 12.1. After the object is made, it is coated with raw lacquer to reinforce the object, and sealed with about 1-5 layers of lacquer that has been mixed with finely ground powder, then sealed with at least one layer of lacquer which has been mixed with fine polishing powder. After hardening, the last layer is smoothed and polished with coarse and fine whetstones, respectively, using water in some cases. Once the object has been polished, the "real" lacquer coating is applied. At least three layers (bottom, middle, and top) of lacquer coating are required. Some sophisticated and very expensive lacquerwares may have dozens of layers of lacquer coating. The object is repeatedly smoothed and polished after each lacquer coating layer is applied, with the final coating layer of the best quality lacquer and more careful smoothing and polishing to finish the process.

The base material should have good compatibility with lacquer, without any variation and/or shrinkage. Wood, leather, paper, ceramics, fabrics, and metals are commonly used. The surface coating is very important during the lacquer coating process, because the base coating of the object must ensure the best adhesion between the object and subsequent lacquer coats. To better understand the lacquer process, we give an example. Figure 12.2 is a typical lacquer process that shows the untreated object (Figure 12.2a), after surface treatment (Figure 12.2b), and then after lacquer coating to create a lacquerware bowl (Figure 12.2c).

In Myanmar, many lacquerware objects are woven from bamboo. After surface treatment, they are usually coated with several layers of lacquer, and finally, paste gold foil to form a full gold lacquerware, as shown in Figure 12.3. From right to left are the bamboo object, first surface treatment, second surface treatment, first layer lacquer coating, second layer lacquer coating, third layer lacquer coating, and final

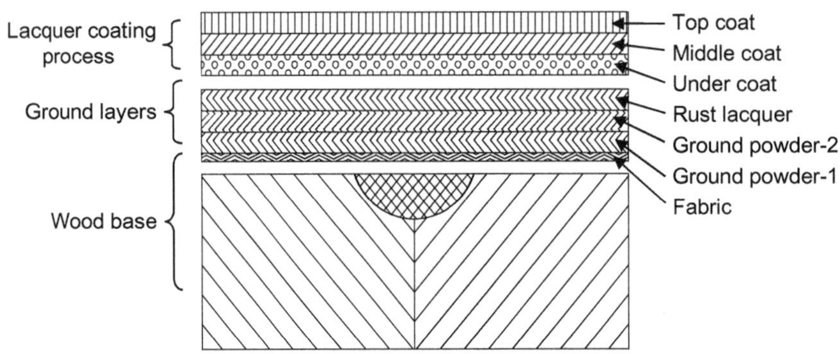

FIGURE 12.1

Lacquer coating process.

(a) (b) (c)

FIGURE 12.2

Object (a), surface treatment (b), and lacquer coating to create a lacquerware bowl (c).

FIGURE 12.3

Myanmar bamboo lacquer process.

paste gold foil lacquerware, respectively. Producing lacquerwares is time consuming, but it can be effectively organized as a cottage industry. In fact, in Japan, lacquerwares are mainly produced in a lacquerware factory, while in Myanmar, the family is the main unit of production.

Another Southeast Asian country, Laos, also has its own lacquer culture and technology in its history and modern life, especially in Luang Prabang. Luang Prabang is a city located in north central Laos, where the Nam Khan River meets the Mekong River about 425 km (264 mi) north of Vientiane, the Laotian capital. Luang Prabang was registered on the World Heritage List in 1995. The ancient capital of the Lane Xang kingdom has historically been a vibrant center for Buddhist culture and arts. Today, the splendid temples and everyday rituals of the people continue to display the living spirit of Buddhism that is the hallmark of Luang Prabang, and lacquerwares are very popular with the Luang Prabang donors.

Lacquer was also used for decorative items during the royal period, for example, lacquer-coated baskets, tobacco boxes, and the interiors and exteriors of temples as well as the furniture. Lacquer is also used for lacquer appliqué. Lacquer appliqué is a traditional technique used to decorate the exterior of temples, for example, the decorative molding on gables, eaves, columns, porches, and purlins. It is also used to decorate items inside the temple, such as the surface of altars and the platforms on

FIGURE 12.4

Lao lacquer process, from object (left) to finished art (right).

which Buddha images and the king sit. Lacquer was applied to the column pillars of the Royal Palace building. In ancient times, the local people used lacquer to decorate important buildings, and then apply gold leaf, making lacquer appliqué molding flower patterns to decorate the walls and altars. Lacquer is used to apply colored glass to tell the history of the Buddha, legends of Phavet (Reincarnation of Buddha Vet), and other stories on the walls of temples.

Lao lacquer coating processes are almost the one same as those of Myanmar, as shown in Figure 12.4. The object is first treated with a kind of lacquer mixed with weed ash several times (usually three), then polished to a gloss, and painted with a good-quality lacquer sometime mixed with a drying oil in order to produce a gloss. Finally, paste gold foil to make various designs patterns and a gold foil lacquerware is obtained.

Because lacquer is very useful for decorative work, recently the Luang Prabang local people have used lacquer to decorate guesthouses, hotels, and houses.

12.3 LACQUER IN INDUSTRY

The purpose of coating is the protection and decoration of objects. In the case of lacquer, of the phrase "lacquer technology" is usually used to suggest the technical content and artistic beauty of the lacquer process and lacquer object. Although lacquer is an ultradurable coating that can withstand storage for thousands of years, its application in industry is very limit. The reason can be attributed to the negative factors of lacquer such as it is "allergenic," "expensive," "slow drying," and "has weak light resistance." However, due to its excellent rust, heat, chemical, and water resistance, research on how to use lacquer in industry is in progress. Lacquer has been used as insulating [11] and anticorrosive [12] materials in modern Japan. Recently, in order to examine how lacquer film changes when used outdoors, Oyabu *et al.* reported use of a lacquer refined by a three-roll mill to paint a whole car. The painting procedures

FIGURE 12.5

Car painted with lacquer sap.

used acrylic urethane resin as the Shitachi coating, then red lacquer was used as the middle coating, and finally, translucent lacquer was used as the top coating, as shown in Figure 12.5 [13]. The car was driven for about 1000 days from 7 June 2003 to 19 February 2006, and the mileage was about 6000 km. The gloss maintenance and decreases of film thickness from 7 June 2003 to 19 February 2006 are shown in Figures 12.6 and 12.7, respectively.

It can be observed from Figure 12.6 that an obvious change in film gloss occurred from the rainy season to summer, and subsequently decreased slightly during autumn to winter, with no clear difference in the flat roof and perpendicular doors. The car has been driven about 1000 days during three rainy seasons to summer, but the final of gloss value was over 70, suggesting strong water and light resistance.

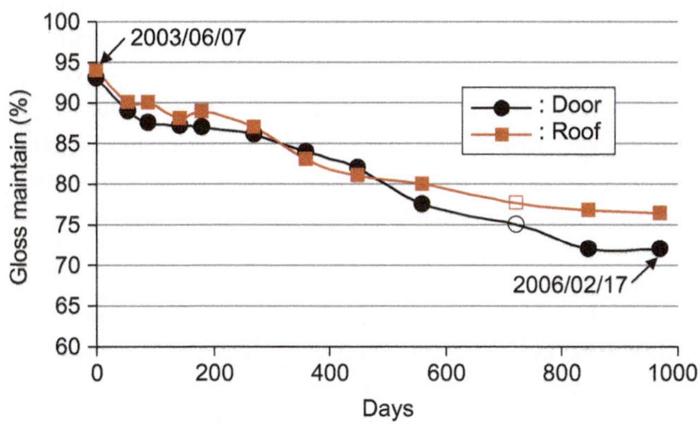

FIGURE 12.6

Gloss maintenance of lacquer film.

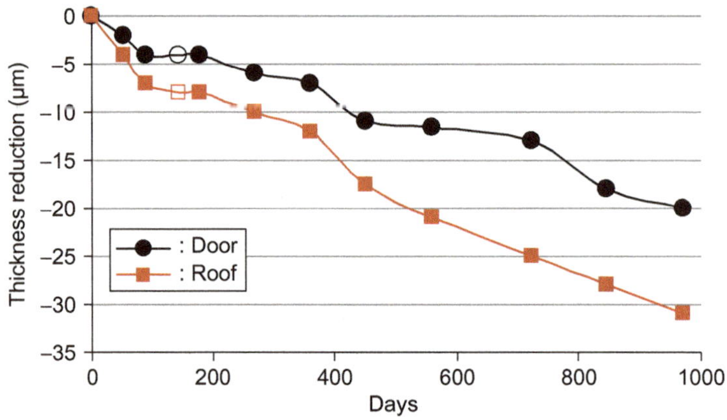

FIGURE 12.7

Change of thickness of lacquer film.

The film thickness also decreased greatly during the rainy season to summer and decreased slightly during the autumn to spring, in parallel with the change in gloss (Figure 12.7). The surface of the lacquer film was degraded by strong rain and ultraviolet light in the rainy and summer seasons, when the saccharides and glycoproteins which are embedded in the film became exposed and were washed away by water or blown away by wind causing the decrease in film thickness.

We have reported the application of hybrid lacquer as a chrome-free corrosion-protective coating [14], as a continuous research project. Recently we used the Myanmar lacquer, which is plentiful and cheap and black instead of the Chinese lacquer to develop a corrosion-protective coating. The Myanmar lacquer was first hybridized with silicate MS51 by the following detailed process: 10 g Myanmar lacquer was mixed with 0.5 g of MKC silicate MS51, in the ratio of 100:5 wt%, and stirred for about 60 min. Both the blank Myanmar and hybrid lacquer were diluted with 20 ml of toluene before application as a protective paint because of the high viscosity. SPCC steel disks (1.0 × 30 × 70 mm) were immersed into the hybrid lacquer, drained of excess lacquer, and dried at 190 °C for 50 min in a drying oven (Isuzu Motors Limited Factory, analog control system ANS-111S, 260 °C max, Japan), to obtain one-layer lacquer-coated SPCC steel disks. A repeat immersion into the lacquer plus drying produce disks coated with two layers. Changes in gloss and color, pencil hardness, and thickness were examined, and crosscut, boiling water immersion, salt solution immersion, and promotion salt solution immersion tests of the lacquer coating film were carried out.

The changes in gloss and color, thickness, and pencil hardness are summarized in Table 12.1. It can be observed that the hybrid lacquer film has a slightly higher gloss than the blank lacquer film, and the two-layer coating has a lower L^*, a^*, and b^* than the one-layer coating, suggesting that two-layer film is darker than the one-layer film. According to the Commission Internationale d'Éclairage (CIE) system, L^*, a^*, and b^* mean lightness, (+red)/(−green) and (+yellow)/(−blue), respectively.

Table 12.1 Changes of gloss and color of lacquer film coating on SPCC steel disks

Entry number	Lacquer name	Coating layer	Gloss (at 60°)	Thickness (μm)	Pencil hardness	Lightness and color		
						*L**	*a**	*b**
1	Blank	1	68	8	2H	2.29	4.91	2.11
2	Hybrid	1	77	7	4H	0.91	0.02	0.06
3	Blank	2	88	15	H	1.20	0.88	0.23
4	Hybrid	2	93	13	3H	0.93	−0.05	−0.02

The thickness of the one-layer film is about 7-8 μm, and that of the two-layer film is about 13-15 μm; both are less than 20 μm. The one-layer film has higher pencil hardness than the two-layer film, but the reason was not clear.

The crosscut results are shown in Figure 12.8. In an area of 1 cm^2, 100 small boxes were drawn with a knife on a piece of tape, the tape was applied to a lacquer-coated object, and then peeled off. The adhesive ability of the lacquer was evaluated according to how many small boxes were stripped off. The entry numbers are the same in Table 12.1. Almost no small boxes were stripped, implying that both natural and hybrid Myanmar lacquers have a good adhesion to the SPCC steel disk. However, after the SPCC steel disks coated with lacquer film were immersed in boiling water for 60 min, the blank natural lacquer films (entries 1 and 3) were completely stripped off, while the hybrid lacquer film showed very strong water resistance as shown in Figure 12.9.

The SPCC steel disks coated with lacquer were immersed into 3.5 wt% salt solution at 20 °C for 1 week, and the results are shown in Figure 12.10. After being immersed in the salt solution for 72 h, some blisters arose from the corners of the natural

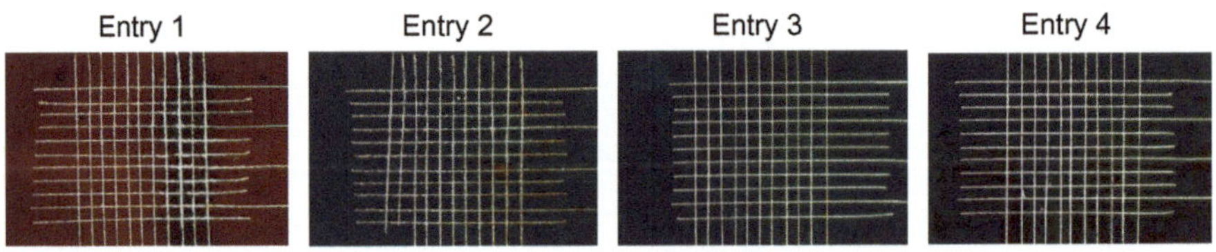

FIGURE 12.8

Results of crosscut test of lacquer films.

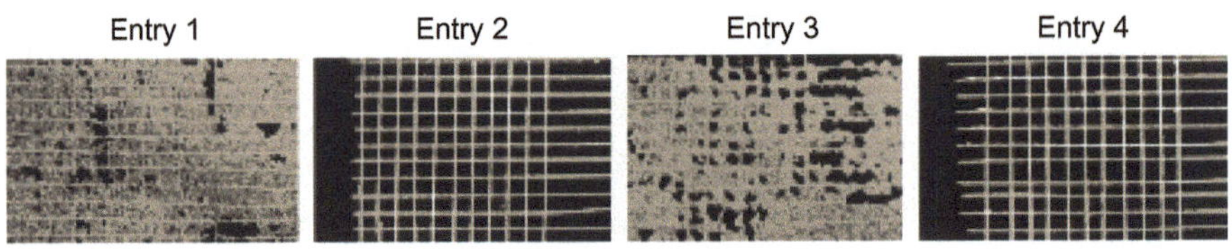

FIGURE 12.9

Results of boiling water immersion test of lacquer films.

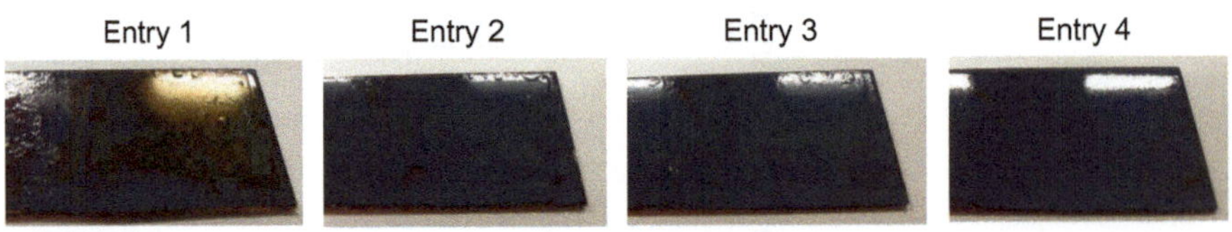

FIGURE 12.10

Results of salt solution immersion test of lacquer films.

FIGURE 12.11

Nuts and bolts of a motorcycle painted with antirust lacquer coating.

Myanmar one-layer lacquer films and rust appeared after 100 h (entry 1). Seven days later, the disks coated with natural two-layer (entry 2) and hybrid one-layer (entry 3) lacquer exhibited a little rust from the corners. There is no any change in the two-layer hybrid lacquer film even when immersed in salt solution for 1 week (entry 4).

The salt solution immersion test also showed the same tendency results as the salt solution test. Although the film thickness is thicker, it was found that the two-layer coating had less corrosion than the one-layer coating, and adding silicate further improved the anticorrosion property of lacquer film. This chrome-free lacquer coating is being tested for application on motorcycles and automobiles as a black antirust coating for nuts and bolts, as shown in Figure 12.11.

In this book, the authors have described recent advances, including their own original research, on the analysis of lacquer sap components, structure, and biological activities of lacquer polysaccharides, structure of lacquer glycoprotein, synthesis of urushiol derivatives, and modifications of lacquer sap such as repeat of kurome, hybridization, and microdispersion. Lacquer is the only natural product that dries by enzymatic polymerization. From the viewpoint of energy conservation and environmental protection, natural and artificial lacquers are promising as paints. Not only does lacquer form a dense film without a solvent at ambient temperature and pressure, but it also is a planted resource crop that can produce an ecolacquer paint. Instead of producing paints in a factory, paints grown in a field may be feasible in the future.

REFERENCES

[1] A. Hutermann, C. Mai, A. Kharazipour, Modification of lignin for the production of new compounded materials, Appl. Microbiol. Biotechnol. 55 (2001) 387.

[2] http://www.ndcnc.gov.cn/datalib/2002/Ware/DL/DL-172153.

[3] http://inoues.net/ruins/torihama.html.

[4] http://www.daily-tohoku.co.jp/kikaku/tyouki_kikaku/jomon/jomon_22.htm.

[5] http://ja.wikipedia.org/wiki/%E6%BC%86%E5%99%A8.

[6] T. Miyakoshi, K. Nagase, T. Yoshida, Progress of Lacquer Chemistry, IPC, Japan, 1999, p. 43.

[7] T. Terada, K. Oda, H. Oyabu, T. Asami, Lacquer—The Science and Practice, Rikou Publisher, Tokyo, Japan, 1999, pp. 25-26.

[8] F. Zhang, Chinese Lacquer: Technology and Conservation, Science Publisher, Beijing, China, 2010.

[9] S. Bhirasri, Thai Lacquer Works, seventh ed., Fine Arts Department Publisher, Bangkok, Thailand, 2006.

[10] S. Fraser, Burmese lacquerware, first ed., Sirivatana Publisher, Bangkok, Thailand, 1985.

[11] H. Kiyomisu, N. Inai, Urushi industry, Denka 3 (1939) 7.

[12] Y. Ohashi, Study on fried urushi, Kagaku to Kogyo 279 (1953).

[13] H. Oyabu, N. Ando, Explore the possibility of urushi, Paint Technol. 1 (2010) 126 (in Japanese).

[14] T. Honda, R. Lu, T. Miyakoshi, Chrome-free corrosion protective coating based on lacquer hybridized with silicate, Prog. Org. Coat. 56 (2006) 279.

Appendix

1 IDENTIFICATION METHODS OF LACQUER IN CHINA

NATIONAL STANDARD OF THE PEOPLE'S REPUBLIC OF CHINA (GB/T 14703-2008)

Foreword: This standard is a revision of the GB/T 14703-1993. The main technical differences between GB/T 14703-2008 and GB/T 14703-1993 are as follows.

1. Range

This standard specifies lacquer-related terms including definitions, quality requirements, test methods, packaging, storage, and transportation. It is a guideline to lacquer production and sales.

2. Normative reference documents

The following documents become part of this standard by reference to this standard. Dated references and subsequent amendments or revisions documents do not apply to this standard. However, use of the latest version of this standard is encouraged according to the agreement reached by parties to this standard. For documents without dated references, the latest edition is suitable for this standard.

GB/T 3186: Sampling methods for colored and pure lacquers.

GB 9273: Marking test method of lacquer film.

GB/T 9750: Packaging marks of lacquer.

3. Term and definition

The following terms and definitions apply to this standard:

3.1 Color

Lacquer is a liquid emulsion with color.

3.2 Color change

Stirring stationary lacquer will cause the color to change due to reaction with air.

Note: Lacquer color change is due to the reaction of urushiol with oxygen in the air catalyzed by laccase to form urushiol quinone.

3.3 Odor

Lacquer emits a smell that can be detected by the human olfactory sense.

3.4 Tiny granules

Milky-white particles like broken rice grains exist in lacquer sap. They are oval in shape with different sizes.

3.5 Sand-like lines

Stirring and crushing the tiny granules forms a white line dependent on the movement of lacquer sap. A light brown sandy line outside the white line is also found to move at the same time.

Note: If stirred too much, the granular phenomenon will gradually disappear, but gradually recurs after static.

3.6 Stretch drawing property

After the lacquer sap is stirred with a paint paddle, the paddle is inserted into the sap, then withdrawn, and the sap will flow down in a long thin silky strip. The paint strip should crack and shrink rapidly as hooked when a certain tension is exceeded.

3.7 Lacquer dregs

The dregs are a sawdust-like mass and natural lacquer conjunctiva in lacquer sap during the tapping process.

3.8 Special steelyard

A special steelyard balance made of purple copper (1 g, for 10 points).

3.9 Nonvolatile content

This means the nonvolatile matter left after removing the moisture and volatile materials.

3.10 Heating loss

The moisture and volatile matter are evaporated at a prescribed temperature, and the evaporation loss is calculated by the difference.

3.11 Nitrogenous matter and gum

These are mainly glycoproteins and polysaccharides.

4. Technical quality

The technical quality requirements for lacquer are summarized in Table A1.

5. Sampling method

Stir lacquer sap (250 g) in a clean, dry, wide-mouth bottle, then cover with a lid. Label with the inspection number, inspection authority, product name, product origin, batch number, quantity, and production date.

6. Physical and chemical testing methods

6.1 Determination of total urushiol concentration

Method and principle: A reaction between urushiol, ferric chloride, and triethanolamine (TEA) produces ternary TEA-Fe^{3+}-urushiol complexes, which have a stable blue-purple color and strong absorption at 625 nm.

6.1.1. Instrument and reagents

(1) UV-visible spectrophotometer.
(2) Volumetric flasks: 25 ml, 50 ml, 100 ml.
(3) Small beakers: 25 ml, 50 ml.
(4) Pipettes: 1 ml, 5 ml, 10 ml.
(5) Filter paper: 7 cm diameter medium speed qualitative filter paper.
(6) Glass funnel: 4 cm diameter.
(7) Anhydrous ethanol: Analytical grade.
(8) Triethanolamine solution: Dissolve 1.6% analytical grade TEA in ethanol.

Table A1 Lacquer technical quality index

Item		Quality index		
		First grade	**Second grade**	**Third grade**
Sensory test	Color	From milky-white to dark-brown emulsion		
	Color darking	Gradually change from milky-white to deep coffee color		Distinctly color change
	Odor	Strong sour smell	Common sour smell	Weak sour or no smell
	Tiny granule	Very obvious	Obvious	No obvious
	Stretch drawing property	Long filaments, fast shrinkage	Short filaments, slow shrinkage	Short thick filaments, no shrinkage
	Dreg content (%)	≤3	≤4	≤5
Physical and chemical properties	Nonvolatile matter content (%)	≥73	≥68	≥63
	Total urushiol (%)	≥65	≥58	≥50
	Heating loss (%)	≤27	≤32	≤37
	Nitrogenous and gum (%)	6-14		
	Touch-free time (h)	4	3	3

(9) Ferric chloride solution: Dissolve analytical grade ferric chloride in 0.1% ethanol solution which contains six crystallization water.

(10) Mixed triethanolamine and ferric chloride solution: Place 25 ml 1.6% triethanolamine solution and 25 ml 0.1% ferric chloride solution in a 100-ml volumetric flask and dilute with ethanol to 100 ml.

(11) Standard solutions: dissolve 0.0500 g saturated urushiol in anhydrous ethanol in a 25-ml volumetric flask, and dilute it with anhydrous ethane to 25 ml. Pipette 10 ml into a 50-ml volumetric flask and dilute with anhydrous ethane to 50 ml, giving a solution concentration of 0.4 mg/ml.

Note: Normative synthesis of saturated urushiol is described in Appendix A.

6.1.2. Determination steps

(1) Sampling method: Pierce the lacquer film with a glass rod, then insert another glass rod and stir; remove the glass rods and let the lacquer sap flow into a beaker that was previously weighed.

(2) Working curve: Place standard solutions of saturated urushiol (0.4 mg, 1.0, 2.0, 3.0, 4.0, and 5.0 mg/ml) into 25-ml volumetric flasks; then add 10 ml triethanolamine-ferric chloride solution and dilute with anhydrous ethanol to

25 ml. After 15 min, absorbance is measured at 625 nm using a 1-cm quartz cell and reagent solution as the blank. The volume of the standard solution based on saturated urushiol (ml) as the abscissa and the corresponding absorbance as the ordinate is plotted for a working curve.

(3) Determination: Place 0.2-0.3 g lacquer sap into a 25-ml beaker, and then quickly add anhydrous ethanol to dissolve it, and filter it into a 100-ml volumetric flask. Wash the filter residue several times with small amounts of anhydrous ethanol until the filtrate is colorless, and then dilute with anhydrous ethanol to 100 ml. Pipette 1 ml of this solution into a 25-ml volumetric flask, then add 10 ml triethanolamine-ferric chloride solution, dilute with anhydrous ethanol to 25 ml, and shake. After 15 min, measure absorbance at 625 nm using a 1-cm quartz cell and reagent solution as the blank. The corresponding urushiol concentration is calculated by the following formula:

$$U = \frac{d \times V \times 100}{m \times K} \tag{1}$$

U is the percentage of urushiol (%); d is the concentration of standard saturated urushiol solution (mg/ml); V is the volume detected by absorbance from the working curve (ml); m is the sample weight (mg); and K is the sampling proportion (1/100).

Allowable deviation: In this standard range, the parallel test results permit deviation of not more than 0.50%, and the result is an average of all results.

6.2 Determination of heating loss, nitrogenous matter, and gum

6.2.1. Instruments and reagents

(1) Drying oven.
(2) Glass desiccator.
(3) Beakers (50 ml).
(4) Qualitative filter paper, glass funnel.
(5) Analytical grade anhydrous ethanol.

6.2.2. Determinations steps

(1) Sampling method: Similar to the measurement of total urushiol method.

(2) Heating loss: Weigh 1.0-1.3 g raw lacquer sap, place it in a 50-ml beaker, and distribute the sap evenly on the bottom of the beaker. Heat the beaker in a drying oven at 105-110 °C for 1 h, immediately place it in a glass desiccator to cool to room temperature, and weigh. The result is calculated according to the following formula:

$$W = \frac{m_1 - m_2}{m} \times 100 \tag{2}$$

W is the heating loss (%); m_1 is the weight of raw lacquer sample and beaker before drying (g); m_2 is the weight of raw lacquer sample and beaker after drying (g); and m is the sample weight (g).

Allowable deviation: In this standard range, the parallel test results permit a deviation of not more than 0.50%, and the result is an average of all results.

(3) Determination of nitrogenous matter and gum: Add 25 ml anhydrous ethanol to the heating loss test sample and stir with a glass rod. After several minutes, filter with a filter paper that has been dried and weighed. Wash the filter paper several times with a small amount of anhydrous ethanol, until the filtrate is colorless, and heat the filter paper in an oven at 105-110 °C for 1 h. Then, immediately place it in the glass desiccator to cool to room temperature, and weigh it. The result is calculated according to the following formula:

$$S = \frac{m_2 - m_1}{m} \times 100 \tag{3}$$

S is the percentage of nitrogenous matter and gum; m_1 is the filter paper weight before drying (g); m_2 is the weight of filter paper and residues after drying (g); and m is the sample weight (g).

Allowable deviation: In this standard range, the parallel test results permit a deviation of not more than 0.50%, and the result is an average of all results.

6.3 Determination of touch-free drying time

The touch-free drying is carried out in a constant temperature and humidity box according to GB/T9273 standard. The temperature is set at 23 ± 1 °C and relative humidity at 80 ± 5%.

6.4 Preparation of lacquer film

Spread lacquer sap on the glass with an applicator to a wet film thickness of about 100 μm.

7. Traditional testing methods

Traditional testing methods are used mainly for rapid testing at the work site. Methods for the identification of foreign matter in raw lacquer sap can be seen in Appendix B.

7.1. Tools

Special steelyard balance, alcohol lamp, and lacquer-checking board.

7.2. Steelyard fraction testing method

Weigh 1 g dreg-free lacquer sap with a steelyard balance, and slowly heat it with an alcohol lamp. It first produces a big bubble, then small bubbles, and then a flocked-velvet flower. Finally, when an oil-like hole appears in the middle of flocked-velvet flower, rapidly rotate the steelyard, until bubble disappears and smoke appears, and then weigh it.

7.3. Methods for testing color, smell, and tiny granules

(1) Color: Stir the lacquer sap and observe its color.

(2) Tiny granules: Observe stirred lacquer sap using the lacquer-checking board.

(3) Smell: Smell lacquer sap using the lacquer-checking board.

7.4. Color darkening

Insert the lacquer-checking board to the bottom of the lacquer storage barrel and stir, and the lacquer color will change due to oxidation in the air. Good-quality lacquer should change gradually from a shallow to deep color, not change color suddenly, and will have a luster.

7.5. Stretch drawing property

Insert the lacquer-checking board into lacquer sap and then lift it allow the sap to drain naturally while observing the silk-like line and retraction force.

8. Testing rules

8.1. General inspection can be carried out according to the traditional method (Section 7). When the test results are inconsistent, the physical and chemical properties can be tested again according to this standard.

8.2. The lacquer buyer can examine a lacquer sample according to this standard. If the lacquer quality does not meet the technical requirements of this standard, it can be resampled and tested again according to GB/T3186. If the retest results still do not comply with the standard technical requirements, the lacquer can be considered a substandard product, and acquirer has the right to refuse to purchase or receive it.

8.3. When the lacquer is tested according to GB/T3186 standards, samples should be divided into two parts, one sealed for reference purpose and another for testing.

8.4. Both buyer and supplier should examine product packaging, amount of product, and product marks for verification. If the packing leaks, quantity discrepancies are found, or the mark does not meet the requirements, the buyer should promptly notify the relevant processing departments.

8.5. When there is a question about lacquer quality, the lacquer should be tested by a product supervision and arbitration authority.

9. Mark

Lacquer product should be marked according to GB/T9750.

10. Packaging, storage, and transport

10.1. Packaging

(1) Container: Double-bottom and double-lid wood storage barrel, 25 or 50 kg.

(2) Before filling with lacquer sap, the container joints and inner wall must be sealed in order to make sure it does not leak.

(3) After filled with lacquer sap, tightly cover the inner and outer lids, and label the container with the name, size, gross weight, net weight, place of origin, and date.

10.2. Storage

(1) Lacquer should be stored in a cool and humid warehouse with the temperature kept between 0 and 25 °C. If the warehouse temperature is too high in summer, it should be cooled to prevent the storage barrel from cracking and leaking lacquer.

(2) Lacquer shall not be stored in open air; it should be stored in a special warehouse. If no special warehouse exists, lacquer should be stored separately from other matter. It is strictly prohibited to store with acid, alkali, and salts.

(3) The effective storage period for lacquer sap stored under conditions (1) and (2) is 1 year from the production date. Samples exceeding the storage period of 1 year should be tested again according to this standard, but it still can be used if it meets the quality requirements.

10.3. Transport

When lacquer is transported, avoid exposure to rain and sunlight.

Appendix A: Synthesis of saturated urushiol

1. Instruments and reagents

1.1. Eudiometer, 200 ml.

1.2. Hydrogen cylinder.

1.3. Hydrogen bottle, 250-ml conical flask.

1.4. Magnetic stirrer.

1.5. Nickel-aluminum alloy: powder containing 40-50% nickel.

1.6. Sodium hydroxide (chemical grade).

1.7. Petroleum ether (analytical grade).

2. Experiment

2.1. Preparation of catalyst

Slowly add 54 g powdered nickel-aluminum alloy to 500 ml 35% sodium hydroxide solution in a beaker and stir until bubbles appear and the temperature increases. Put the beaker in an ice-water bath to prevent the temperature from exceeding 50 °C. After the nickel-aluminum alloy is added, continue to stir for 30 min. Then heat in a boiling water bath for 30 min to obtain the catalyst.

2.2. Preparation of saturated urushiol

Wash the catalyst prepared above with distilled water to neutralize it, then wash three times with anhydrous ethanol, and store in anhydrous ethanol until use. Add the catalyst (about half of lacquer sap) to 20 ml anhydrous ethanol in a hydrogen bottle, and adding a certain amount of lacquer sap, then stir and introduce hydrogen to cause the hydrogenation reaction. After filtering to remove the catalyst and gum, the reaction solution is evaporated under low pressure, and a brown crude saturated urushiol is obtained.

2.3. Purification of saturated urushiol

Dissolve crude saturated urushiol in the proper amount of benzene, and then perform chromatography in a silicone column eluting with benzene. Collect various fractions according to different eluate colors. Remove benzene by evaporation, and recrystallize the solid with petroleum ether to obtain a white saturated urushiol. The melting point of pure saturated urushiol is 58-59 °C (concentration is not less than 98%). If the melting point is not 58-59 °C, purification by recrystallization is necessary.

Appendix B: Identification method of foreign matter contained in lacquer sap

1. Sensory identification

1.1. Film properties

The film of pure lacquer sap is hard, black, finely wrinkled, and elastic. If water is added, the lacquer sap the film is flat, unwrinkled, thick, brittle, and lacks luster. Lacquer sap with semidry oil and/or nondry oil does not form a film or forms a thin film that is soft, sticky, and dark. The film of lacquer sap with sugar shows irregular wrinkles and a yellow-brown foam. The film of lacquer sap with water in which plant roots, stems, and leaves (including herbs) have been boiled is thick and rough, hard and inelastic, and some films have yellow-brown film. When lacquer sap is mixed with mineral oil (especially a lubricating oil), oil can be observed on the sap surfaces, and the sap do not form a film.

1.2. Color and luster

In general, after precipitation for 5-7 days, pure lacquer sap should form three layers: top oil, yellow medium, and bottom powder, respectively. Lacquer sap mixed with foreign materials such as honey and lubricating oil, should divide into 2-4 irregular layers, and the colors are often black-blue, dark-gray, gray, and red-yellow. However, lacquer sap collected from a lacquer tree grown on the Yunnan-Guizhou Plateau of China is an exception—it is yellow-gray.

1.3. Serofluid water

When lacquer sap is static, the serofluid water sinks to the bottom, and generally constitutes around 1%. Sap tapped from Damu (big trees) lacquer trees contains more serofluid water than that from Xiaomu (small trees), and the more serofluid water, the better the sap quality. In general, lacquer sap mixed with foreign matter does not have serofluid water.

1.4. Color darkening

The color of pure lacquer sap usually changes from lighter to darker in clear steps. Lacquer sap mixed with more water changes color more slowly, and the color is unclear. Lacquer sap mixed with water boiled with plant roots, stems, and leaves changes color quickly, and the color stops at red and does not change to black. The color of lacquer sap mixed with mineral oil, vegetable oil, sugar, and some others, changes very slowly, and the luster is bad.

1.5. Tiny granules

The more obvious the line of tiny granules is in lacquer sap, the better the quality of the sap. There are more tiny granules in Damu sap and fewer in Xiaomu sap. When sap is mixed with more foreign matter, the tiny granules disappear and the sand line is less obvious. When lacquer is mixed with certain foreign materials, it may also show tiny granules and a sand line in a short time, and it is important to be careful about fake tiny granules when testing.

1.6. Stretch drawing property

The silk line of good-quality lacquer sap is slender and long, has a uniform speed, a strong contraction force after breaking, and looks obviously like a hook. Lacquer sap mixed with water including boiling water from tea, lacquer tree leaves, and other plants is a highly viscous liquid, the silk line is short and thick, contracts poorly, and forms a heap if dropped on the surface of a liquid. Lacquer sap mixed with mineral oil and/or vegetable oil is thinner, the silk line is not elastic, and its contraction without a hook-like head.

1.7. Smell

Lacquer sap has a sour smell. Mixing with foreign matter gives the sap different odors. Mixing with more water makes the odor more sour after long storage. Mixing with mineral oil gives it a mineral oil smell.

1.8. Lacquer dregs

Dregs contained in lacquer sap are divided into two kinds, original and pseudo dregs. The original dregs mainly refer to the sawdust wood and sap natural film, and the pseudo dregs are the intentionally included foreign materials such as silt, iron filings, mineral powder, starch, and other debris. The original dregs are almost homogenously dispersed in the sap, and when packing the sap, it is easy to separate sap and dregs. On the other hand, the pseudo dregs sink almost to the bottom of vessel, and when packing the sap, it is difficult to separate sap and dregs. In the testing process, check not only the sap surface, but also check the vessel bottom to identify original or pseudo dregs.

1.9. Drying properties

Drying characteristics are a most important aspect of lacquer quality. In general, Chinese lacquer can achieve touch-free dry within 3-6h at room temperature and relative humidity. To test it, wrap absorbent cotton around one end of a chopstick and then gently tap lacquer film. If no cotton adheres to the lacquer, it is touch-free dry. Lacquer mixed with foreign materials usually does no dry quickly.

2. Special steelyard balance observation

A special steelyard balance can not only directly measure the lacquer scores (specification), but also identify pure sap or sap mixed with other foreign matter. The special steelyard measurement of pure lacquer sap shows "smoke that appears, then a bubble that disappears," and the dish is clear without any heterogeneous precipitates. If "smoke appears, but the bubble does not disappear," and a precipitate appears on the dish bottom, it can be said that the sap was mixed with foreign matter.

When lacquer sap is mixed with kerosene, a white smoke persists and kerosene can be smelled during the special steelyard testing process. When sap is mixed with vegetable oil, small bubbles first appear in the dish, and then smoke appears. The small bubbles do not cease, and they overflow the dish. When sap is mixed with sugar, first black bubbles appear, then black smoke, and finally white smoke continuously comes out, and the sap overflows the dish. The dish bottom is dirty, and the lacquer sap becomes black.

When lacquer sap is mixed with resin (rosin) during the special steelyard testing process, it first produces a white foam and overflows the dish, and white smoke constantly comes out, small foam does not cease, and the sap changes to black and adheres to the dish. Lacquer sap mixed with water also produces a clear dish bottom, so it is necessary to be careful during testing such lacquer sap. In the inspection process, the following steps are necessary: (1) observation of the color, stretch drawing property, and tiny granules; (2) smelling the sap; (3) heating in the special steelyard to calculate each fraction; and (4) testing the drying time. Any kind of foreign matter mixed with the lacquer sap should change the color, stretch drawing property, tiny granules, smell, and drying time. A correct conclusion will be reached as long as the inspection procedures are followed strictly. If the lacquer sap cannot be judged immediately, let the sap sit for several days, and then observe the changes in properties such as color, tiny granules, smell, and drying time.

2 IDENTIFICATION OF LACQUER FROM JAPAN

JAPANESE INDUSTRIAL STANDARD (JIS) K-5950 (K IS JAPANESE "KAGAKU," MEANING CHEMISTRY) FOR REFINED RHUS LACQUER

1. Range

   This standard is for refined lacquer and its inspection methods.
   Reference standards:
   JIS K-1409: About the use of TiO_2 in textile chemistry.
   JIS K-8101: Ethyl alcohol (99.5%), ethanol (99.5%) (reagents).
   JIS K-8987: Anhydrous sodium sulfate (reagents).
   JIS K-3503: Glassware for chemical analysis.
2. Meaning of terms

   2.1. Refined lacquer: Raw lacquer appropriately processed according to the application.

(1) Raw lacquer is sap tapped from a lacquer tree after removing foreign matter. Grade 1 is the highest quality sap, mainly used in lacquer art, the bottom coat in high-quality lacquerware, and top coat for polishing. Sap below grade 2 is mainly used as the bottom coat, blowing coat, wood preservative, rust prevention of metal, and dyeing stencils.

(2) Nashiji lacquer is prepared from a high-quality lacquer sap, mixed with yellow dyes and pigments, such as orpiment to give a yellow color. It has high transparency because it is subjected to nayashi and kurome (a refinement process, heating and stirring). It is manly used as cover coating on gold, silver, and tin powder, and is also used to represent the wood grain abrasive coating.

(3) Transparent roiro lacquer is prepared from a good-quality lacquer sap, mixed with various dyes and pigments to make colored lacquer sap, and mainly used to represent the wood grain abrasive coating.

(4) Transparent gloss lacquer is prepared from a good-quality lacquer sap, mixed with suitable auxiliary agents, and used mainly as the top coat without polishing or used in various colored lacquers.

(5) Transparent hakushita lacquer is prepared from a good-quality lacquer sap mixed with drying oil or natural resin, and mainly used to attach gold, silver, and tin foil.

(6) Transparent intermediate coating lacquer is mainly used as intermediate coating material.

(7) Transparent no-gloss lacquer is mainly used to eliminate lacquer gloss.

(8) Black roiro lacquer is prepared from a good-quality lacquer sap mixed with black matter and mainly used as top coat.

(9) Black gloss lacquer is mainly used as black top coat.

(10) Black hakushita lacquer is mainly used as primer to attach gold, silver, and tin foil.

(11) Black intermediate coating lacquer is mainly used as intermediate coat.

(12) Transparent nongloss lacquer is used mainly to eliminate black lacquer gloss, and it is almost always mixed with nongloss materials.

2.2. Raw lacquer sap: Refers to sap tapped from the trunk of a lacquer tree. (Exposed to air, it will form a film. The main ingredient is a polyvalent phenol containing two hydroxyl groups.)

2.3. Nayashi is a lacquer refining process of stirring and kneading in order to increase film gloss.

2.4. Kurome is a lacquer refining process of stirring, kneading, and heating in order to decrease the water concentration and diameter of water drops in lacquer sap. Kurome lacquer is also mixed with suitable auxiliary agents for different uses.

3. Type: Refined lacquer types are as follows;

(1) Raw lacquer: Grades 1, 2, 3, and 4

(2) Transparent lacquer: Nashiji lacquer grades 1 and 2

Transparent roiro lacquer grades 1 and 2

Transparent gloss lacquer grades 1, 2, 3, and 4

Transparent hakushita lacquer grades 1 and 2

Transparent intermediate lacquer grades 1 and 2

Transparent nongloss lacquer grades 1 and 2

(3) Black lacquer

Black roiro lacquer grades 1 and 2

Black gloss lacquer grades 1, 2, 3, and 4

Black hakushita lacquer grades 1 and 2

Black intermediate lacquer grades 1 and 2

Black nongloss lacquer grades 1 and 2

4. Quality

4.1. Raw lacquer is the sap tapped from a lacquer tree after removal of foreign matter. Its quality must comply with the parameters summarized in Table A2.

4.2. Transparent lacquer is lacquer sap that treated by nayashi and kurome processes to remove solid foreign matter. Its quality must comply with the parameters summarized in Table A3.

4.3. Black lacquer is prepared from a good-quality lacquer sap by nayashi and kurome processes, mixed with iron powder or ferric hydroxide, then filtered to remove the solid matter. Its quality must comply with the parameters summarized in Table A4.

5. Sampling method

Randomly select one or several 100 kg buckets of refined lacquer, evenly stir it, and take out 100 g each from different buckets or different sites of the same bucket, and mix them well together. Put 50 g into a 60-ml wide-mouth bottle which complies with the JSK 3503 (glassware for chemical analysis) standard, and then close the cover and send it to the test lab.

6. Test method

6.1. Test conditions: Samples used for drying time, film, transparency, and polishing tests should be dried at 20 °C and 80% relative humidity, and samples used for the hardness test should be dried at 20 °C and 70% relative humidity. The samples not only need to comply with this standard, but also meet other test requirements if necessary.

Table A2 Parameters of raw lacquer

Types	Items					
			Analysis test			
	Drying time (h)	Hardness time (h)	Heating loss (%)	Titration (ml)	Nitrogenous compounds (%)	Gum (%)
Raw lacquer grade 1	<6	<5	<28	>2.5	1.5-5.0	<7.0
Raw lacquer grade 2	<8	<7	<30	>2.4	1.5-5.0	<10.0
Raw lacquer grade 3	<10	<9	<35	>2.2	1.5-5.0	<13.0
Raw lacquer grade 4	<12	<10	<38	>1.7	1.5-5.0	<18.0

Table A3 Parameters of transparency lacquer

Types	Items					
					Analysis test	
	Film test	Transparence test	Drying time (h)	Polishing test	Heating loss (%)	Titration (ml)
Nashiji grade 1	Qualified	Qualified	<18	Qualified	3-6	>3.0
Nashiji grade 2	Qualified	Qualified	<18	Qualified	3-6	>2.7
T-R grade 1	Qualified	Qualified	<10	Qualified	3-6	>3.2
T-R grade 2	Qualified	Qualified	<10	Qualified	3-6	>2.9
T-G grade 1	Qualified	Qualified	<10	–	3-6	>2.6
T-G grade 2	Qualified	Qualified	<10	–	3-6	>2.4
T-G grade 3	Qualified	Qualified	<12	–	3-6	>2.1
T-G grade 4	Qualified	Qualified	<12	–	3-6	>1.9
T-H grade 1	Qualified	Qualified	<10	–	3-6	>2.6
T-H grade 2	Qualified	Qualified	<10	–	3-6	>2.4
T-I grade 1	Qualified	Qualified	<8	–	3-6	>2.4
T-I grade 2	Qualified	Qualified	<8	–	3-6	>2.2
T-N grade 1	Qualified	Qualified	<10	–	3-6	>2.6
T-N grade 2	Qualified	Qualified	<10	–	3-6	>2.4

T-R: transparency roiro lacquer; T-G: transparency gloss lacquer; T-H: transparency hakushita lacquer; T-I: transparency intermediate lacquer; T-N: transparency nongloss lacquer.

Table A4 Parameters of black lacquer

Types	Items					
					Analysis test	
	Film test	Black test	Drying time (h)	Polishing test	Heating loss (%)	Titration (ml)
B-R grade 1	Qualified	Qualified	<10	Qualified	3-6	>3.2
B-R grade 2	Qualified	Qualified	<10	Qualified	3-6	>2.9
B-G grade 1	Qualified	Qualified	<10	–	3-6	>2.6
B-G grade 2	Qualified	Qualified	<10	–	3-6	>2.4
B-G grade 3	Qualified	Qualified	<12	–	3-6	>2.1
B-G grade 4	Qualified	Qualified	<12	–	3-6	>1.9
B-H grade 1	Qualified	Qualified	<10	–	3-6	>2.6
B-H grade 2	Qualified	Qualified	<10	–	3-6	>2.4
B-I grade 1	Qualified	Qualified	<8	–	3-6	>2.4
B-I grade 2	Qualified	Qualified	<8	–	3-6	>2.2
B-N grade 1	Qualified	Qualified	<10	–	3-6	>2.6
B-N grade 2	Qualified	Qualified	<10	–	3-6	>2.4

B-R: black roiro lacquer; B-G: black gloss lacquer; B-H: black hakushita lacquer; B-I: black intermediate lacquer; B-N: black nongloss lacquer.

6.2. Drying time: Paint about 0.6g lacquer sap on a clear glass plate (165×60mm) on 1 dm^2 with a lacquer brush[1] as much as possible the same thickness. Let the film dry flat according to Section 6.1 and record the drying time.

6.3. Hardness time. (3.1 g lacquer sap: 4.6 g tonoko). Mix 4.6 g tonoko[2] with 2.3 g water and knead with a spatula; then add 3.1 g lacquer sap and evenly knead it. A clean wooden board[3] is coated with about 3 g this mixture sample in 1 dm^2 by a lacquer brush (see footnote 1) as nearly the same thickness as possible, and allowed to dry. After reaching the longest drying time of this standard, touch the film surface with a fingernail; if the nail does not leave a clear trace, the sample is considered hard.

6.4. Film test: The film obtained according to Section 6.2 should be compared with the standard film. If the variation in the gloss and surface smoothness are not greater than the standard film deviation, little oil seeps out, and no small holes are found in the film surfaces, the sample is considered to be satisfactory.

6.5. Transparency test: The test sample is prepared in the ratio of sap/TiO_2=1:0.8. TiO_2 complies with JIS K 1409 (the use of TiO_2 in textile chemistry). The lacquer sap and TiO_2 are kneaded together, and then film is painted on a clear glass plate (165×60mm) with a lacquer brush as much as possible the same thickness. Place the film face down and let it dry flat according to Section 6.1. The standard film was also prepared at the same condition. After drying for 24h, check the film color. If sample is whiter than the standard film, it is considered satisfactory.

6.6. Black test: The film should be prepared according to Section 6.2 and compared with the standard film. If it is blacker than the standard film, it is considered satisfactory.

6.7. Polishing test: Coat sample film and standard film on frosted glass sheets with a lacquer brush, and then dry according to Section 6.1. After drying for 72h, the film surface is first polished with water and suruga-charcoal,[4] second with water and roiro charcoal,[5] and third with rapeseed oil and tonoko, until the charcoal grinding traces completely disappear when checked with a cotton cloth. Then the oil (if present) which appears on the film surface is removed by a finger with tonoko. If the gloss of the film is not bad compared with the standard film and it lacks other shortcomings, it is considered satisfactory.

---

[1] *Note*: The lacquer brush used in this test is made from human hair, width about 30mm.

[2] *Note*: Tonoko is a powder made from ash, or soil, clay, and loess. Yamanashi Tonoko was used in this test (Yamanashi is in Kyoto, Japan).

[3] *Note*: The wooden board used in this test is made from cypress or *Cercidiphyllum japonicum* wood, and is 165×60×7mm.

[4] *Note*: Suruga-charcoal is a kind of charcoal made from tung tree oil. It is used with a straight-line reciprocating movement on the film surface.

[5] *Note*: Roiro charcoal is a kind of charcoal made from *Styrax japonica*, *Lagerstroemia indica*, or *Camellia japonica* tree. It is used with a circular movement on the film surface.

6.8. Heating loss: Weigh 1 g lacquer sap and put it in a 25-ml preweighed beaker. Stir and heat in a water bath until the bubbles completely disappear. Place the beaker in a drying oven at 105-110 °C for 30 min; then, immediately place the beaker in the glass desiccator cool it to room temperature, and weigh it. The heating loss (%) is calculated according to the following formula:

$$\text{Heating loss}(\%) = \frac{\text{Weight loss}(g)}{\text{Sample weight}(g)} \times 100$$

6.9. Titration

6.9.1. Preparation of samples (raw and transparent lacquer): After removing water according to Section 6.8, add 20 ml anhydrous alcohol,[6] stir and static to precipitate the insoluble materials, and filter the supernatant by pouring it through a preweighed filter paper. The operation is repeated until the filtrate is colorless. Then the filter paper is washed with anhydrous alcohol, and filtrate and washings are combined, adding anhydrous alcohol to 100 ml, as a titration test. The insoluble material on the filter paper can be used for quantitative determinations of nitrogenous material and gum.

6.9.2. Preparation of samples (black lacquer): After removing water according to Section 6.8, dissolve the sample dissolved in 20 ml ether, and drop into concentrated HCl to remove the blackness and precipitate the insoluble material. Filter the supernatant through filter paper into a separatory funnel, which has small amount of water in advance. Repeat the operation several times until the filtrate is colorless, then wash it with ether. Wash the ether with water several times, and add anhydrous sodium sulfate (JIS K 8987) to remove the water. Filter it in a conical flask with filter paper, evaporate the ether, and add ethyl alcohol to 100 ml.

6.9.3. Test method: Place 20 ml of alcoholic solution of the raw, transparent, and black lacquer samples, prepared as described in Sections 6.9.1 and 6.9.2, in a glass tube,[7] add 10 ml anhydrous ethanol, and a minimum amount of N/4 hydroxide barium solution, cover with a lid and shake. Hold the glass tube vertically until the upper 1-2 mm becomes transparent, then immediately observe the color by transmitted light. If the supernatant shows only a very slightly red color, it is considered that the titration determinations did not reach the prescribed number.

6.10. Nitrogenous matters: Wash the insoluble matter in anhydrous alcohol obtained in Section 6.9 with hot water, and dry in an oven at 105-110 °C till constant in weight, and weight the residue on

---

[6] *Note*: JIS K 8101: ethyl alcohol (99.5%) and ethanol (99.5%) (reagents).

[7] *Note*: Glass tube: inside diameter 15 mm, height 400 mm, capacity 60 ml with cover.

the filter paper. The concentration of nitrogenous matter (%) is calculated according to the following formula:

$$\text{Nitrogenous matter}(\%) = \frac{\text{Residue}(\text{g})}{\text{Sample weight}(\text{g})} \times 100$$

6.11. Gum: Evaporate the washing liquid obtained in Section 6.10, and dry it in an oven at 105-110 °C till constant weight and weigh the residue. The gum concentration is calculated according to the following formula:

$$\text{Gum}(\%) = \frac{\text{Residue}(\text{g})}{\text{Sample weight}(\text{g})} \times 100$$

REFERENCES

Relationship between titration and urushiol percentage or laccol percentage: The titration volume reflects the refined lacquer quality, and 1 ml titration is equivalent to 2.415% urushiol (or laccol). The percentage of urushiol or laccol can be calculated from the titration volume. Table A5 shows the titration volumes of 0.6-3.8 ml and the equivalent urushiol (or laccol) percentages.

Table A5 Relationship between titration and urushiol percentage or laccol percentage

Titration (ml)	Urushiol or laccol (%)	Titration (ml)	Urushiol or laccol (%)	Titration (ml)	Urushiol or laccol (%)
0.6	14	1.7	41	2.8	68
0.7	17	1.8	43	2.9	70
0.8	19	1.9	46	3.0	72
0.9	22	2.0	48	3.1	75
1.0	24	2.1	51	3.2	77
1.1	27	2.2	53	3.3	80
1.2	29	2.3	56	3.4	82
1.3	31	2.4	58	3.5	85
1.4	34	2.5	60	3.6	87
1.5	36	2.6	63	3.7	89
1.6	39	2.7	65	3.8	91

Index

Note: Page numbers followed by *f* indicate figures and *t* indicate tables.

E

F

G

H

V

W

X

Y

Kolibri 360 GmbH
Ferdinand-Jühlke-Straße 7
99095 Erfurt, Deutschland
produktsicherheit@kolibri360.de